Neutrinos and Explosive Events in the Universe

NATO Science Series

A Series presenting the results of scientific meetings supported under the NATO Science Programme.

The Series is published by IOS Press, Amsterdam, and Springer (formerly Kluwer Academic Publishers) in conjunction with the NATO Public Diplomacy Division.

Sub-Series

I. Life and Behavioural Sciences	IOS Press
II. Mathematics, Physics and Chemistry	Springer (formerly Kluwer Academic Publishers)
III. Computer and Systems Science	IOS Press
IV. Earth and Environmental Sciences	Springer (formerly Kluwer Academic Publishers)

The NATO Science Series continues the series of books published formerly as the NATO ASI Series.

The NATO Science Programme offers support for collaboration in civil science between scientists of countries of the Euro-Atlantic Partnership Council. The types of scientific meeting generally supported are "Advanced Study Institutes" and "Advanced Research Workshops", and the NATO Science Series collects together the results of these meetings. The meetings are co-organized by scientists from NATO countries and scientists from NATO's Partner countries — countries of the CIS and Central and Eastern Europe.

Advanced Study Institutes are high-level tutorial courses offering in-depth study of latest advances in a field.
Advanced Research Workshops are expert meetings aimed at critical assessment of a field, and identification of directions for future action.

As a consequence of the restructuring of the NATO Science Programme in 1999, the NATO Science Series was re-organized to the four sub-series noted above. Please consult the following web sites for information on previous volumes published in the Series.

http://www.nato.int/science
http://www.springeronline.com
http://www.iospress.nl

Neutrinos and Explosive Events in the Universe

edited by

Maurice M. Shapiro

University of Maryland,
College Park, U.S.A.

Todor Stanev

Bartol Research Institute,
University of Delaware, Newark, U.S.A.

and

John P. Wefel

Louisiana State University,
Baton Rouge, U.S.A.

 Springer

Published in cooperation with NATO Public Diplomacy Division

Proceedings of the NATO Advanced Study Institute on
Neutrinos and Explosive Events in the Universe

A C.I.P. Catalogue record for this book is available from the Library of Congress.

ISBN-10 1-4020-3747-3 (PB)
ISBN-13 978-1-4020-3747-4 (PB)
ISBN-10 1-4020-3746-5 (HB)
ISBN-13 978-1-4020-3746-7 (HB)
ISBN-10 1-4020-3748-1 (e-book)
ISBN-13 978-1-4020-3748-1 (e-book)

Published by Springer,
P.O. Box 17, 3300 AA Dordrecht, The Netherlands.

www.springeronline.com

Printed on acid-free paper

Printed in the Netherlands.

*For Maurice M. Shapiro,
who started this school, and
without whom none of this
would have been possible.*

T.S.&J.P.W.

Contents

Preface

"Neutrinos and Explosive Events in the Universe" brought together experts from diverse disciplines to offer a detailed view of the exciting new work in this part of High Energy Astrophysics. Sponsored by NATO as an Advanced Study Institute, and coordinated under the auspices of the International School of Cosmic Ray Astrophysics (14th biennial course), the ASI featured a full program of lectures and discussion in the ambiance of the Ettore Majorana Centre in Erice, Italy, including visits to the local Dirac and Chalonge museum collections as well as a view of the cultural heritage of southern Sicily. Enrichment presentations on results from the Spitzer Infrared Space Telescope and the Origin of Complexity complemented the program.

This course was the best attended in the almost 30 year history of the School with 121 participants from 22 countries. The program provided a rich experience, both introductory and advanced, to fascinating areas of observational Astrophysics Neutrino Astronomy, High Energy Gamma Ray Astronomy, Particle Astrophysics and the objects most likely responsible for the signals Explosions and related phenomena, ranging from Supernovae to Black Holes to the Big Bang. Contained in this NATO Science Series volume is a summative formulation of the physics and astrophysics of this newly emerging research area that already has been, and will continue to be, an important contributor to understanding our high energy universe. The volume is suitable for students and advanced researchers wanting a current picture of high energy astrophysics either for personal edification or for use in a course of study.

A highly successful ASI requires the combined effort of many individuals, foremost of whom are the Lecturers who give of their time and expertise in formal presentations, informal discussions and through their contributions to this volume. We salute them! The ASI was co-directed by J.P. Wefel and V.S. Ptuskin, assisted by M.M. Shapiro (Director of the School), T.S. Stanev, A. Smith and an exceptional organizing committee. Without the assistance of NATO, this ASI would not have been possible, and we thank F. Pedrazzini and A. Trapp in the Collaborative Programmes Section, and the members of the NATO Science committee, for help and support. The exceptional facilities of the Ettore Majorana Centre provided the setting for this ASI, and we acknowledge Centre Director, A. Zichichi, plus Fiorella, Pino, Alessandro, Alberto, and a host of others who contributed to the ASIs success. We are also grateful to the Sicilian Regional Government, the Italian Ministry of Education, the US National Science Foundation and all of the institutes, universities and government agencies who helped to support the participants. Special thanks are due to G. Sutton and A. Eads for their work on the organization, programme and manuscripts and to A. Kersbergen for her gracious help with this volume.

M.M. Shapiro, V.S. Ptuskin, T.S. Stanev and J.P. Wefel

I

CONTEMPORARY CHALLENGES IN ASTROPHYSICS AND COSMOLOGY

OUTSTANDING PROBLEMS IN PARTICLE ASTROPHYSICS

Thomas K. Gaisser

Bartol Research Institute, University of Delaware, Newark, DE 19716, USA

gaisser@bartol.udel.edu

*Work supported in part by the U.S. Department of Energy under DE-FG02 91ER40626

Abstract   The general features of the cosmic-ray spectrum have been known for a long time. Although the basic approaches to understanding cosmic-ray propagation and acceleration have also been well understood for many years, there are several questions of great interest that motivate the current intense experimental activity in the field. If the energy-dependence of the secondary to primary ratio of galactic cosmic rays is as steep as observed, why is the flux of PeV particles so nearly isotropic? Can all antiprotons and positrons be explained as secondaries or is there some contribution from exotic sources? What is the maximum energy of cosmic accelerators? Is the "knee" of the cosmic-ray spectrum an effect of propagation or does it perhaps reflect the upper limit of galactic acceleration processes? Are gamma-ray burst sources (GRBs) and/or active galactic nuclei (AGN) accelerators of ultra-high-energy cosmic rays (UHECR) as well as sources of high-energy photons? Are GRBs and/or AGNs also sources of high-energy neutrinos? If there are indeed particles with energies greater than the cutoff expected from propagation through the microwave background radiation, what are their sources? The purpose of this lecture is to introduce the main topics of the School and to relate the theoretical questions to the experiments that can answer them.

Keywords: High Energy Cosmic Rays

Introduction

It is appropriate to begin this lecture with a diagram from the review of Shapiro & Silberberg, 1970, which compares the abundances of elements in the cosmic radiation with solar system abundances. This classic measurement is one of the foundations of cosmic-ray physics. The elements lithium, beryllium and boron are quite abundant among cosmic rays even though they constitute only a tiny fraction of the material in the solar system and the interstellar medium. This fact is understood largely as the result of spallation of the

3

M. M. Shapiro et al. (eds.), Neutrinos and Explosive Events in the Universe, 3–31.
© 2005 *Springer. Printed in the Netherlands.*

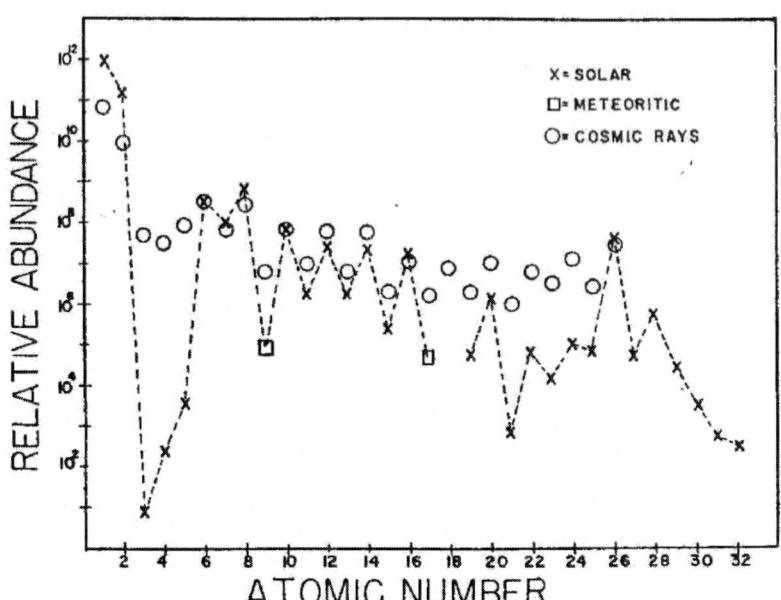

Figure 1.   Comparison of cosmic-ray abundances with abundances of nuclei in the solar system (Shapiro & Silberberg, 1970).

"primary" nuclei carbon and oxygen during their propagation in the interstellar medium. The idea is that the protons and nuclei are accelerated from the interstellar medium and/or from the gas in or near their sources. The composition of the "primary" nuclei (defined as those initially accelerated and injected into the interstellar medium with high energy) reflects the combination of nuclei present in the material that is accelerated, which contains negligible amounts of "secondary" elements such as Li, Be, B. These secondary nuclei are fragmentation products of heavier primary nuclei.

In this picture, the amount of secondary nuclei is a measure of the characteristic time for propagation of cosmic rays before they escape from the galaxy into inter-galactic space. A simplified version of the diffusion equation that relates the observed abundances and spectra to initial values is

$$\frac{N_i(E)}{\tau_{esc}(E)} = Q_i(E) - \left(\beta c n_H \sigma_i + \frac{1}{\gamma \tau_i}\right) N_i(E) + \beta c n_H \sum_{k \geq i} \sigma_{k \to i} N_k(E). \quad (1)$$

Here $N_i(E)$ is the spatial density of cosmic-ray nuclei of mass i, and n_H is the number density of target nuclei (mostly hydrogen) in the interstellar medium, $Q_i(E)$ is the number of primary nuclei of type i accelerated per cm^3 per second, and σ_i and $\sigma_{k \to i}$ are respectively the total and partial cross sections for interactions of cosmic-ray nuclei with the gas in the interstellar medium. The second term on the r.h.s. of Eq. 1 represents losses due to interactions with cross section σ_i and decay for unstable nuclei with lifetime τ_i. The energy per nucleon, E, remains constant to a good approximation in the transition from parent nuclei to nuclear spallation products, which move with velocity βc and Lorentz factor $\gamma = E/m_p$.

A crude first estimate of the characteristic diffusion time τ_{esc} can be made by neglecting propagation losses for a primary nucleus P and assuming that $Q_S = 0$ for a secondary nucleus S. If we also neglect collision losses by the secondary nucleus after it is produced, then the solution of Eq. 1 is

$$n_H \tau_{esc} = \frac{1}{\beta c \sigma_{P \to S}} \frac{N_S}{N_P}. \tag{2}$$

Since the density in the disk of the galaxy is of order one particle per cm^3 and the typical partial cross section for a light nucleus will be of order 100 mb, the scale for the characteristic time is $\sim 10^7$ years. The full analysis requires self-consistent solution of the coupled equations for all species accounting for all loss terms, as described in John Wefel's lecture in this volume (Wefel, 2005). A nice overview of propagation models is given by Jones *et al.*, 2001.

The thickness of the disk of the galaxy is of order 300 pc = 1000 light years, which is much shorter than the characteristic propagation time of 10 million years. The explanation is that the charged particles are trapped in the turbulent magnetized plasma of the interstellar medium and only diffuse slowly away from the disk, which is assumed to be where the sources are located. Measurements of the ratio of unstable to stable secondary nuclei (especially $^{10}Be/^9Be$) are used to determine τ_{esc} independently of the product $n_H \tau_{esc}$ and hence to constrain further the models of cosmic-ray propagation.

Another important fact is that the ratio N_S/N_P is observed to decrease with energy. From Eq. 2 this implies that τ_{esc} also decreases. Simple power-law fits to ratios like B/C give

$$\tau_{esc} \propto E^{-\delta}, \tag{3}$$

with $\delta \approx 0.6$. This behavior has important consequences for the source spectrum. To see this, consider an abundant, light primary nucleus such as hydrogen or helium. They are sufficiently abundant so that feed-down from heavier nuclei can be neglected and their cross sections are small enough so that energy losses in the interstellar medium can also be neglected in a first approximation. Then Eq. 1 reduces to

$$Q(E) = N(E)/\tau_{esc} \approx N(E) \times E^\delta. \tag{4}$$

The local energy-density spectrum of cosmic rays is related to the observed flux ϕ(particles per cm^2 per GeV per second per steradian) by

$$N(E) \;=\; \frac{4\pi}{c}\,\phi(E). \tag{5}$$

Since $\phi(E) \approx E^{-\alpha}$ with $\alpha \approx 2.7$ the inference is that the cosmic accelerators are characterized by a power law with spectral index $\alpha_S \approx 2.1$. This value is close to the spectral index for first order acceleration by strong shocks in the test-particle approximation (Ostrowski, 2005).

If the time for diffusion out of the galaxy continues to be described by $\delta \approx 0.6$ to very high energy there is a problem: as $c\tau_{esc}$ decreases and approaches galactic scales, the cosmic-ray fluxes should become significantly anisotropic, which is not observed. One possibility is that the observed energy-dependence at low energy is due to a combination of "reacceleration" (Seo & Ptuskin, 1994, Heinbach & Simon, 1995) by weak shocks in the interstellar medium after initial acceleration by strong shocks. In such models, the high-energy behavior of diffusion is typically described by a slower energy dependence with $\delta \approx 0.3$. If so, the source spectral index would be steeper, approximately $\alpha_S = 2.4$. In any case, in more realistic non-linear treatments of acceleration by strong shocks, the spectrum has some curvature, being steeper at low energy and harder at the high energy end of the spectrum (Berezhko & Ellison, 1999). What we observe may be some kind of average over many sources, each of which is somewhat different in shape and maximum energy.

The assumption underlying the discussion above is that the sources accelerating cosmic rays are in the disk of the galaxy and that the energy density in cosmic rays observed locally is typical of other regions of the galactic disk. If so, the total power P_{CR} required to maintain the cosmic radiation in equilibrium may be obtained by integrating Eq. 4 over energy and space. The result is

$$P_{CR} \;=\; \int d^3x \int Q(E)\,dE \;=\; V_G\,\frac{4\pi}{c}\int \phi(E)\,/\,\tau_{esc}(E)\,dE. \tag{6}$$

Using the observed spectrum and the value of τ_{esc} explained above, one finds numerically

$$P_{CR} \;\sim\; 10^{41}\,\mathrm{erg/s}. \tag{7}$$

The kinetic energy of an expanding supernova remnant is initially of order 10^{51} erg, excluding the neutrinos, which carry away most of the energy released but do not disturb the interstellar medium. There are perhaps 3 supernovae per century, which gives $P_{SN} \sim 10^{42}$ erg/s as an estimate of the power which is dissipated in the interstellar medium by means of strong shocks driven by supernova ejecta. These are just the ingredients needed for acceleration of galactic cosmic rays. I return to the subject of cosmic-ray acceleration in §3 below.

Secondary cosmic-rays

Depending on the context, the term *secondary cosmic rays* can refer either to particles produced by interactions of primary cosmic rays with the interstellar gas or to particles produced by interactions of cosmic rays in the Earth's atmosphere. The production mechanisms are similar, and there are some common features.

We have already seen one example, the secondary nuclei produced by occasional interactions of primary cosmic-ray nuclei during their propagation in the interstellar medium. In that case, to a good approximation, the energy per nucleon of the secondary nucleus is the same as that of the parent nucleus. The reason is that the nuclear fragments are only spectators to any production of secondary pions that may occur in the collisions. In general, however, the production term (last term of Eq. 1) will involve an integration over the energy of the parent particle. The main process to consider is production of pions by interaction of protons with a target nucleus. The production spectrum of the pions is

$$d\phi_\pi(E_\pi) = \frac{dX}{\lambda_p} \int_{E_\pi}^{\infty} \phi_p(E_p) \frac{dn_\pi(E_\pi, E_p)}{dE_\pi} dE_p. \tag{8}$$

Here $\lambda_p = A\,m_p/\sigma_{pA}^{\text{inel}}$ is the interaction length of protons in a medium consisting of nuclei of mass A, and $dX = \rho\,d\ell$ is the differential element of mass traversed in distance $d\ell$ in a medium of density ρ. It is often a useful approximation at high energy (i.e. energy $\gg$ particle masses) to assume a scaling form for the dimensionless production spectrum:

$$E_p \frac{dn_\pi(E_\pi, E_p)}{dE_\pi} = \frac{dn_\pi(\xi)}{d\xi}. \tag{9}$$

The scaling variable is $\xi = E_\pi/E_p$, and Eq. 1 becomes

$$d\phi_\pi(E_\pi) = \frac{dX}{\lambda_p} \int_0^1 \phi_p\left(\frac{E_\pi}{\xi}\right) \frac{dn_\pi(\xi)}{d\xi} \frac{d\xi}{\xi} \rightarrow \frac{dX}{\lambda_p} K\, E_\pi^{-\alpha}\, Z_{p\to\pi}(\alpha). \tag{10}$$

The last step on the r.h.s. of Eq. 10 follows when the parent spectrum is a power law in energy ($\phi_p(E) = K\,E^{-\alpha}$). In that case, in the high-energy scaling approximation

$$d\phi_\pi(E_\pi) \rightarrow \frac{dX}{\lambda_p} Z_{p\to\pi} \times \phi_p(E_\pi), \tag{11}$$

i.e. the energy spectrum of secondaries has the same power behavior as the primaries scaled down by a factor

$$Z_{p\to\pi}(\alpha) = \int_0^1 \xi^{\alpha-1} \frac{dn_\pi}{d\xi} d\xi. \tag{12}$$

The spectrum-weighted moment $Z_{p \to \pi}(\alpha)$ depends both on the physics of production of the secondary pion and on the value of the differential spectral index α.

Galactic secondaries

Diffuse gamma-rays. Gamma-ray emission from the disk of the galaxy is a powerful probe of the model of cosmic-ray origin and propagation as well as of the structure of the galaxy. Unlike charged cosmic rays, secondary photons propagate in straight lines. Since the galaxy is transparent for γ-rays of most energies, it is possible to search for concentrations of primary cosmic-ray activity from the map of the γ-ray sky after subtracting point sources. For example, if cosmic-ray acceleration is correlated with regions of higher density such as star-forming regions where supernovae are more frequent, then one would expect a quadratic enhancement of secondary production because of the spatial correlation between primary flux and target density.

The baseline calculation is to assume that the intensity observed locally at Earth is representative of the distribution everywhere in the disk of the galaxy. One can then look for interesting variation superimposed on this baseline. Following the analysis of Eqs. [9,10,11,12], the average number of neutral pions produced per GeV per unit volume in the interstellar medium is

$$q_\pi = 4\pi \, n_H \, \sigma_{pH}^{\text{inel}} \, Z_{p \to \pi} \cdot \phi_p(E_\pi). \tag{13}$$

Next this expression has to be convolved with the distribution of photons produced in $\pi^0 \to \gamma\gamma$. In the rest frame of the parent pion, each photon has $E_\gamma = m_\pi/2 = 70 \text{ MeV}$. The angular distribution is isotropic, so

$$\frac{dn_\gamma}{d\Omega^*} = \frac{1}{2\pi} \frac{dn_\gamma}{d\cos\theta^*} = \frac{1}{2\pi}, \tag{14}$$

where θ^* is the polar angle of the photon along the direction of motion of the parent pion but evaluated in the rest frame of the pion. For decay in flight of a pion with Lorentz factor γ and velocity βc, the energy of each of the resulting photons is

$$E_\gamma = \gamma \frac{m_\pi}{2} (1 + \beta \cos\theta^*) \tag{15}$$

with $\cos\theta_1^* = -\cos\theta_2^*$ for the two photons. Changing variables in Eq. 14 then gives

$$\frac{dn_\gamma}{dE_\gamma} = \frac{2}{\beta\gamma m_\pi}. \tag{16}$$

For $E_\pi > E_\gamma \gg m_\pi/2$ the convolution of the distribution (16) with the production spectrum of neutral pions (13) gives

$$q_\gamma(E_\gamma) \approx 4\pi\, n_H\, \sigma_{pH}^{inel} \times \frac{2}{\alpha}\, Z_{p\to\pi^\circ}(\alpha)\, \phi_p(E_\gamma). \tag{17}$$

Numerically, in the approximation of uniform cosmic-ray density and uniform gas density in the interstellar medium, the observed gamma-ray flux would be

$$\frac{\phi_\gamma(E_\gamma)}{\phi_p(E_\gamma)} \approx 3 \times 10^{-6} \left(\frac{n_H}{\text{cm}^3}\right) \left(\frac{r_{max}(b, \phi_\ell)}{1 kpc}\right). \tag{18}$$

Here $r_{max}(b, \phi_\ell)$ is the distance in a direction $\{b, \phi_\ell\}$ to the effective edge of the galactic disk, where b and ϕ_ℓ are galactic latitude and longitude. The effective distance is defined with respect to an equivalent disk of uniform density. Eq. 18 compares well in order of magnitude with the measured intensity of GeV photons by Egret (Hunter *et al.*, 1997). The derivation of Eq. 18 is grossly oversimplified compared to the actual model calculation made in the paper of Hunter *et al.*, 1997 of the diffuse, galactic gamma-radiation. The reader is urged to consult that paper to understand the impressive level of detail at which the data are understood. It is also interesting to compare Eq. 18 to the TeV diffuse flux measured by Milagro (Goodman, 2005).

The implication of Eq. 17 is that for $E_\gamma \gg 70$ MeV the diffuse gamma-ray spectrum should have the same power law behavior as the proton spectrum, $\alpha \approx 2.7$. What is observed, however, is that the spectrum of gamma-rays from the inner galaxy is harder than this, having a power-law behavior of approximately $E_\gamma^{-2.4}$ (Hunter *et al.*, 1997). This is currently not fully understood. One possibility is that the cosmic-ray spectrum producing the gamma rays is harder than observed locally near Earth (Hunter *et al.*, 1997).

Cosmic-ray electrons also contribute to the diffuse gamma-radiation by bremsstrahlung and by inverse Compton scattering. Fitting the observed spectrum requires a complete model of propagation that includes all contributions (Hunter *et al.*, 1997). The distinguishing feature of π^0-decay photons is a kinematic peak at $E_\gamma = m_\pi/2$. The origin of this feature can be seen in Eq. 15 from which the limits on E_γ for any given Lorentz factor of the parent pion are

$$\sqrt{\frac{1-\beta}{1+\beta}}\frac{m_\pi}{2} < E_\gamma < \sqrt{\frac{1+\beta}{1-\beta}}\frac{m_\pi}{2}.$$

The distribution dn_γ/dE_γ is flat between these limits for each γ and is always centered around $\ln(m_\pi/2)$ when plotted as a function $\ln(E_\gamma)$. The individual contributions for parent pions of various energies always overlap at $\ln(E_\gamma) = m_\pi/2$, so the full distribution from any spectrum always peaks at this value (Stecker, 1971).

Antiprotons and positrons.   Antiprotons and positrons are of special interest because an excess over what is expected from production by protons during
propagation could reflect an exotic process such as evaporation of primordial
black holes or decay of exotic relic particles (Bottino et al., 1998). At a more
practical level, they are important because they are secondaries of the dominant
proton component of the cosmic radiation. As a consequence their spectra and
abundances provide an independent constraint on models of cosmic-ray propagation (Moskalenko et al., 1998).

Secondary antiprotons have a kinematic feature analogous to that in π^0-
decay gamma rays but at a higher energy related to the nucleon mass. In this
case the feature is related to the high threshold for production of a nucleon-
antinucleon pair in a proton-proton collision. This kinematic feature is observed in the data (Orito et al., 2000), and suggests that an exotic component
of antiprotons is not required. Antiproton fluxes are consistent with the basic
model of cosmic-ray propagation described in the Introduction.

Positrons are produced in the chain

$$p \rightarrow \pi^+ \rightarrow \mu^+ \rightarrow e^+.$$

Secondary electrons are produced in the charge conjugate process, but their
number in the GeV range is an order of magnitude lower than primary electrons
(i.e. electrons accelerated as cosmic rays). Because of radiative processes, the
spectra of positrons and electrons are more complex to interpret than high-
energy secondary γ-rays (Moskalenko & Strong, 1998). The measured intensity of positrons appears to be consistent with secondary origin (DuVernois et
al., 2001).

γ-rays and ν from young supernova remnants.   The same equations that
govern production of secondaries in the interstellar medium also apply to production in gas concentrations near the sources. For example, a supernova
exploding into a dense region of the interstellar medium (Berezhko & Völk,
2000) or into the gas generated by the strong pre-supernova wind of a massive
progenitor star (Berezhko, Pühlhofer & Völk, 2003) would produce secondary
photons that could show up as point sources.

Indeed, for many years, observation of π^0-decay γ-rays from the vicinity of
shocks around young supernova remnants (SNR) has been considered a crucial
test of the supernova model of cosmic-ray origin (Drury et al., 1994; Buckley
et al., 1998). Note, however, that a sufficiently dense target is required. Moreover, it is difficult to distinguish photons from π^0 decay from photons originating in radiative processes of electrons (Gaisser, Protheroe & Stanev, 1998).
There are two signatures: at low energy, observation of a shoulder reflecting
the π^0 peak at 70 MeV would be conclusive. At higher energy one has to de-

pend on the hardness and shape of the spectrum for evidence of hadronic origin of the photons. (See Drury *et al.*, 2001 for a current review.)

Observation of high-energy neutrinos would be strong evidence for acceleration of a primary beam of nucleons because such neutrinos are produced in hadronic interactions. Expected fluxes are low (Gaisser, Halzen & Stanev, 1995), so large detectors are needed (Montaruli, 2003, Migneco, 2005).

Atmospheric secondaries

Production of secondary cosmic rays and γ-rays in the interstellar medium generally involves less than one interaction per primary. In the language of accelerators, this is the thin-target regime. In contrast, the depth of the atmosphere is more than ten hadronic interaction lengths, so we have a thick target to deal with. The relevant cascade equation is

$$\frac{\mathrm{d}N_i(E,X)}{\mathrm{d}X} = -\left(\frac{N_i(E,X)}{\lambda_i(E)} + \frac{N_i(E,X)}{d_i(E)}\right) \qquad (19)$$
$$+ \sum_i \int_E^\infty \frac{N_k(E',X)}{\lambda_k(E')} \frac{F_{k\to i}(E,E')}{E} \mathrm{d}E',$$

where

$$F_{k\to i} = \frac{1}{\sigma_k} E \frac{\mathrm{d}\sigma_{k\to i}}{\mathrm{d}E}.$$

The equation describes the longitudinal development of the components of the atmospheric cascade in terms of slant depth ($\mathrm{d}X = \rho\,\mathrm{d}\ell$) along the direction of the cascade.

The loss terms on the r.h.s. of Eq. 19 represent interactions and decay, in analogy to Eq. 1. Here

$$d_i = \rho\,\gamma\,c\,\tau_i = \rho\,\frac{E_i\tau_i c}{m_i c^2} \qquad (20)$$

is the Lorentz dilated decay length of particle i in g/cm^3. The expression $\lambda_i = d_i$ defines a critical energy below which decay is more important than re-interaction. Because the density of the atmosphere varies with altitude, it is conventional to define the critical energy at the depth of cascade maximum (Gaisser, 1990). For pions the critical energy in the terrestrial atmosphere is $\epsilon_\pi = 115$ GeV, while $\epsilon_K^\pm = 850$ GeV. In astrophysical settings, the density is usually low enough so that decay always dominates over hadronic interactions. An intermediate case of some interest is production of secondary cosmic rays in the solar chromosphere, where the scale height is larger than in the Earth's atmosphere so that decay remains dominant for another order of magnitude (Seckel, Stanev & Gaisser, 1991).

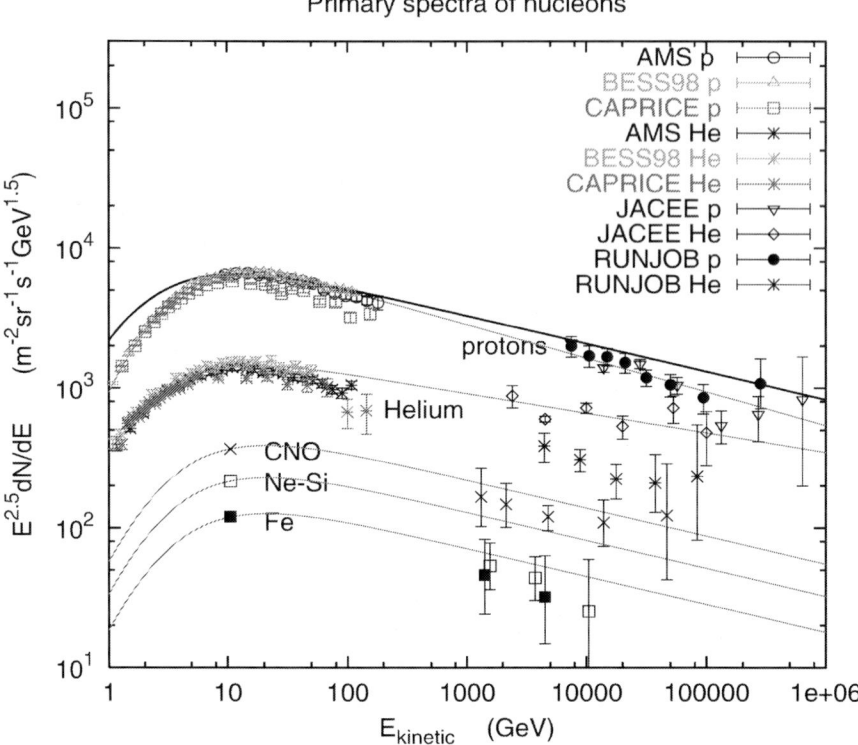

Figure 2. The flux of nucleons. The heavy black line shows the numerical form of Eq. 22. The lighter lines show extrapolations of fits (Gaisser, Honda, Lipari & Stanev, 2001) to measurements of protons, helium and three heavier groups below 100 GeV/nucleon.

The same set of cascade equations (see Eq. 19) governs air showers and uncorrelated fluxes of particles in the atmosphere. The boundary condition for an air shower initiated by a primary of mass A and total energy E_0 is

$$N(X)|_{X=0} = A\,\delta(E - E_0/A) \tag{21}$$

and $N(0) = 0$ for all other particles. This approximation, in which a nucleus is treated as consisting of independently interacting nucleons, is called the superposition approximation. In practice in Monte Carlo solutions of the cascade equation it is straightforward to remove this approximation given a model of nuclear fragmentation, (e.g. Battistoni, *et al.*, 1997).

The other important boundary condition is that for uncorrelated fluxes in the atmosphere:

$$N(X)|_{X=0} = \phi_p(E) = \approx 1.7 \times 10^4\, E^{-2.7}\ (\mathrm{GeV\,cm^2 s\,sr})^{-1}. \tag{22}$$

The numerical approximation is for the flux of all nucleons summed over the five major nuclear groups shown in Fig. 2. In Eq. 22, E is total energy per nucleon. This numerical approximation is shown as the heavy solid line in Fig. 2. Its curvature at low energy is just a consequence of plotting the power law in total energy per nucleon as a function of kinetic energy per nucleon. Only a subset of available data is shown in Fig. 2. Data from the magnetic spectrometers BESS98 (Sanuki, *et al.*, 2000) and AMS (Alcarez *et al.*, 2000a) for protons are indistinguishable on the plot, although the two experiments differ somewhat for helium (Alcarez *et al.*, 2000b). Data from the CAPRICE spectrometer (Boezio *et al.*, 1999) are 15-20% lower than BESS98 above 10 GeV/nucleon. The higher energy data on the plot are from balloon-borne calorimeters, which are subject to larger systematic errors because not all energy is sampled in the calorimeter. Proton and helium data from the electronic calorimeter ATIC (preliminary, Ahn *et al.*, 2004) and from the JACEE emulsion chamber (Asakimori *et al.*, 1998) are shown. Data of the RUNJOB emulsion chamber (Apanasenko *et al.*, 2001; Furukawa *et al.*, 2003) are shown for five groups of nuclei (protons and helium, CNO, Ne-Si and Fe) above 1000 GeV. The fits to CNO, Ne-Si and Fe are normalized at 10.6 GeV/nucleon to data of Engelmann *et al.*, 1990.

Uncorrelated fluxes in the atmosphere. The simplest physical example to illustrate the solution of Eq. 22 is to calculate the vertical spectrum of nucleons as a function of depth in the atmosphere. Nucleons are stable compared to the transit time through the atmosphere, so only losses due to interactions are important in the cascade equation 19. In the approximation of scaling, the dimensionless distribution $F_{N \to N}(E, E') \to F(\xi)$ as in Eq. 9, with $\xi = E/E'$. Eq. 19 becomes

$$\frac{\mathrm{d}N(E, X)}{\mathrm{d}X} = -\frac{N(E, X)}{\lambda} + \frac{1}{\lambda_N} \int_0^1 N\left(\frac{E}{\xi}, X\right) F(\xi) \frac{\mathrm{d}\xi}{\xi^2}. \quad (23)$$

The dependence on energy and depth can be factorized, and the solution of Eq. 23 is

$$N(E, X) = K e^{-X/\Lambda_N} \times E^{-\alpha}, \quad (24)$$

where the boundary condition 22 is satisfied if $K = 1.7 \times 10^4$ and $\alpha = 2.7$. The attenuation length is related to the interaction length λ by

$$\Lambda_N = \frac{\lambda_n}{1 - Z_{NN}}, \quad (25)$$

where $Z_{NN} = \int_0^1 \xi^{\alpha-2} F_{N \to N}(\xi) \, \mathrm{d}\xi \approx 0.3$ is the spectrum-weighted moment for nucleons, analogous to Eq. 12 for pions.

The solution outlined above has several obvious approximations such as neglect of nucleon-anti-nucleon production and neglect of energy-dependence

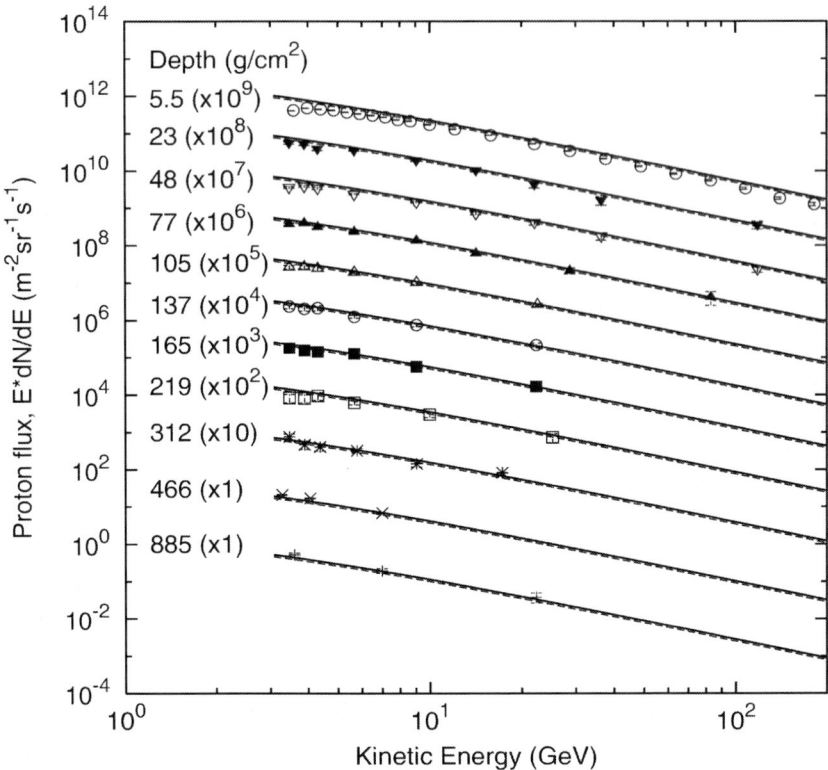

Figure 3.   The flux of protons. The data are from the measurements of Mocchiutti, 2003 with the CAPRICE detector.

of the cross sections, but it nevertheless gives a reasonable representation of measurements of the spectrum of protons at various atmospheric depths, as shown in Fig. 3. For comparison with measurements of protons, the solution of Eq. 24 for all nucleons must be modified to remove neutrons, which increase slightly as a fraction of the total flux of nucleons with increasing depth in the atmosphere. The correction, as described in Gaisser, 1990, is included in the calculations shown in Fig. 3.

Another benchmark measurement of secondary cosmic rays in the atmosphere is the flux of muons. The main source of muons is from decay of charged pions. There is also a small contribution from decay of charged kaons, which becomes somewhat more important at high energy. At very high energy the muon energy spectrum becomes one power steeper than the parent spectrum of nucleons as a consequence of the extra power of E_π in the ratio λ_π/d_π, which represents the decreasing probability of decay relative to re-interaction

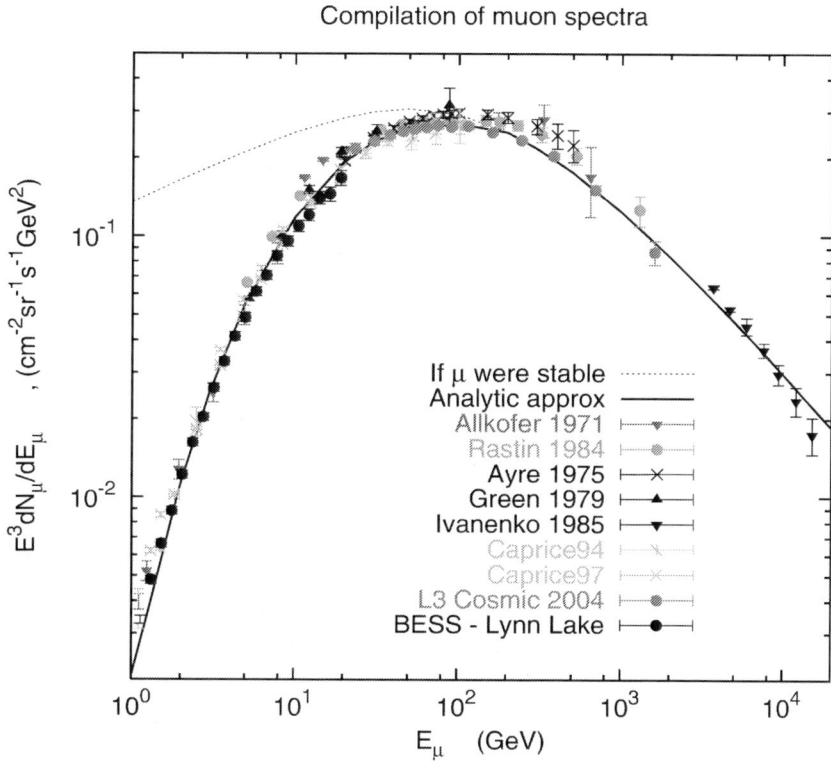

Figure 4. Summary of measurements of the vertical muon intensity at the ground. The solid line shows an analytic calculation Gaisser, 2002. The dotted line shows the spectrum in the absence of decay and energy loss, or equivalently the muon production spectrum integrated over the atmosphere.

for charged pions at high energy (see Eqs. 19 and 20). For $E < \epsilon_\pi$ essentially all pions decay, and the muon production spectrum has the same power behavior as the parent pion and grandparent nucleon spectrum ($\alpha \approx 2.7$). At low energy, however, muon energy-loss and decay become important, and the muon spectrum at the ground falls increasingly below the production spectrum. To account for all the complications one generally resorts to Monte Carlo calculations. However, analytic approximations (Lipari, 1993) of these effects are also possible. The full line in fig. 4 shows one such calculation (Gaisser, 2002), which uses as input the simple power-law primary spectrum of Eq. 22. This simple result compares relatively well to various measurements in several energy ranges. The recent data are from CAPRICE (Kremer *et al.*, 1999), L3-Cosmic (Achard *et al.*, 2004) and BESS (Motoki *et al.*, 2003). One can also see the level of relative systematic uncertainties between different measurements.

Although I will not discuss the subject here, The most important secondary cosmic-ray flux is the atmospheric neutrino beam because of the discovery of neutrino oscillations by Super-Kamiokande (Fukuda *et* al., 1998). The experimental situation is reviewed by Kajita & Totsuka, 2001 and Jung *et* al., 2001, and the calculations by Gaisser & Honda, 2002.

Air Showers.   Above about 100 TeV cascades generated by individual primary nuclei have a big enough footprint deep in the atmosphere to trigger an array of widely spaced detectors on the ground. The threshold energy may be somewhat lower for closely spaced detectors, especially at high altitude. The threshold also may be made much higher by separating the detectors by a large distance. Examples of the latter are the Akeno Giant Air Shower Array (AGASA) (Takeda *et* al., 2003) and the surface detector of the Auger Project (Watson *et* al., 2004). Such ground arrays work by looking for coincidences in an appropriate time window, then reconstructing the primary direction and energy from the timing pattern of the hits and the size of the signals in the detectors. There are large fluctuations from shower to shower which complicate the interpretation of the data. The air shower technique is used at very high energy where the flux is too low to accumulate meaningful statistics with detectors carried aloft by balloons or spacecraft. The dividing line at present is approximately 100 TeV.

Because of the complicated cascade of interactions that intervenes between the primary cosmic-ray nucleus incident on the atmosphere and the sparse data on the ground, Monte Carlo simulations are used to interpret the data. The other important reason for the necessity of a Monte Carlo generation of showers is that the detectors only sample a tiny fraction of the particles in the shower. Simulation of the response of an air shower detector to showers is therefore crucial. The standard, fully stochastic, four-dimensional air shower generator is CORSIKA (Heck & Knapp, 2003). A cascade generator of similar scope and design is AIRES (Sciutto, 2001). A fast, one-dimensional cascade generator (Alvarez-Muñiz *et* al., 2002) that uses libraries of pre-generated subshowers at intermediate energies inside cascades is useful for analysis of ultra-high energy showers measured by fluorescence detectors for which knowledge of the lateral distribution is less important. The three-dimensional hybrid generator SENECA (Drescher & Farrar, 2003) uses stochastic Monte Carlo methods for the high-energy part of the shower and at the detector level, but saves time by numerically integrating the cascade equations for intermediate energies. The FLUKA program (Fassò *et* al., 2001) is a general code for transport and interaction of particles through detectors of various types, including a layered representation of the atmosphere. The FLUKA interaction model (Battistoni *et* al., 2004) is built into the code and cannot be replaced by a different event generator.

There is a variety of hadronic event generators on the market (Bopp *et al.*, 2004, Kalmykov *et al.*, 1997, Engel *et al.*, 2001, Werner, 1993, Bossard *et al.*, 2001), which can be called by cascade programs like CORSIKA or AIRES to generate showers. Because the event generators are based on interpolations between measurements with accelerators at specific points in phase space and because the energies involved require extrapolations several orders of magnitude beyond those accessible with present accelerators, different hadronic event generators give different results for observables in air showers. We will see examples of this in the discussion of air showers below.

An air shower detector essentially uses the atmosphere as a calorimeter. Each shower dissipates a large fraction of its energy as it passes through the atmosphere, which is sampled in some way by the detector. It is therefore customary to plot the energy spectrum in the air shower regime by total energy per particle rather than by energy per nucleon as at low energy. In the lower energy regime the identity of each primary nucleus can be determined as it passes through the detector on a balloon or spacecraft. With air showers one has to depend on Monte Carlo simulations to relate what is measured to the primary energy and to the mass of the primary particle. The resulting energy assignments typically have uncertainties $\Delta E / E \sim$20-30%. Primary mass is often quoted as an average value for a sample of events in each energy bin, or at best as a relative fraction of a small number of groups of elements.

Acceleration

A detailed review of the theory of particle acceleration by astrophysical shocks is given in the lectures of Ostrowski, 2005. The main feature necessary for understanding the implications of air shower data for origin of high-energy cosmic rays is the concept of maximum energy. Diffusive, first-order shock acceleration works by virtue of the fact that particles gain an amount of energy $\Delta E \propto E$ at each cycle, where a cycle consists of a particle passing from the upstream (unshocked) region to the downstream region and back. At each cycle, there is a probability that the particle is lost downstream and does not return to the shock. Higher energy particles are those that remain longer in the vicinity of the shock and have time to achieve high energy.

After a time T the maximum energy achieved is

$$E_{max} \sim Ze\beta_s \times B \times T V_s, \qquad (26)$$

where $\beta_s = V_s/c$ refers to the velocity of the shock. This result is an upper limit in that it assumes a minimal diffusion length equal to the gyroradius of a particle of charge Ze in the magnetic fields behind and ahead of the shock. Using numbers typical of Type II supernovae exploding in the average interstellar medium gives $E_{max} \sim Z \times 100$ TeV (Lagage & Cesarsky, 1983). More recent estimates give a maximum energy larger by as much as an order of magnitude or more for some types of supernovae (Berezhko, 1996).

The nuclear charge, Z, appears in Eq. 26 because acceleration depends on the interaction of the particles being accelerated with the moving magnetic fields. Particles with the same gyroradius behave in the same way. Thus the appropriate variable to characterize acceleration is magnetic rigidity, $R = pc/Ze \approx E_{tot}/Ze$, where p is the total momentum of the particle. Diffusive propagation also depends on magnetic fields and hence on rigidity. For both acceleration and propagation, therefore, if there is a feature characterized by a critical rigidity, R^*, then the corresponding critical energy per particle is $E^* = Z \times R^*$.

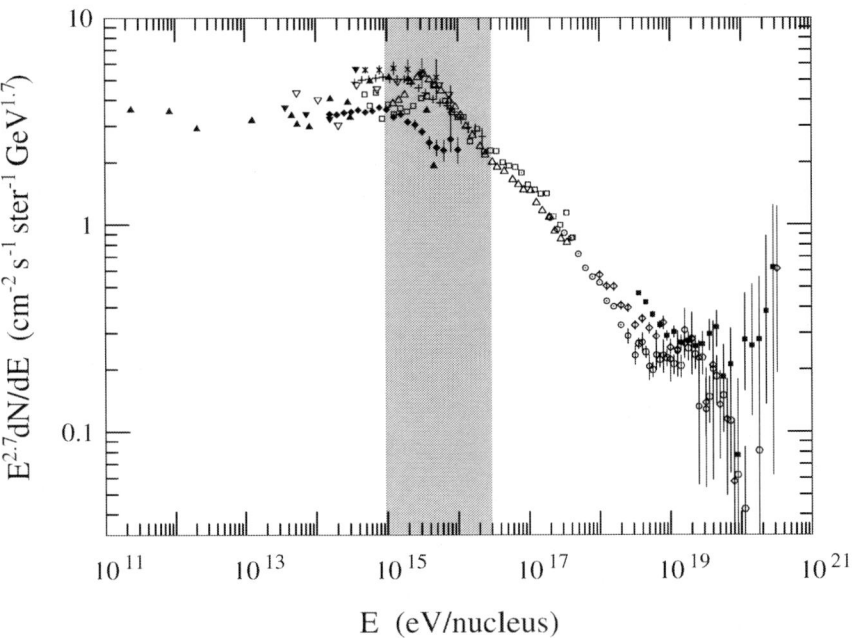

Figure 5. High-energy cosmic-ray spectrum. References to the data are given in (Gaisser & Stanev, 2004). The shaded region indicates a factor of 30 in total energy (see text).

The anatomy of the cosmic-ray spectrum

The knee of the spectrum is the steepening that occurs above 10^{15} eV, as shown in Fig. 5, while the ankle is the hardening around 3×10^{18} eV. One possibility is that the knee is associated with the upper limit of acceleration by galactic supernovae, while the ankle is associated with the onset of an extra-galactic population that is less intense but has a harder spectrum that dominates at sufficiently high energy. The general idea that the knee may signal the end of the population of particles produced in the Galaxy is an old one that I will trace in the next two subsections.

The knee

If the knee is a consequence of galactic cosmic accelerators reaching their limiting energy, then there are consequences for energy-dependence of the composition that can be used to check the idea. This follows from the form of Eq. 26. Consider first the simplest case in which all galactic accelerators are identical. Then $E_{max} = Z \times R^*$, where R^* characterizes the maximum rigidity. When the particles are classified by total energy per nucleus, protons will cut off first at $E_{max} = e R^*$, helium at $E_{max} = 2 e R^*$, etc. Peters, 1961 described this cycle of composition change and pointed out the consequences for composition in a plot like that reproduced here as Fig. 6. Since the abundant elements from protons to the iron group cover a factor of 30 in Z, the "Peters cycle" should occupy a similar range of total energy.

Because the observed spectrum does not abruptly stop, Peters hypothesized a new population of particles coming in with a somewhat harder spectrum, as indicated by the line in Fig. 6. Comparison with the data in Fig. 5 shows that reality is more complicated. The shaded area indicates the factor of 30 for a Peters cycle assuming $R^* = 10^{15}$ eV. What is observed is that the steepened spectrum above the knee continues on smoothly for more than an order of magnitude in energy before any sign of a hardening. Even postulating a significant contribution from elements heavier than iron (up to uranium, Hörandel, 2004) cannot explain the smooth continuation all the way up to the ankle.

One possibility is that most galactic accelerators cut off around a rigidity of perhaps 10^{15} eV, but a few accelerate particles to much higher energy and account for the region between the knee and the ankle (Erlykin & Wolfendale, 2001). This scenario would be a generalization of Peters' model. Its signature would be a sequence of composition cycles alternating between light and heavy dominance as the different components from each source cut off. As emphasized by Axford, 1994, however, the problem with this type of model is that it requires a fine-tuning of the high-energy spectra so that they rise to join smoothly at the knee then steepen to fit the data to $\sim 10^{18}$ eV. As a consequence, several models have been proposed in which the lower-energy accelerators ($E < 10^{15}$ eV) inject seed particles into another process that accelerates them to higher energy. In this way the spectrum above the knee is naturally continuous with the lower energy region. One such possibility is acceleration by interaction with multiple supernova shocks in a cluster of supernovae (Ax-

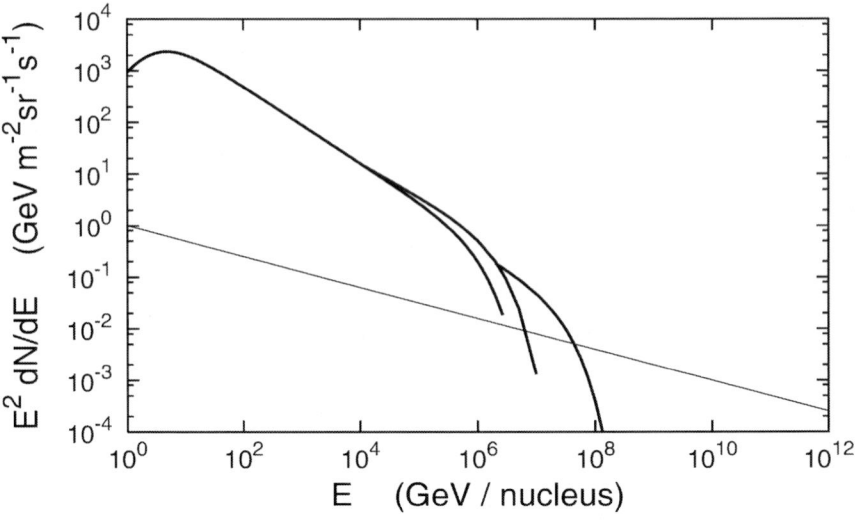

Figure 6.   Recreation of a figure from Peters, 1961 showing a sequence of cutoffs for ions with successively higher charge when classified by total energy per particle. The line shows a hypothetical second component with a harder spectrum than the lower energy component. (See text for discussion).

ford, 1994). Another possibility (Völk & Zirakashvili, 2003) is acceleration by a termination shock in the galactic wind (Jokipii & Morfill, 1987).

The fine tuning problem (i.e. to achieve a smooth spectrum with a sequence of sources with different maxima) was actually clearly recognized by Peters, 1961 in his original statement of this idea. He correctly pointed out, however, that since the cutoff is a function of rigidity while the events are classified by a quantity close (but not equal) to total energy, the underlying discontinuities are smoothed out to some extent. An interesting question to ask in this context is what power at the source would be required to fill in the spectrum from the knee to the ankle. The answer depends on what is assumed for the spectrum of the sources and the energy dependence of propagation in this energy region. Reasonable assumptions (e.g. $Q(E) \propto E^{-2}$ and $\tau_{esc} \propto E^{-\delta}$ with $\delta \approx 0.3$) lead to an estimate of $\sim 2 \times 10^{39}$ erg/s, less than 10% of the total power requirement for all galactic cosmic-rays. For comparison, the micro-quasar SS433 at 3 kpc distance has a jet power estimated as 10^{39} erg/s (Distefano *et al.*, 2002).

Another possibility is that the steepening of the spectrum at the knee is a result of a change in properties of diffusion in the interstellar medium such that above a certain critical rigidity the characteristic propagation time τ_{esc} decreases more rapidly with energy. If the underlying acceleration process were featureless, then the relative composition as a function of total energy per particle would change smoothly, with the proton spectrum steepening first by 0.3, followed by successively heavier nuclei. It is interesting that this possibility was also explicitly recognized by Peters, 1961.

A good understanding of the composition would go a long way toward clarifying what is going on in the knee region and beyond. A recent summary of direct measurements of various nuclei shows no sign of a rigidity-dependent composition change up to the highest energies accessible ($\sim 10^{14}$ eV/nucleus) (Battiston, 2004). The change associated with the knee is in the air shower regime. Because of the indirect nature of EAS measurements, however, the composition is difficult to determine unambiguously. The composition has to be determined from measurements of ratios of different components of air showers at the ground. For example, a heavy nucleus like iron generates a shower with a higher ratio of muons to electrons than a proton shower of the same energy. Swordy *et al.*, 2002 reviewed all available measurements of the composition at the knee. A plot of mean log mass ($\langle \ln(A) \rangle$) (Ahrens *et al.*, 2004b) shows no clear pattern when all results are plotted together. The best indication at present comes from the KASCADE experiment (Roth, 2003), which shows clear evidence for a "Peters cycle", the systematic steepening first of hydrogen, then of helium, then CNO and finally the iron group. The transition occurs over an energy range from approximately 10^{15} eV to 3×10^{16} eV, as expected, but the experiment runs out of statistics by 10^{17} eV, so the data do not yet discriminate among the various possibilities for explaining the spectrum between the knee and the ankle.

The ankle

Above some sufficiently high energy it seems likely that the cosmic rays will be of extra-galactic origin. A proton of energy 10^{18} eV has a gyroradius of a kiloparsec in a typical galactic magnetic field, which is larger than the thickness of the disk of the Galaxy. Given constraints from observed isotropy of particles with $E \sim 10^{19}$ eV, where the corresponding proton gyroradius is comparable to the full extent of the Galaxy, the usual assumption is that particles above 10^{19} originate outside our galaxy. There is a suggestion of an anisotropy just around 10^{18} eV from the central regions of the galaxy (Hayashida *et al.*, 1999) that may be due to neutrons, which survive for a mean pathlength of ~ 10 kpc at this energy, and could therefore reach us from the galactic center. If so, this would suggest that at least some fraction of the cosmic rays around 10^{18} eV are still of galactic origin. An interesting discussion of possible implications for correlated production of antineutrinos is given by Anchordoqui *et al.*, 2004a. Crocker *et al.*, (2004) relate the possible neutrons to gamma-rays detected from the galactic center. They give a detailed discussion of evidence from a variety of sources for cosmic-ray acceleration up to the ankle in the center of the Milky Way Galaxy. The questions of whether there is a transition from galactic to extragalactic cosmic rays and at what energy such a transition occurs are clearly of great interest.

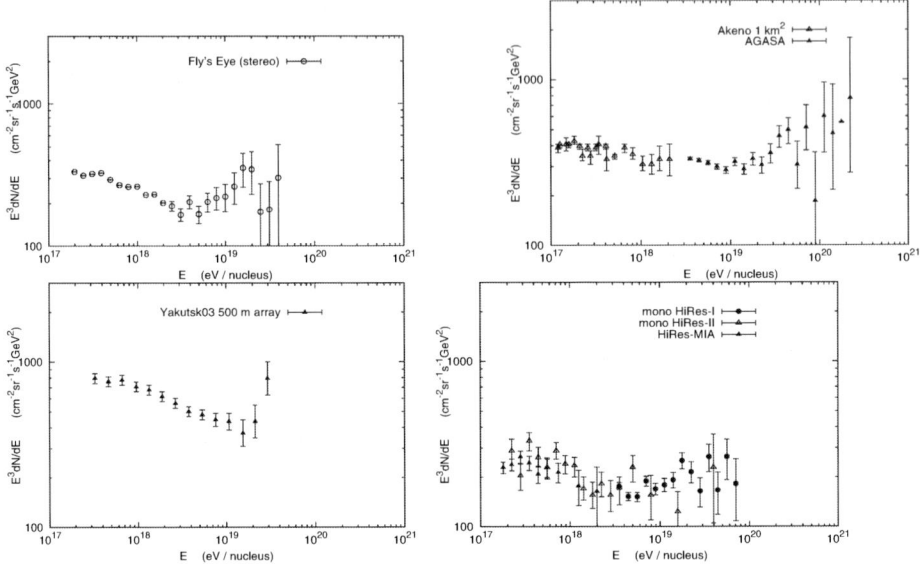

Figure 7.   Four measurements of the spectrum above 10^{**} eV.

As background for a discussion of the significance of the ankle and how it should be interpreted, it is helpful to look at the spectrum from individual measurements separately so that systematic differences between measurements do not obscure any details that may be present. Results from four groups are shown in Fig. 7: Fly's Eye stereo (Bird, *et al.*, 1993); Akeno (Nagano *et al.*, 1992) and AGASA (Takeda *et al.*, 2003); Yakutsk (Glushkov *et al.*, 2003); and the measurements made with Hi-Res (Abassi *et al.*, 2004 and Abu-Zayyad *et al.*, 2001). In all these measurements there is a suggestion of a steepening just below 10^{18} eV, which is sometimes referred to as the "second knee." The ankle appears as a saddle-like shape with its low point between 3×10^{18} and 10^{19} eV, depending on the experiment. One could fit the saddle with a final Peters cycle starting just below 10^{18} eV and an extragalactic component crossing as in Fig. 6 to contribute to the ankle in the overall spectrum (Bahcall & Waxman, 2003). Alternatively, it is possible to make a model in which extragalactic cosmic rays account for the entire observed flux down to 10^{18} eV (Berezinsky *et al.*, 2004) or even lower (Bergman, 2004). The difference lies in the assumptions made for the spectrum and cosmological evolution of the sources. I will return to this issue in the next section.

First it is interesting to ask what data on primary composition may tell us about the changing populations of particles above 10^{17} eV. In this energy range the composition is measured by the energy dependence of the position of shower maximum, X_{max}. An air shower consists of a superposition of electromagnetic cascades initiated by photons from decay of π^0 particles produced by hadronic interactions along the core of the shower as it passes through the atmosphere. Most of the energy of the shower is dissipated by ionization losses of the low-energy electrons and positrons in these subshowers. The composite shower reaches a maximum number of particles (typically 0.7 particles per GeV of primary energy) and then decreases as the individual photons fall below the critical energy for pair production. Because each nucleus of mass A and total energy E_0 essentially generates A subshowers each of energy E_0/A the depth of maximum depends on E_0/A. Since cascade penetration increases logarithmically with energy,

$$X_{max} = \lambda_{ER} \log(E_0/A) + C, \tag{27}$$

where λ_{ER} is a parameter (the "elongation rate") that depends on the underlying properties of hadronic interactions in the cascade.

Fig. 8 shows results of measurements with the Fly's Eye stereo detector (Bird, *et al.*, 1993) compared to measurements of HiRes (Abu-Zayyad *et al.*, 2001; Archbold & Sokolsky, 2003). A weak inference about composition can be made by comparing to the results of simulations, two of which are shown in the figure. Both calculations use CORSIKA (Heck & Knapp, 2003) with two different interaction models (Kalmykov *et al.*, 1997, Engel *et al.*, 2001). The measurement with the Fly's Eye Stereo detector suggests a transition from a large fraction of heavies below 10^{18} eV to a larger fraction of protons by 10^{19} eV (how much larger depending on which interaction model is chosen). The coincidence of the change of composition from heavier toward lighter) around $\sim 3 \times 10^{18}$ eV with the ankle feature in the Fly's Eye data at the same energy led to the suggestion of a transition from galactic to extragalactic cosmic rays as a possible explanation (Bird, *et al.*, 1993). This interpretation would favor a model like that of Bahcall & Waxman, 2003. The more recent HiRes data set, however, shows the transition from heavier toward lighter beginning at 10^{17} eV and complete by 10^{18} eV, consistent with the models of Berezinsky *et al.*, 2004 and Bergman, 2004.

Because of the uncertainties in the interaction models above accelerator energies coupled with statistical and systematic limitations of the experiments, the primary composition as a function of energy above 10^{17} eV remains an open question (Watson, 2004).

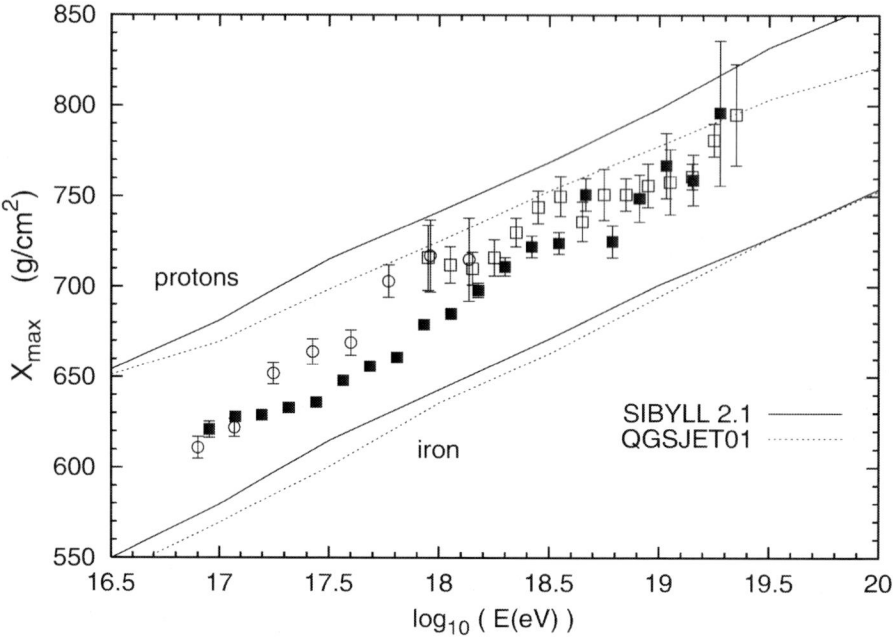

Figure 8.   Plot of data on mean depth of maximum vs energy. Filled squares are data from the stereo Fly's Eye (Bird, *et* al., 1993). Open symbols show the data of HiRes (squares) (Archbold & Sokolsky, 2003) and HiRes prototype with MIA (circles) (Abu-Zayyad *et* al., 2001). This figure also appears in Ref. (Gaisser & Stanev, 2005).

Highest energy cosmic rays

Protons lose energy by three loss processes during propagation through the cosmos. Red-shift losses (adiabatic losses due to expansion of the Universe), which apply to all particles, become important when the distance scales are comparable to the Hubble distance $\approx$ 4Gpc. Protons of sufficiently high energy also interact with the microwave background and lose energy to electron-positron pair production (for $E > \sim 10^{18}$ eV) and to photopion production (for $E > \sim 5 \times 10^{19}$ eV). The corresponding attenuation lengths (for reducing energy by a factor $1/e$) are $\lambda_{e^+ \, e^-} \sim 1$ Gpc and $\lambda_\pi \sim 15$ Mpc, respectively. The photo-pion process leads to the expectation of a suppression of the flux above 5×10^{19} eV unless the sources are within a few tens of Mpc. The suppression is referred to as the GZK cutoff (or GZK feature) in recognition of the authors of the two papers, Greisen, 1966 and Zatsepin & Kuz'min, 1966 who first pointed out the effect shortly after the discovery of the microwave background radiation.

The actual shape of the spectrum at Earth after accounting for these three loss processes depends on what is assumed

- for the spatial distribution of sources,

- for the spectrum of accelerated particles at the sources, and

- for the possible evolution of activity of the sources on cosmological time scales.

A classic calculation is that of Berezinsky & Grigorieva, 1988, in which the energy-loss equation is integrated numerically. This approach neglects effects of fluctuations, which may be noticeable in certain circumstances. A recent example of a Monte Carlo propagation calculation of cosmological propagation is by Stanev *et al.*, 2000, which contains comparisons with other calculations. Figure 9 from Abassi *et al.*, 2002 shows an example of a calculated cosmologically evolved spectrum compared to data of HiRes (Abassi *et al.*, 2004) and AGASA (Takeda *et al.*, 2003).

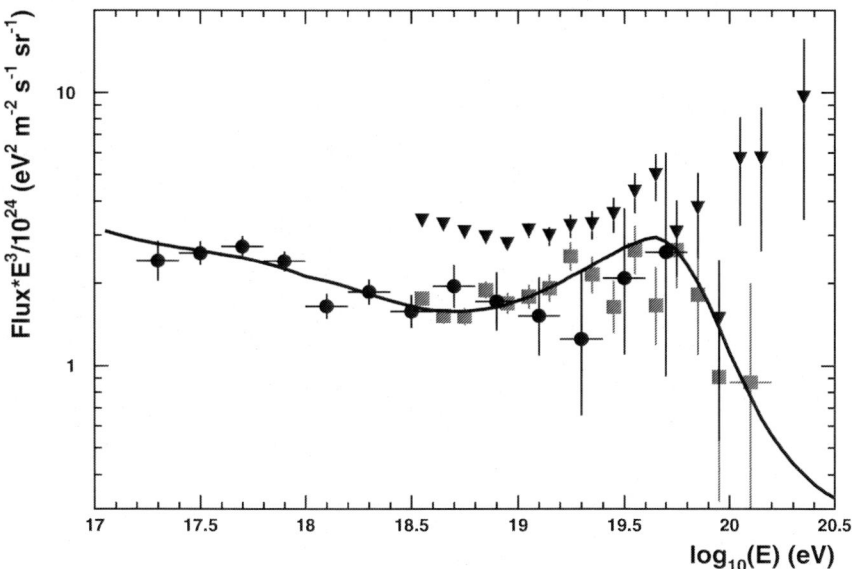

Figure 9. The ultra-high-energy cosmic-ray spectrum from a paper by Abassi *et al.*, 2002. The points with the fitted curve are from the HiRes fluorescence detectors (Abassi *et al.*, 2004), while the higher set of points are the AGASA spectrum (Takeda *et al.*, 2003). See text for a discussion of the curve.

The calculation illustrated in Fig. 9 is for a uniform distribution of sources out to large red shifts. The "pile-up" (which is amplified by plotting $E^3 \times$ differential flux) is populated by particles that have fallen below the GZK energy. The saddle or ankle feature on the $E^3 dN / dE$ plot is a consequence of energy losses to pair production. The data below 10^{18} eV have been fit in this example by adding an assumed galactic contribution, which is not shown separately here (but see Bergman, 2004 for a discussion). As pointed out long ago by Hillas, 1968 (see also Hillas, 1974), the degree to which the extragalactic flux contributes at low energy (for example below 10^{18} eV) can be adjusted by making different assumptions about increased activity at large redshift, z > 1. Such high redshift sources do not contribute at high energy because of adiabatic losses, but they can contribute at low energy if they were sufficiently abundant. On the other hand, it is also possible to construct a model in which the extragalactic sources only contribute significantly at very high energy (e.g. above 10^{19} eV), as in Bahcall & Waxman, 2003.

The question of the energy above which the extragalactic component accounts for all of the observed spectrum is important because it is directly related to the amount of power needed to supply the extragalactic component. An estimate of the power needed to supply the extragalactic cosmic rays is obtained by the replacements $\tau_{esc} \rightarrow \tau_H$ and $\phi(E) \rightarrow \phi_{EG}(E)$ in Eq. 6, where $\tau_H \approx 1.4 \times 10^{10}$ yrs is the Hubble time and ϕ_{EG} is the extragalactic component of the total cosmic-ray flux. The result depends on where the EG-component is normalized to the total flux and on how it is extrapolated to low energy. A minimal estimate follows from taking the hardest likely spectral shape (differential index $\alpha = -2$) and normalizing at 10^{19} eV. This leads to a total power requirement of $P_{EG} = 10^{37}$ erg/Mpc^3/s. If the rate of GRBs is 1000 per year, then they would need to produce 2×10^{52} erg/burst in accelerated cosmic rays to satisfy this requirement. The density of AGNs is $\sim 10^{-7} / \mathrm{Mpc}^3$ (Peebles, 1993), so the corresponding requirement would be 10^{44} erg/s per AGN in accelerated particles. Normalizing the extragalactic component at 10^{18} eV increases the power requirement by a factor of 10, and assuming a softer spectral index increases it even more. These are very rough estimates just meant to illustrate the relation between the intensity of extragalactic cosmic rays and the power needed to produce them. A proper treatment requires an integration over redshift, accounting for the spatial distribution of sources and their evolution over cosmic time (Stanev, 2003).

In addition to being able to satisfy the power requirement, the sources also have to be able to accelerate particles to $\sim 10^{20}$ eV. The requirements on size and magnetic field in the sources are essentially given by setting $E_{\mathrm{max}} = 10^{20}$ eV in Eq. 26 and solving for the product of $B \times T V_s$ (or $B \times R$). The resulting constraint on the sources is the famous plot of Hillas, 1984. Sites with a sufficiently large $B \times R$ included active galactic nuclei (AGN) (Berezinsky *et* al., 2004) and the termination shocks of giant radio galaxies (Biermann & Strittmatter, 1987). Since 1984 two more potential candidates have emerged, jets of gamma-ray bursts (GRB Waxman, 1995, Vietri, 1995) and magnetars.

Neutrino telescopes may make a decisive contribution to identifying ultra-high energy cosmic accelerators by finding point sources of TeV neutrinos and associating them with known objects, for example with gamma-ray sources such as AGN or GRB. Even the detection of a diffuse flux of high-energy, extraterrestrial neutrinos, or the determination of good limits can constrain models of cosmic-ray origin under the assumption that comparable amounts of energy go into cosmic-rays and into neutrinos produced by the cosmic-rays at the sources (Anchordoqui *et* al., 2004b).

As is well-known, the AGASA measurement (Takeda *et* al., 2003) shows the spectrum continuing beyond the GZK energy without a suppression, while the HiRes measurement (Abassi *et* al., 2004) is consistent with the GZK shape. The discrepancy between the two experiments is not as great as it appears in Fig. 9. The difference is amplified quadratically when the differential spectrum is multiplied by E^3 in the plot. A 20-30% shift in energy assignment (downward for AGASA or upward for HiRes) brings the two data sets into agreement below 5×10^{19} eV. A quantitative estimate of the statistical significance of the difference above 5×10^{19} eV is ambiguous because it depends on what is assumed for the true spectrum. The difference is generally considered to be less than three sigma, however (Olinto, 2004). One of the first goals of the Auger project (Klages, 2005) is to resolve the question of whether the spectrum extends significantly beyond the GZK energy. If not, then higher statistics accumulated by Auger may be used to constrain models of extra-galactic sources by making precise measurements both above and below the GZK energy. If the spectrum does continue beyond the cutoff then identification of specific, cosmologically nearby sources from the directions of the ultra-high energy events would be likely.

Conclusion and outlook

While the questions surrounding the highest energy particles are clearly of the greatest importance, there are several important open questions at all energies. One is the transition, if it exists, from galactic to extra-galactic cosmic rays. Proposed experiments associated with the telescope array (Arai *et* al., 2003) may be optimized to provide good coverage down to 10^{17} eV (Thompson, 2004).

Understanding the knee of the spectrum remains an outstanding problem in cosmic-ray astrophysics. KASCADE-Grande (Haungs *et* al., 2003) will cover the energy range from below the knee to 10^{18} eV with a multi-component air shower array at sea level. The IceCube detector at the South Pole (Ahrens *et* al., 2004a) will have a kilometer square surface component, IceTop (Gaisser *et al.*, 2003; Stanev, *et al.*, 2005), 1.4 km above the top of the cubic kilometer neutrino telescope. The whole constitutes a novel, large three-dimensional air shower detector with coverage of the cosmic-ray spectrum from below the knee to 10^{18} eV. In addition to its main mission of neutrino astronomy, Ice-Cube therefore also has the potential to make important contributions to related cosmic-ray physics. The high altitude of the surface (9300 m.a.s.l.) allows good primary energy resolution with the possibility of determining relative importance of the primary mass groups from the ratio of size at the surface to muon signal in the deep detector. With a coverage extending to 10^{18} eV, these detectors also have the potential to clarify the location of a transition to cosmic-rays of extragalactic origin.

While the knee of the spectrum remains in the realm of air shower experiments for the time being, there is much activity aimed at extending direct measurements of the primary spectrum and composition to reach the knee from below. The detectors include Advanced Thin Ionization Calorimeter (ATIC, Wefel *et al.*, 2003); Transition Radiation Array for Cosmic Energetic Radiation (TRACER, Müller, 2005); Cosmic Ray Energetics and Mass (CREAM, Seo *et al.*, 2003) and Advanced Cosmic-ray Composition Experiment for Space Station (ACCESS, http://www.atic.umd.edu/access.html). While ACCESS is to be flown in space, ATIC, TRACER and CREAM all take advantage of NASA's long-duration balloon program which is regularly achieving flights of two to four weeks in Antarctica. These experiments also have the opportunity to extend measurements of the ratio of secondary to primary nuclei to much higher energy and hence to resolve the related questions about the average source spectral index and about the isotropy of galactic cosmic rays.

References

Abassi, R.U. *et al.* (2002). astro-ph/0208301.

Abassi, R.U. *et al.* (2004). *Phys. Rev. Letters*, **92**:151101.

Abu-Zayyad, T. *et al.* (2001). *Ap.J.*, **557**:686.

Achard, P. *et al.* (2004). *Phys. Lett. B*, **598**:15.

Ahn H. for the ATIC Collaboration (2004). COSPAR (Paris) talk.

Ahrens, J. *et al.* (2004a). *Astropart. Phys.*, **20**:507.

Ahrens, J. *et al.* (2004b). *Astropart. Phys.*, **21**:565.

Alcarez, J. *et al.* (2000a) *Phys. Lett. B*, **490**:27.

Alcarez, J. *et al.* (2000b) *Phys. Lett. B*, **494**:193.

Alvarez-Muñiz, J. *et al.* (2002). *Phys. Rev. D*, **66**:033011.

Allkofer, O.C., Carstensen, K. & Dau, W.D. (1971). *Phys. Lett. B*, **36**:425.

Anchordoqui, L.A., Goldberg, H., Halzen, F. & Weiler, T.J. (2004a). *Phys. Lett. B*, **593**:42.

Anchordoqui, L.A., Goldberg, H., Halzen, F. & Weiler, T.J. (2004b). hep-ph/0410003.

Apanasenko, A.V. *et al.* (2001). *Astropart. Phys.*, **16**:13.

Arai, Y. *et* al. (2003). in *Proc.* 28th Int. Cosmic Ray Conf. (Tsukuba, ed. T. Kajita *et* al., Universal Academy Press), **2**:1025.

Archbold, G. & Sokolsky, P. (2003). in *Proc.* 28th Int. Cosmic Ray Conf. (ed. T. Kajita *et* al., Universal Academy Press) **1**:405.

Asakimori, K. *et* al. (1998). *Ap.J.*, **502**:278.

Axford, W.I. (1994). *Ap.J.* Suppl., **90**:937.

Ayre, C.A. *et* al. (1975). *J.* Phys. G, **1**:584.

Bahcall, J.N. & Waxman, E. (2003). *Phys.* Lett. B, **556**:1.

Battiston, R. (2004). in *Frontiers of Cosmic Ray Science* (ed. T. Kajita *et* al., Universal Academy Press):229.

Battistoni, G., Forti, C., Ranft, J. & Roesler, S. (1997). *Astropart.* Phys., **7**:49.

Battistoni, G. *et* al. (2004). astro-ph/0412178

Berezhko, E.G. (1996). *Astropart.* Phys., **5**:367.

Berezhko, E.G. & Ellison, D.C. (1999). *Ap.J.*, **526**:385.

Berezhko, E.G. & Völk, H.J. (2000). *Astropart.* Phys., **14**:201.

Berezhko, E.G., Pühlhofer, G. & Völk, H.J. (2003). *Astron.* Astrophys., **400**:971.

Berezinsky, V.S. & Grigorieva, S.I. (1988). *Astron.* Astrophys., **199**:1.

Berezinsky, V., Gazizov, A. & Grigorieva, S. astro-ph/0410650.

Bergman, D. (2004). astro-ph/0407244.

Biermann, P. & Strittmatter, P.A. (1987). *Ap.J.*, **322**:643.

Bird, D.J. *et* al. (1993). *Phys.* Rev. Letters, **71**:3401.

Boezio, M *et* al. (1999). *Ap.J.*, **518**:457.

Bopp, F.W., Ranft, J., Engel,, R. & Roesler, S. (2004). astro-ph/0410027 and references therein.

Bossard, G. *et* al. (2001). *Phys.* Rev. D, **63**:054030.

Bottino, A. *et* al. (1998). *Phys.* Rev. D, **58**:123503.

Buckley, J.H. *et* al. (1998). *Astron.* Astrophys., **329**:639.

Crocker, R.M. *et* al. (2004). astro-ph/0408183.

Distefano, C., Guetta, D., Waxman, E. & Levinson, A. *Ap.J.*, **575**:378.

Drescher, H.-J. & Farrar, G. (2003). *Phys.* Rev. D, **67**:116001.

Drury, L.O'C., Aharonian, F.A. & Völk, H.J. (1994). *Astron.* Astrophys., **287**:959.

Drury, L.O'C. *et* al. (2001). *Ap.* Sci. Rev., **99**:329.

DuVernois, M.A. *et* al. (2001). *Ap.J.*, **559**:296.

Engel, R., Gaisser, T.K. & Stanev, Todor (2001). *Proc.* 27th Int. Cosmic Ray Conf. (ed. K.-H. Kampert, G. Heinzelmann & C. Spiering, Copernicus Gesellschaft, Hamburg) **2**:431.

Engelmann, J.J. *et* al. (1990). *Astron.* Astrophys., **233**:96.

Erlykin, A.D. & Wolfendale, A.W. (2001). *J.* Phys. G, **27**:1005.

Fassò, A., Ferrari, A., Ranft, J. & Sala, P.R. (2001). in *Proc.* MonteCarlo 2000, Lisbon (ed. A. Kling *et* al., Springer-Verlag, Berlin):955.

Fukuda, Y. *et* al. (1998). *Phys.* Rev. Letters, **81**:1562.

Furukawa, M. *et* al. (2003). in *Proc.* 28th Int. Cosmic Ray Conf. (Tsukuba, ed. T. Kajita *et* al., Universal Academy Press), **4**:1837.

Gaisser, T.K. (1990). *Cosmic Rays and Particle Physics* (Cambridge University Press).

Gaisser, T.K., Halzen, F. & Stanev, Todor (1995). *Phys.* Reports, **258**:173.

Gaisser, T.K. Protheroe, R.J. & Stanev, Todor (1998). *Ap.J.*, **492**:219.

Gaisser, T.K., Honda, M., Lipari, P. & Stanev, Todor (2001). *Proc.* 27th Int. Cosmic Ray Conf. (Hamburg) **5**:1643.

Gaisser, T.K. (2002). *Astropart. Phys.*, **16**:285.

Gaisser, T.K. & Honda, M. (2002). *Ann. Revs. Nucl. Part. Sci.*, **52**:153.

Gaisser, T.K. *et al.* (2003). in *Proc.* 28th Int. Cosmic Ray Conf. (Tsukuba, ed. T. Kajita *et al.*, Universal Academy Press), **2**:1117.

Gaisser, T.K. & Stanev, Todor (2004). in S. Eidelman *et al.*, Reviews of Particle Properties, Physics Letters B, **592**:228.

Gaisser, T.K. & Stanev, Todor (2005). *Nucl. Phys.* A:(to be published).

Glushkov, A.V. *et al.* (2003). in *Proc.* 28th Int. Cosmic Ray Conf. (Tsukuba, ed. T. Kajita *et al.*, Universal Academy Press), **1**:389.

Goodman, Jordan (2005). This volume.

Green, P.J. *et al.* (1979). *Phys. Rev.* D, **20**:1598.

Greisen, K. (1966). *Phys. Rev. Letters*, **16**:748.

Hayashida, N. *et al.* (1999). *Astropart. Phys.*, **10**:303.

Heck, D. & Knapp, J. (2003). *Extensive Air Shower simulation with CORSIKA: A user's Guide*, (V 6.020, FZK Report, February 18, 2003). http://www-ik.fzk.de/corsika/

Heinbach, U. & Simon, M. (1995). *Ap.J.*, **441**:209.

Hillas, A.M. (1968). *Canadian J. Phys.*, **46**:S623.

Hillas, A.M. (1974). *Phil. Trans. R. Soc. Lond.* A, **277**:413.

Hillas, A.M. (1984). *Ann. Revs. Astron. Astrophys.*, **22**:425.

Hörandel, J.R. (2004). *Astropart. Phys.*, **21**:241.

Haungs, A. *et al.* (2003). in *Proc.* 28th Int. Cosmic Ray Conf. (Tsukuba, ed. T. Kajita *et al.*, Universal Academy Press), **2**:985.

Hunter, S.D. *et al.* (1997). *Ap.J.*, **481**:205.

Ivanenko, I.P. *et al.* (1985). in *Proc.* 19th Int. Cosmic Ray Conf., La Jolla (NASA Conf. Publ. No 2376) **8**:21

Jokipii, J.R. & Morfill, G.E. (1987). *Ap.J.*, **312**:170.

Jones, F.C., Lukasiak, A., Ptuskin, V. & Webber, W. (2001). *Ap.J.*, **547**:246.

Jung, C.K., Kajita, T., Mann, T. & McGrew, C. (2001). *Ann. Rev. Nucl. Part. Sci.*, **51**:451.

Kajita, T. & Totsuka, Y. (2001). *Revs. Mod. Phys.*, **73**:85.

Kalmykov, N.N., Ostapchenko, S.S. & Pavlov, A.I. (1997). *Nucl. Phys.* B (Proc. Suppl.), **52**:17.

Klages, H. (2005). This volume.

Kremer, J. *et al.* (1999). *Phys. Rev. Letters*, **83**:4241.

Lagage, P.O. & Cesarsky, C.J. (1983). *Astron. Astrophys.*, **118**:223 and **125**:249.

Lipari, P. (1993). *Astropart. Phys.*, **1**:195.

Migneco, E. (2005). This volume.

Mocchiutti, E. (2003). *Atmospheric and Interstellar Cosmic Rays Measured with the CAPRICE98 Experiment* (Thesis, KTH, Stockholm)

Montaruli, T. (2003). astro-ph/0312558 (*Nucl. Phys.* B (Suppl.), to be published).

Moskalenko, I.V. & Strong, A.W. (1998). *Ap.J.*, **493**:694.

Moskalenko, I.V., Strong, A.W., Ormes, J.F. & Potgieter, M.S. (2002). *Ap.J.*, **565**:280.

Motoki, M. *et al.* (2003). *Astropart. Phys.*, **19**:113.

Müller, D. (2005). This volume and http://tracer.uchicago.edu

Nagano, M. *et al.* (1992). *J. Phys.* G, **18**:423.

Olinto, A. (2004). in *Frontiers of Cosmic Ray Science* (ed. T. Kajita *et al.*, Universal Academy Press):299.

Orito, S. *et* al. (2000). *Phys.* Rev. Letters, **84**:1078.

Ostrowski, M. (2005). This volume.

Peebles, P.J.E. (1993). *Principles of Physical Cosmology* (Princeton University Press).

Peters, B. (1961). *N*uovo Cimento, **XXII**:800.

Rastin, R.C. (1984). *J.* Phys. G, **10**:1609.

Roth, M. *et* al. (2003). in *Proc.* 28th Int. Cosmic Ray Conf. (Tsukuba, ed. T. Kajita *et* al., Universal Academy Press), **1**:139.

Sanuki, T. *et* al. (2000). *Ap.*J., **545**:1135.

Sciutto, S.J. (2001). *Proc.* 27th Int. Cosmic Ray Conf. (ed. K.-H. Kampert, G. Heinzelmann & C. Spiering, Copernicus Gesellschaft, Hamburg) **1**:237. http://www.fisica.unlp.edu.ar/auger/aires/

Seckel, D. Stanev Todor & Gaisser, T.K. (1991). *Ap.*J., **382**:651.

Seo, E.S. & Ptuskin, V.S. (1994). *Ap.*J., **431**:705.

Seo, E.-S. *et* al. (2003). in *Proc.* 28th Int. Cosmic Ray Conf. (Tsukuba, ed. T. Kajita *et* al., Universal Academy Press), **4**:2101. http://cosmicray.umd.edu/cream/cream.html

Shapiro, Maurice M. & Silberberg, Rein (1970). *Ann.* Revs. Nucl. Sci., **20**:323.

Stanev, Todor *et* al. (2000). *Phys.* Rev. D, **62**:093005.

Stanev, Todor (2003). *H*igh Energy Cosmic Rays (Springer Verlag, Berlin).

Stanev, T. for the IceCube Collaboration (2005). astro-ph/0501046.

Stecker, F.W. (1971). *C*osmic Gamma Rays (NASA Scientific and Technical Information Office, NASA SP-249).

Swordy, S. *et* al. (2002). *A*stropart. Phys., **18**:129.

Takeda, M. *et* al. (2003). *A*stropart. Phys., **19**:447.

Thompson, G. (2004). Talk given at Leeds Workshop on Ultra-High Energy Cosmic Rays, 22 July.

Vietri, M. (1995). *Ap.*J., **453**:883.

Völk, H.J. & Zirakashvili, V.N. (2003). in *Proc.* 28th Int. Cosmic Ray Conf. (Tsukuba, ed. T. Kajita *et* al., Universal Academy Press), **4**:2031.

Watson, A.A. *et* al. (2004). Nucl. Inst. Methods A, **523**:50.

Watson, A.A. (2004). astro-ph/0410514.

Waxman, E. (1995). *Phys.* Rev. Letters, **75**:386.

Wefel, John *et* al. (2003). in *Proc.* 28th Int. Cosmic Ray Conf. (Tsukuba, ed. T. Kajita *et* al., Universal Academy Press), **4**:1849.

Wefel, John (2005). This volume.

Werner, K. (1993). *Phys.* Rep., **232**:87.

Zatsepin, G.T. & Kuz'min, V.A. (1966). *J*ETP Letters, **4**:78.

CONSTRUCTING THE COSMOS, FROM SUNBOATS TO SUPERCLUSTERS

Virginia Trimble
Department of Physics and Astronomy, University of Maryland,
College Park, MD 20742, USA
vtrimble@uci.edu

Abstract   Our current understanding of the nature and origins of the large scale structure
of the Universe has arison from the very long sequence of earlier understanding,
with size and time scales and complexity increasing more or less monotonically.
This history is explored beginning with ideas from pre-literate societies and ending with some fine tuning of the current "consensus cosmology".

Introduction

Three coherent themes appear as we examine changing notions of what the
world as a whole is like: Expanding horizons, Hierarchial levels of structure,
and Increasing mediocrity.

Expanding horizons: To most pre-literate and immediately postliterate societies (not all), the world was the Earth and it was a size that could be circum-
navigated by humans in a lifetime and by a god in a day. Examples include
Apollo and his chariot, the traditional Indian structure of tortoises supported by
elephants supported by tortoises (etc.), and the ancient Egyptian earth god Geb
overarched by the sky goddess Nut, supported by the air god Shu. Around this,
the sun god Ra sailed once a day, changing between his day and night-boats at
the horizon, thus accounting for the slight slowing of the apparent motion of
the sun at the horizons, which we now attribute to differential refraction in the
atmosphere.

In contrast, the modern universe is at least 10^{10} LY (Light Year) in extent
by direct observation, arguably at least 10^{15} LY in extent from the absence of
gradients, and (because once inflation gets started it is not so very anxious to
stop) could be $10^{10^{\cdot\cdot\cdot}}$ LY in extent, the size at which replications occur, so
somewhere else you are also reading this paper, and yet somewhere else I am
writing it. Statements about recent cosmological ideas that are not otherwise
credited here come from a meeting "Concordant Cosmology and Beyond" held

33

M. M. Shapiro et al. (eds.), Neutrinos and Explosive Events in the Universe, 33–45.
© 2005 *Springer. Printed in the Netherlands.*

in Cambridge, UK during the first week of the Erice summer school. At least one respectable scientist thought each was likely to be correct, at least for the 30 minute duration of his talk. Yes, the Cambridge meeting, like the summer school, had only one senior woman speaker.

Time scales have expanded correspondingly. Most mythical worlds (that of Genesis, of the Australian aborigines and the North American Indians, though not that of the Asian Indians) had endured from 100 to 1000 generations since creation by some higher authority, with the expected future duration comparable. In contrast, the time since our universe was very different (hot and dense) is known to be 13.7 Gyr with considerable precision (Bennett et al. 2003). Its future life expectancy could be infinite and is, at minimum, at least 50 Gyr to a big crunch (if the negative pressure stuff decays away) or a big rip (if the negative pressure stuff takes over completely). Some brane world cosmologies and others (discussed by P. Steinhardt in Cambridge) are cyclic, with finite time between bangs but infinite total duration.

Hierarchical levels: Early, earth-centered worlds were widely held to be unique (Trimble 2004 for details and references). A heliocentric solar system soon suggested the possibility of other planetary systems orbiting other stars. That stars are genuinely clustered in some cases (and in bound pairs in others) was shown statistically by John Michell in 1767. Then came other galaxies like the Milky Way and the first grouping of galaxies, the Local Group (these both from Edwin Hubble in 1923-24 and 1936 respectively). All astronomers active in the field now recognize that galaxies are clustered and the clusters, often to be found in sheet-like and filamentary superclusters around extended voids, although until rather recently there were defenders of the idea that the distribution of galaxies in space is really quite homogeneous and the structures we see are due to differential interstellar absorption. In between, Fritz Zwicky (1961-68) long held the view that there were clusters but no superclusters, and from this deduced that the range of the gravitational force was not much more than 10 Mpc, and the mass of the graviton was 10^{-63} g.

Are there levels of structure and organization still larger than the 100 - 200 Mpc aggregates seen in modern extensive surveys of galaxy redshifts? DeVaucouleurs (1970) felt strongly that there should be, on the grounds that it would be very strange that he should be writing just as the largest structures had been mapped and just as estimates of the age of the universe stopped increasing with time. Strange, but, it seems, true, and the preponderence of current data strongly suggest that supercluster structure joins smoothly onto fluctuations in the X-ray background and those, in turn, join onto the very small fluctuations of the cosmic microwave background (Jones et al. 2004).

Decreasing centrality; increasing mediocrity: It gradually became clear over the centuries that we are at the center of neither the solar system nor the Milky Way Galaxy. Neither we, nor anyone else, can be at the center of a general relativistic universe, whether finite or infinite (as noted by Digges before 1600). In addition, there are enormous numbers of stars very much like the Sun and enormous numbers of galaxies very much like the Milky Way. At least 5-10% of sun-like stars have one or more planets, and it could be all of them have planets, though current search techniques can reveal only massive (Jupiter-like) planets in relatively short period orbits (Butler et al. 2003 and references therein). Are there other universes, in the sense of other four (or more) dimensional space-times with which we cannot communicate, even in principle? That there are is one of the implications of selfreproducing, eternal, chaotic inflation (Linde 1996) and of some of the others of the roughly 125 "flavors" of inflation (from anisotropic brane to TeV-scale hybrid) compiled by Paul Shellard. Uniqueness is also possible.

To what extent are these three themes played out? Additional levels seem to have been exhausted, i.e. we see the largest structures in the observable universe. The temporal and spatial extent of that universe hovers between absolutely enormous and infinite, and no observation on the near horizon can settle the issue. The two are, however, conceptually different. If merely "enormous", then a non-trivial portion is sampled by existing and future data; if "infinite" then our sample will always be a set of measure zero of the total, and so perhaps not at all representative. The same would be true if we are one of an infinite number of (finite) universes, though not if there are "only" the 104 of the Calabi-Yau manifolds mentioned by P. Steinhardt at Cambridge. As for mediocrity, no very earth-like planes have yet been found nor can they be with existing techniques employed from the ground (Trimble 2004 and other papers in the volume), but we know in principle what is required and how to recognize another Earth if it should turn up. A spectrum with the red absorption edge produced by chlorophyll (seen in integrated earth light reflected from the moon) is perhaps more than we can expect, or perhaps not. Observations that might reveal the existence of other universes are rather difficult to come by, but it is not unreasonable to look for signs of the kinds of inflation that lead us to expect them.

Despite these fairly coherent three themes, it will turn out not to be possible to tell the story of the discovery of the large scale structure and evolution of the cosmos in a single, linear tale, because, to take only one example, Charlier is looking for hierarchical clustering of galaxies and Lundmark is measuring the distances to a few nearby ones by the same time (1910-23) that Shapley, van Maanen, and others are denying the very existence of such galaxies.

OUTWARD STEPS

The following are brief verbal descriptions of major stages from earthsized and earth-centered to really big and centerless. Most can also be presented as images, and were, taken from standard texts (Jaki 1972, Whitney 1971, Struve and Zebergs 1967, Berendzen et al. 1976, Peebles 1993, Harrison 2000).

1. The size and shape of the Earth. Arguments for a spherical Earth were known to the Greeks, and its size measured by Eratosthenes (c. 200 BCE) by the classic method of examining shadow lengths at noon at two different latitudes whose linear separation was known. The accuracy of his answer depends on the somewhat uncertain length of the Roman stadium. Sphericity was sometimes forgotten, for instance in the drawings around 1150 CE by Hildegaard of Bingen (who also put hail and lightning further from the Earth than the inner planets and fixed stars). The overall structure was pineapple shaped. The earth was re-rounded by Thomas Aquinas in his circa 1250 CE synthesis of Greek philosophy (mostly Aristotle) and medieval church doctrine, and the round shape carried over into Protestantism by Martin Luther and others.

2. The distance to the moon. Hipparchus in 130 BCE or thereabouts used the curve of the terrestrial shadow across the moon during eclipses to estimate this in units of earth radii and got roughly the right answer. A geometrical method for determining the ratio of sun distance to moon distance failed, because the lunar orbit is not a circle.

3. Non-centrality of the Earth. Aristarchus and other Greeks considered this possibility and were aware that stellar parallax was a definitive test. They couldn't see it and so mostly opted for earth-centered. Copernicus, publishing almost post-humously in 1543, chose the sun as a more suitable center, though the absence of parallax distressed potential followers like Tycho (who invented his own compromise system in the late 1500s). Observational confirmation of "eppur se muovere" came only in 1729 with the measurement of aberration of starlight by James Bradley, though by then the scholarly community had long since been converted by the moons of Jupiter, the phases of Venus, and much else.

4. The orb of star. Copernicus put all his stars in a thin shell outside the orbit of Saturn, and not far outside. An infinite distribution was favored by Nicholas of Cusa (1450), Thomas Digges (1576), Giordano Bruno (before 1600), and Rene Descartes (1636). Reversion to a finite orb came from Kepler (1620) and Newton (1660), who, however, later considered an infinite distribution and worried about the gravitational equivalent of Olbers' Paradox. He expected Divine providence to take care of the stability of the total system, perhaps by occasionally sacrificing small portions of it to collapse. William Gilbert's 1603 stellar distribution was again declared infinite in his text, though drawn as finite (presumably owing to the high cost of 17th century paper and parchment). In addition, his stars were not all the same intrinsic brightness, as indeed real stars are not. Descartes's infinite universe was tessellated into vortices with squared-off corners, each dominated by a central star and with the potential for formation of planetary systems in the churning vortices. Otto von Guenicke (discoverer of terrestrial vacuums, with the help of an evacuated iron sphere and two horses) returned again to a finite starry cosmos surrounded by an infinite void.

5. *The shape of the Milky Way.* That white (milky!) band was obviously known to the Greeks, and its resolution into stars was achieved, at the latest, in 1609 by Galileo. Yet it is not until 1734 that Swedenborg provided a drawing with the stars other than spherically or uniformly distributed. He had in mind a sort of magnetic dipole shape (inspired by the earth's field, recognized by Gilbert in 1600) and envisioned an infinite number of other such systems in the universe. A later (1750) drawing by Thomas Wright of Durham is most often reproduced as the first version of the Milky Way as a disk. One version of the drawing looks quite modern (his edge-on galaxy), while the other (the galaxy face on) has sort of an eye-of-god appearance. In either case, he supposed there to be an infinite number of such systems. William Herschel, with a way of estimating stellar distances, was able to be somewhat quantitative about the axis ratio of the disk, about four to one (not actually as extreme as the truth), and also recorded the Cygnus rift as an uneven edge to the disk, which we now attribute to interstellar absorption of starlight. His best known drawing dates from 1785.

6. *The distances to the stars.* If the stars are suns, they must be enormously further away. John Michell, the first to be quantitative about this said millions of times in 1767. "At least," we would now say. William Herschel refined Michell's method to his "star gauging" and concluded that the disk was a kilo-light-year or so across. Firm numbers for stellar distances came only with the first measurements of parallax, almost simultaneously in 1838 by Bessel (of the functions), Struve (the middle one of a dynasty), and Henderson (working from South Africa and so able to choose Alpha Centauri for his campaign).

7. *Position of the Milky Way in the scheme of things.* From William Herschel to Jacobus Kapteyn (1920), nearly every authority had a Milky Way of stars with the sun at or very near its center, sandwiched in a finite "realm of the nebulae". William Huggins, who showed that many nebulae are truly diffuse gas and not unresolved star clusters, concurred around 1860, as did Harlow Shapley in 1920. The Third Earl of Rosse, with his homemade telescope (well, they all were in those days) was the first to resolve the spiral arms of a few of these nebulae, beginning with M51. Doubters seem to have been few, though Simon Newcomb (first president of the American Astronomical Society), who otherwise has rather bad press, asked rhetorically in 1901 whether we might be victims of some fallacy (like that afflicting Ptolemy) in thinking ourselves at the center of the Milky Way. He also asked some very modern-sounding questions about the extent and duration of the universe, the form and extent of the stars of large proper motion, which would have passed through Herschel's Milky Way in only a million years (and sorting that one out required

both an understanding of star formation and the discovery of dark matter!). Cornelius Easton in 1900 drew a perfectly charming spiral Milky Way with the sun solemnly at the center of all the circles, but the center of the spiral pattern far off to one side toward Cygnus. Several contemporaneous drawings (Eddington; Alfred Russel Wallace) have a heliocentric galaxy with a ring of stars outside the main ellipsoid. This is probably Gould's belt.

8. Non-centrality of the Sun. The Curtis-Shapley debate (Curtis 1921, Shapley 1921) centered on this and on the distance scale of the galaxy. Curtis said small, sun-centered while Shapley said big, with the sun something like 20 kpc from the center, based on his distance scale for globular clusters, derived in turn from the apparent brightness of the RR Lyrae stars in them. Shapley was pretty much the winner on this one, and he has been compared with Copernicus for moving us away from the center.

9. Some nebulae are actually other galaxies, island universes, and such. Herschel held at various times to both viewpoints – nebula independent, or nebulae part of the Milky Way. Michell in his 1767 paper had already said that if those fuzzy things were as big as the Milky Way, then the brightest stars in them should appear at V=13.8 (too faint for him to see). Herschel, Curtis, and Lundmark were "external galaxy people" up to about 1923. Shapley, van Maanen, and others were "everything inside" people. The issue was definitely settled by Edwin P. Hubble in 1923-24, when he found and measured light curves for Cepheid variables in NGC 6822, M31, and, in due course, other galaxies.

Recognition of the rotation of the Milky Way by Bertil Lindblad and Jan Oort (who is generally assigned the lion's share of the credit) belongs also to the mid 1920s.

10. Existence of interstellar dust and absorption. Somewhat confusingly, the stuff that was responsible for the misconceptions by Herschel, Kapteyn, and all, was not recognized until after the distance scale and other galaxy issues had been settled. The possibility of interstellar obscuration had been debated for decades; its existence was settled by Trumpler (1930) who noticed that distant star clusters looked fainter than they should if all were about the same size and light fell off only as $1/R^2$. A peacemaker, it seems, he drew a Milky Way centered where Shapley's globular clusters said it should be, but with a "Kapteyn universe" within the dusty disk, centered at the sun, and cut off where obscuration reached a few magnitudes at a couple of kpc away from us. It took some years for the community to re-evaluate the distance scales chosen by Hubble and Shapley in light of this obscuration, and the eventual effect was to make the galaxy rather smaller (8.5 vs. 20 kpc to the center) and other galaxies more distant (by a factor 7-10) than the pioneers had found.

11. Nebulae are paired and clustered. Non-randomness of fuzzy things in the sky was remarked upon by both the Herschels, and anyone with access to Messier's catalog could have seen it. His catalog of things that were not comets (his only real interest) contained 107 fuzzy objects (though only 6 were known to Halley in 1715 and the Herschels took the numbers about 103) and if you color code them for spirals, ellipticals, and star clusters, at least the Virgo and Leo clusters pop out at you and perhaps a couple of others. But they were not color-coded for Messier, and spiral arms belong to the 1850s, not the 1750s when he began collecting his non-comets. If you plot them all, the most conspicuous grouping is around Sagitarius and consists of globular clusters near the center of the Milky Way. After the general recognition of external galaxies, the existence of physical pairs was pointed out by Lundmark in 1932 (using the same statistical argument that John Michell had used to demonstrate pairing and clustering of stars; it is about the same as the one to figure out how many people you need in a room for two to be likely to have the same birthday). Lundmark's student Holmberg catalogued a number of pairs the year after Hubble (1936) pointed out that we live in a crowded region, which he dubbed the Local Group (the first group or cluster to be recognized). Hubble maintained that clusters, though the Local Group, Virgo, and Coma certainly existed, were rather rare, while Shapley and Bart Bok, also in the 1930s, claimed that most galaxies were clustered. The issue was resolved, again statistically, by J. Neyman (a well known statistician) and his student Elizabeth Scott at Berkeley in 1953. Clustering is the norm. Carl Charlier, in 1908 and 1920, was the last person to look at this issue before the nature of the nebulae had definitely been settled. He recognized clustering and superclustering and was a firm advocate of a fractal universe (as a solution to Olbers paradox among other virtues).

12. The redshift-distance relation. Lundmark in 1925 and H.P. Robertson and Georges Lemaitre in 1926-28 had attempted to correlate the galactic wavelength displacement measured by Vesto Melvin Slipher (at Lowell Observatory) with estimates of galactic distances but were not taken very seriously. Lematire's value for what we now call the Hubble constant would have been about 600 km/sec/Mpc. Edwin P. Hubble (again) settled the issue for most astronomers with his 1929 velocity distance relation. The velocities were still Slipher's, though Milton Humason was soon adding larger ones from the 100-inch telescope on Mt. Wilson. What Hubble did better were the distances, using variable stars, brightest single stars, and soon whole galaxies as his standard candles. Well, all right, his distances were wrong by a factor somewhere around 7 to 10, and yielded a Hubble constant (k term to him, though one doesn't suppose he really objected to the new name) of $H = 500$-550 km/sec/Mpc, with an estimated error of 10%, as Hubble constant estimated errors have been from that day to this. But, they were selfconsistent, so that the linear relationship could be seen.

13. Higher-order clustering. Superclusters were discernible in the 1953 analysis by J. Neyman, E.L. Scott, and C.D. Shane of the counts of galaxies carried out by Shane and Wirtanen from the Lick Observatory at that time. The Palomar Observatory Schmidt survey plates began appearing about the same time, and George Abell firmly advocated the existence of superclusters from the time of a 1961 meeting on the significance of large velocity dispersions in clusters. The existence of superclusters was equally firmly denied by Fritz Zwicky from 1936 until his death in 1974 and by V.A. Ambartsumian and his associates into the 1990s. This was never quite as clear a contradiction as it sounds. Many of Zwicky's "clusters" had substructure and were, on a modern distance scale, 40 or more Mpc across, that is, not so very different from Abell's superclusters. Are there yet larger structures in a hierarchial or fractal distribution? Swedenborg, Wright, Charlier, Lambert (up to 1920) said yes, often to avoid an Olbers' paradox. So also said Gerard Henri de Vaucouleurs, but evidence spanning from deep redshift surveys, to the X-ray background, to the microwave background says **NO**.

14. Expansion of the extragalactic distance scale. H was once 500; now it is 55-75 km/sec/Mpc (my error bars are larger than some other people's here). Step one was taken, oddly, because Baade could not resolve RR Lyrae stars in Andromeda with the then new 200-inch Palomar Mountain telescope (and David Thackeray could in the Magellanic Clouds with a 74 inch), though Mineur and Behr had got there first, before and during World War II. Additional sorting out led to 250 to 125 to 100 and below, with a prolonged period during which most calculations were done assuming 100, because it made the arithmetic easy. The use of H = 100 km/sec/Mpc is the modern survival of that era of good feeling. There was also a prolonged period of not-so-good feeling, during which hostile camps defended 100 and 50 (see Trimble 1996 for further details).

15. Recognition of large scale velocity deviations from smooth Hubble flow. These are surely to be expected if there are large density fluctuations. The first data had a great deal of scatter (Rubin et al. 1973, 1976), which led to the distrusting name "Rubin-Ford effect". Still another decade later, other groups reported comparable large scale deviations (including one associated with a not-very visible "Great Attractor"), and the Milky Way was shown, from a dipole moment in the 3K background, to be part of such a large scale flow, at about 600 km/sec (Lubin and Villela 1986 was the first public announcement of this). Deviations from Hubble flow are now generally recognized, with the reminder that you need a distance indicator other than redshift if you are to find them.

THE UNIVERSE TODAY

Fairly definitive things can now be said about both the global properties of the universe and the merely large scale. On the global side are the values of the cosmic constants now measured by various combinations of (1) apparent brightness of distant Type Ia supernovae, (2) brightness fluctuations around the sky on various angular scales of the cosmic microwave background radiation, (3) gravitational lensing of galaxies and quasars by (other) galaxies and clusters thereof, (4) calculations of product nuclei expected from the early, hot dense universe ("Big Bang") in comparison with the amounts of these (normal hydrogen, deuterium, helium-3 and helium-4, and lithium-7) found in unprocessed gas, and (5) numbers, masses, and sizes of clusters of galaxies, often in comparison with predictions of what should have been formed by the present moment in a simulated universe with particular initial conditions. These numbers will undoubtedly continue to change slightly as more data are collected and analyzed, particularly later years of measurements by the Wilkinson Microwave Anisotropy Probe (WMAP), but, one hopes, not by more than two or three standard deviations. Table 1 is largely taken from Bennett et al (2003).

The merely large scale structure is typically described as sheets, filaments where these cross, and knots, where those cross, of larger-than-average density of galaxies, bordering under-populated (but not entirely empty) voids. The first large survey of galaxy redshifts that revealed this structure in most eyes is generally described as "the Harvard slice" (de Lapparent et al. 1986), reflecting the inability of the average astronomer to distinguish the various parts of the Center for Astrophysics in Cambridge, Massachusetts. The topology has been variously termed bubbles, foam, and sponge, this last meaning that both the high- and low-density regions connect up with others. The average void size is about 40 Mpc (Hoyle and Vogeley 2002), and the galaxies inside them tend to be rather puny things (Mathis and White 2002).

The most extensive picture of how the galaxies are currently distributed in three dimensions and how this has changed in the recent past will eventually come from the Sloan Digital Sky Survey (SDSS) which has already revealed some evolution, (Dodelson et al. 2002). Interim pictures, particularly nice because they show the filament/sheet/void pattern repeating many times, belong to the 2dF (Two-degree field) redshift survey (Peacock et al. 2001), the Southern Sky Redshift Survey (da Costa et al. 1998), and the Las Campanas redshift survey (Shectman et al. 1996). There are corresponding fluctuations of velocities of galaxies around smooth Hubble flow, at worst "not inconsistent" with the density fluctuations (Zehavi et al. 2002).

Table 1.   Cosmological parameters

Parameter	Value
Hubble's Constant	$H_o = 71 \pm 4$ km/sec/Mpc
Critical (closure density)	$\rho_c = 9.45 \times 10^{-\cdot\cdot}$ g/cm$^\cdot$
	(or energy equivalent)
Total Density	$\Omega = 1.02\ 0.02$ (in units of critical density)
Age of the Universe	t = 13.7 + 0.2 Gyr
Baryon Density	$\Omega_b = 0.044 \pm 0.004$
	(of which 0.01 is critical density)
Matter in all forms	$\Omega_m = 0.27 \pm 0.04$ (of which
	nonbaryonic $\simeq 0.23$)
Dark energy	$\Omega_x = 0.73 \pm 0.04$ (with equation
	of state parameter w less than ~ 0.78)
Neutrino density	$\Omega_n = 0.01(0.006$ for likely neutrino
	masses, 3 light flavors)
Temperature of CMB	T = 2.725 K
Photon density	$N_\gamma = 410\ \gamma$/cm$^\cdot$
Baryon-photon ratio	$\eta = 6.1 \times 10^{-\cdot\cdot}$
Initial fluctuation spectrum	n = 0.833 $\pm$ 0.086 at 0.05 Mpc$^{-\cdot}$,
	steepening to 1.03 $\pm$ 0.04 at 0.002 Mpc$^{-\cdot}$
	(it is not impossible that n = 1 on all scales
	is still correct)
Redshift of decoupling	z = 1089 1 (but thickness $\Delta z \simeq 200$)
Age at decoupling	$t_d = 380$ kyr, with thickness of 120 kyr
Redshift of matter = radiation	z = 3200 $\pm$ 200
Normalization of power spectrum	$\sigma\cdot = 0.84 \pm 0.04$ (0.9 $\pm$ 0.1 in a later paper)
Baryon optical depth	$\tau = 0.17 \pm 0.04$ (that is, only 84.4%
	of the photons liberated at decopling get
	through to us unscattered)

THE UNIVERSE OF THE (NOT TOO DISTANT) FUTURE

Do I expect the main items mentioned in the previous section to change significantly in my lifetime? No. Neither, of course, did Ptolemy or William Herschel, and they were right. Major scientific reorientations always take time. What are some of the things we might reasonably expect?

- Better numbers, and independent numbers from other considerations for the standard parameters, for instance W from X-ray clusters, the tilt (or not) of the primordial density spectrum, and the optical depth back to recombination from later WMAP data. Supernova work should improve the limits on the equation of state parameter, w quite soon.

- Reconciliation of some discrepant numbers, for instance Wm and s8 whose best fits from clusters of galaxies are respectively a bit smaller and a bit larger than the consensus ones.

- Additional alternatives to inflation, with, one hopes, additional predictive power. Inflation tells us that W and n should both be unity and that the fluctuation spectrum should be Gaussian and adiabatic, but it does not tell us either how that W should be divided among the various constituents or what the amplitude of those fluctuations should be. Depending on your attitude toward anthropic considerations, it may or may not interest you to realize that a universe with Wx larger than 0.9 or so would never form galaxies and one with characteristic fluctuations D T/T outside the range 10-6-10-4 would either (at the low end) also never form galaxies or (at the high end) turn everything into big black holes.

- More details of the large scale distribution of galaxies as the full SDSS data set becomes available. Already it and 2dF seem to connect up the current power spectrum of density fluctuations vs. length scale very nicely with the CMB numbers.

- Reconciliation (or perhaps clearer contradictions!) of the details of structures made by x-CDM simulations with those in the real world. The standard problems are generally described as "missing satellites" (the expectation of more substructure in dark matter halos than we see in the luminous stuff) and "core/cusp" (the steeper rise in central density of the simulations than we see in centers of galaxies and clusters). There are perhaps some other issues, like the pair-wise velocity dispersion. All occur on length scales where feedback from what the baryons are doing must be important and has not yet been fully included in the calculations.

- Better understanding of what, when, and where the first lights in the universe were. By z " 6, there were big, metal-rich galaxies, quasars as bright as any that came later, and enough UV photons to keep diffuse intergalactic gas more than 99.9% ionized. It is generally advertised that stars (probably very massive and certainly very metal poor stars) began the re-ionization process and the QSOs (quasi-stellar objects) later maintained it, but "more work is needed". In one particular realization of a universe with Wm = 0.3 and Wx = 0.17 (carried out by Tom Abel), the very first star had a mass near 300 M,, formed in a halo of about 106 M, and left as a remnant a black hole of 50-80 M, which might then have become a seed for an active galactic nucleus. It happened at z = 58!

ACKNOWLEDGEMENTS

Colleagues from Abell to Zwicky have contributed to both communal and my personal understanding of the construction of the cosmos, but a special thank you to Patricia Henning for a most unexpected introduction to the warmer side of Simon Newcomb.

References

Abell, G.O., 1961, AJ 66, 607
Bennett, C.L. et al., 2003, ApJS 148, 1
Berendzen, R. et al, Man Discovers the Galaxies, (NY; Science History Publications)
Butler, R.P. et al., 2003, ApJ 582, 455
Curtis, H.D., 1921, Bull. Natl. Res. Council 2, 194
da Costa, L.N., et al., 1998, MNRAS 299, 425
de Lapparent, V., Geller, M.J. and Huchra, J.P., 1986, ApJ 302, 1.
de Vaucouleurs, G., 1970, Science 167, 136
Dodelson, S. et al., 2002, ApJ 572, 140
Harrison, E.R., 2000, Cosmology, 2nd Edition, Cambridge University Press
Hoyle, F. and Vogeley, M.S., 2002, ApJ 566, 641
Hubble, E.P., 1936, The Realm of the Nebulae, Yale University Press
Jaki, S., 1972, The Milky Way, NY; Science History Publications
Jones, B.J.T., Martinex, V., Saar, E., and Trimble, V., 2004, Rev. Mod. Phys. (in press)
Linde, A, 1996, in B. Zuckerman and M.A. Malkan (eds.), The Origin and Evolution of the
    Universe (Jones and Bartlett) p. 127
Lubin, P. and Villela, T.N., 1985, in B. Madore and R.B. Tully (eds.), Galaxy Distances and
    Deviations from Universal Expansion (Reidel) p. 169
Mathis, H. and White, S.D.M., 2003, MNRAS 337, 1193
Michell, J., 1767, Phil. Trans. Roy. Soc. 57, 234
Newcomb, S., 1906, Sidelights on Astronomy and Kindred Fields of Popular Sciences, NY
    (Harper)
Neyman, J., Scott, E.L., and Shane, C.D., 1953, ApJ 117, 92
Peacock, J.A. et al., 2001, Nature 410, 169
Peebles, P.J.E., 1993, Principles of Physical Cosmology, Princeton University Press
Rubin, V.C. et al., 1973, ApJ 181, L111 & 1976, AJ 81, 769
Shapley, H., 1921, Bull. Natl. Res. Council 2, 171
Shectman, S.A. et al., 1996, ApJ 470, 172
Struve, O. and Zebergs, V., 1967, Astronomy of the 20th Century, NY: Macmillain

Trimble, V., 1996, PASP 108, 107 & 2004, in S.S. Holt and D. Deming (eds.) The Search for Other Worlds, AIP Conf. Proc. 713, p. 3

——, 2004, Bull. Astron. Soc. India, (in press)

Trumpler, R.S., 1930, Lick Obs. Bull. 14, 154 Whitney, C.A., 1971, The Discovery of Our Galaxy, NY: Knopf

Zehavi, I. et al., 2002, ApJ 571, 172

Zwicky, F. et al., 1961-1968, Catalog of Galaxies and Clusters of Galaxies (in six volumes), California Inst. of Technology

RECENT RESULTS FROM THE SPITZER SPACE TELESCOPE: A NEW VIEW OF THE INFRARED UNIVERSE

Giovanni G. Fazio

Harvard-Smithsonian Center for Astrophysics, 60 Garden Street, Cambridge, MA   02138, USA

Abstract   The *Spitzer Space Telescope*, NASA's Great Observatory for infrared exploration, was launched on August 25, 2003, and is returning excellent scientific
data. Combining the intrinsic sensitivity obtained with a cooled telescope in
space and the imaging and spectroscopic power of modern array detectors, huge
gains have been achieved in exploring the infrared universe. This paper describes the *Spitzer Space Telescope* and its focal-plane instruments and summarizes some of the spectacular images and new scientific results that have been
obtained.

Keywords: infrared – galaxy classification – interstellar dust – star formation – interacting
galaxies – active galactic nuclei – planetary nebula

Introduction

Infrared astronomy, which covers the wavelength region from 1 to 1000 μm,
is concerned primarily with the study of relatively cold objects in the universe
with temperatures ranging from a few degrees Kelvin to about 2000 K. One
principal source of infrared radiation in the interstellar medium is dust, which
absorbs optical and ultraviolet radiation and reradiates at infrared wavelengths.
Infrared observations can also penetrate dust-enshrouded sources, which are
invisible to optical and ultraviolet radiation. Observations of the early universe
at infrared wavelengths have the advantage of seeing galaxies whose optical
light has been redshifted into the infrared.

However, when using ground-based telescopes, most of the infrared spectrum in invisible because of atmospheric absorption (Figure 1). Only a few
windows from 1 to 30 μm wavelength are available. Also because of thermal
emission from the atmosphere and from the telescope, the background infrared
radiation is relatively high, making ground-based observations very difficult.

By placing an infrared telescope in space, atmospheric emission and absorption can be eliminated and, in addition, the telescope can be cooled to eliminate

M. M. Shapiro et al. (eds.), Neutrinos and Explosive Events in the Universe, 47–71.

the thermal background radiation. With a cooled telescope in space the infrared background radiation is due only to the scattered and thermal emission from the zodiacal dust in our solar system. The reduction in background radiation that can be achieved is rather dramatic, a factor of 1 million (Figure 1).

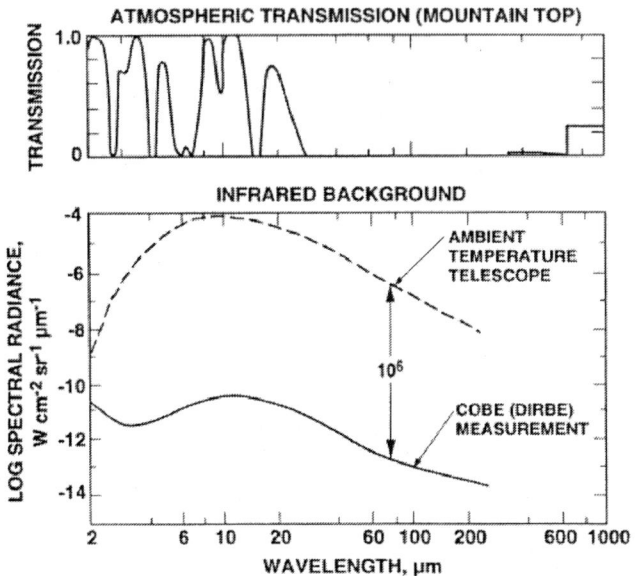

Figure 1. The top figure is a plot of the earth's atmospheric transmission versus infrared wavelength, and the lower figure is a plot the infrared background radiation versus infrared wavelength for a ground-based telescope and helium-cooled space telescope.

The following sections will describe the *Spitzer Space Telescope* and its three focal plane instruments and present a sample of the spectacular images and scientific results that have been produced.

Spitzer Space Telescope

NASA's *Infrared Astronomy Satellite (IRAS)*, which was launched in 1985, consisted of a liquid-helium cooled telescope (60-cm mirror) and produced the first all-sky maps of the infrared universe at 25, 60, and 100 μm wavelength. *IRAS* was followed in 1996 with another cooled telescope in space, the *Infrared Satellite Observatory (ISO)*, an ESA mission, which was a true observatory that could carry out follow-up observations of the *IRAS* sources. In 2003, NASA's *Spitzer Space Telescope*, with an 85-cm mirror, achieved major advances in sensitivity, image quality and field-of-view over *ISO*. Although its mirror was only slightly larger than *ISO's* 60-cm mirror, the use of new, sensitive, and large-area infrared array detectors has permitted this new view of the infrared universe.

The *Spitzer Space Telescope*, NASA's Great Observatory for infrared exploration, was launched on 2003 August 25, into a heliocentric orbit, trailing the Earth (Figure 2). The telescope consists of an 85-cm cryogenically-cooled mirror and three focal-plane instruments, which provide background-limited imaging and spectroscopy covering the spectral region from 3 to 180μm wavelength. Incorporating large-format infrared detector arrays, with the intrinsic sensitivity of a cryogenic telescope, and the high observing efficiency of a solar orbit, Spitzer has already demonstrated dramatic improvements in capability over previous infrared space missions. Following 60 days of In-Orbit-Checkout and 30 days of Science Verification, normal operations began on 2003 December 1. To date, all spacecraft systems continue to operate extremely well (Werner et al. 2004). The Telescope has an expected lifetime of 5 to 6 years. More than 75% of the observing time will be available for General Observers.

Figure 2. Artist's conception of the Spitzer Space Telescope in earth-trailing solar orbit.

Two of the *Spitzer Space Telescope* instruments, the Infrared Array Camera (IRAC; Fazio et al. 2004), and the Multiband Infrared Photometer for Spitzer (MIPS; Rieke et al. 2004) are designed as imaging instruments, although the MIPS also has a spectral energy distribution (SED) channel for very low-resolution spectroscopy. IRAC covers the wavelength region between 3.2 and 9.4 μm, while MIPS covers the region between 21.5 and 175 μm. The third instrument, the Infrared Spectrograph (IRS; Houck et al. 2004) provides low and moderate-resolution spectroscopic capabilities (5 to 40 μm), although it has two small imaging peak-up apertures that can also be used for imaging.

The Infrared Array Camera (IRAC) is a simple four-channel camera that obtains simultaneous broad-band images at 3.6, 4.5, 5.8, and 8.0 μm. Two nearly adjacent 5.2×5.2 arcmin fields of view in the focal plane are viewed by the four channels in pairs (3.6 and 5.8 μm; 4.5 and 8 μm). All four detector arrays in the camera are 256×256 pixels in size (1.22 arcsec pixels), with the two shorter wavelength channels using InSb and the two longer wavelength channels using Si:As IBC detectors. IRAC is a general-purpose, wide-field camera that can be used for a large range of astronomical investigations. In-flight observations with IRAC have already demonstrated that IRAC's sensitivity, pixel size, field of view, and filter selection are excellent for studying numerous topics in galactic and extragalactic astronomy.

MIPS provides the long-wavelength coverage with imaging bands at 24, 70 and 160 μm at a spectral resolution of about 7% and very low resolution spectral energy distribution (SED) spectroscopy from 52 to 100 μm. MIPS uses true detector arrays of 128×128 pixels at 24 μm, 32×32 pixels at 70 μm, and 2×20 pixels at 160 μm, and achieves telescope limited resolution of 6, 18, and 40 arcsec at 24, 70 and 160 μm. For the first time, at these long infrared wavelengths, MIPS can achieve simultaneously high sensitivity, a large imaging field and the maximum possible angular resolution.

IRS consists of four separate spectrograph modules, known by their wavelength coverage and resolution as Short-Low (SL), Short-High (SH), Long-Low (LL) and Long-High (HL). Two Si:As array detectors (128×128 pixels) collect the light into the SL and SH modules., while two Si:Sb array detectors (128×128 pixels) are used in the LL and LH modules. IRS also contains tow peak-up imaging fields with bandpasses centered at 16 μm and 22 μm.

The *Spitzer Space Telescope* is managed by JPL for NASA. Science operations are conducted at the Spitzer Science Center at Caltech, Pasadena, CA.

Additional information on the *Spitzer Space Telescope* and its instruments can be found at the Spitzer Science Center (SSC) web site.[1]

A New View of Galaxy Morphology and Classification

The morphological classification scheme introduced by Hubble (1926, 1936), based on blue-light images, has been modified periodically over the years (e.g., Sandage 1961; de Vaucouleurs 1959; Kormendy 1979; Buta 1995) but remains a fundamental method by which astronomers continue to sort and compare galaxies. While the method has proven very successful over much of the last century, its dependence on fundamental, physical properties intrinsic to the galaxies is indirect due to the complicated emission processes sampled in the B-band.

Using recent observations from the IRAC instrument on the *Spitzer Space Telescope*, the morphological classification scheme using images with resolution and sensitivities similar to those used for traditional optical classifications, but at mid-infrared wavelengths, were achieved (Pahre et al. 2004b). While some nearby galaxies could be resolved by previous infrared space missions operating at these wavelengths, and equal or better spatial resolution can be obtained from the ground, the combined wide-field coverage and sensitive infrared imaging detectors of the IRAC instrument on the *Spitzer Space Telescope* permits us to explore an entirely new region of parameter space for imaging nearby galaxies. These images demonstrate a new approach to galaxy classification based on the properties of the galaxy *interstellar medium* relative to its starlight.

Observations

The galaxies presented here are a small but representative subset of a sample of about 100 scheduled to be observed with Spitzer. These galaxies were taken from a complete sample from Ho et al. (1997). The sample observed to date with *Spitzer* nearly fully spans the classical morphological sequence (only Sa is missing).

The data were taken with the Infrared Array Camera (IRAC; Fazio et al. 2004) on the *Spitzer Space Telescope* during the first five IRAC campaigns of normal operations (2003 December – 2004 April).

The general infrared SED properties of various galaxy components are shown in Figure 2, along with the IRAC filter bandpasses. Starlight is blue in the IRAC bandpasses, because the mid-infrared wavelengths are longer than the peak of the blackbody radiation even for cool M giant stars. (Note that M giants are *bluer* than Vega in the $[3.6] - [4.5]$ color due to CO absorption in the 4.5 μm bandpass.) Warm dust emission appears in lines of polycyclic aromatic hydrocarbon (PAH) molecules in the 5.8 and 8.0 μm bandpasses. Emission from an AGN is *redder* than starlight in the infrared bandpasses.

Infrared Morphologies of Galaxies

Images of the galaxies at 3.6 and 8.0 μm are shown in Figures 3 and 4, organized by optical morphological type. A color representation of the images at 3.6, 4.5, and 8.0 μm is shown at the end of this paper. The 3.6 μm flux samples the unreddened stellar light distribution, while the 8.0 μm flux samples that same starlight (dimmed by a factor of more than four since these wavelengths are on the Rayleigh-Jeans tail) plus emission lines of PAH tracing out warm dust.

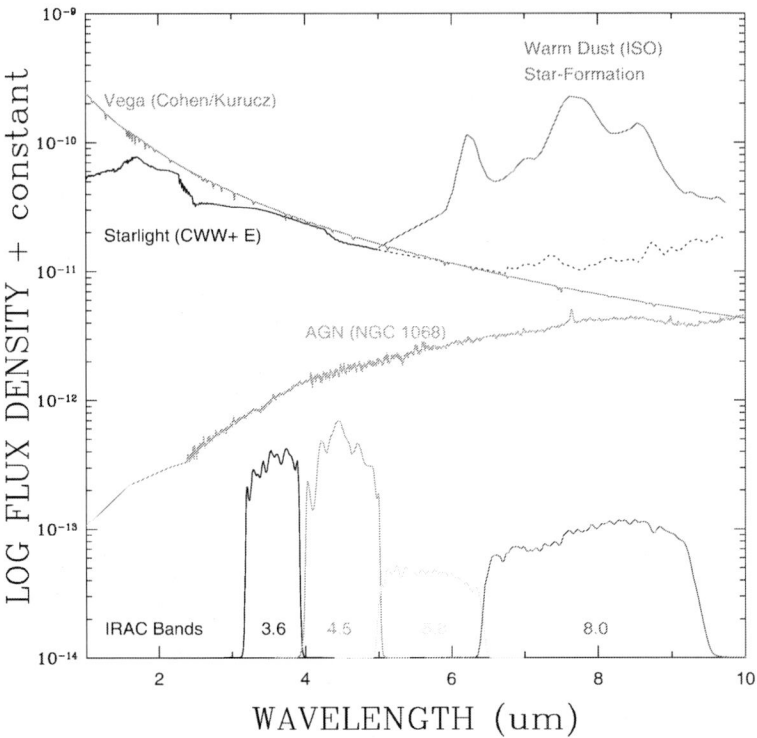

Figure 3. The spectral energy distributions of various galaxy components (starlight, warm dust, and active nuclei) at the IRAC wavelengths. The IRAC bandpasses are shown at the bottom of the figure. Warm dust emitting in the PAH lines, and active nuclei, both appear red in the IRAC bandpasses, while stellar light appears blue. The presence of CO absorption in the 4.5 μm bandpass also results in cool, late-type stars having a bluer color $[3.6] - [4.5]$ than earlier-type stars like Vega.

Comparing Figures 3 and 4, early-type galaxies and the bulges of spiral galaxies are dominated by starlight emission. The disks of late-type galaxies, on the other hand, are dominated by warm dust emission.

Galaxy morphology is more clearly delineated in the dust emission (Figure 4) than it is in the traditional, optical blue light images. The warm dust provides a clean tracer of the reddening-free interstellar medium, which has high contrast and can trace spiral arms all the way to the center of a galaxy. Furthermore, there is a changing ratio of starlight to warm dust emission along the galaxy sequence. This is the basis of a new method of classifying galaxies in the infrared at $3.2 < \lambda < 9.4$ μm using Spitzer images. This is shown graphically in the color figure appearing at the end of the volume (Figure 18), where the starlight is color-coded as blue and the warm dust as red.

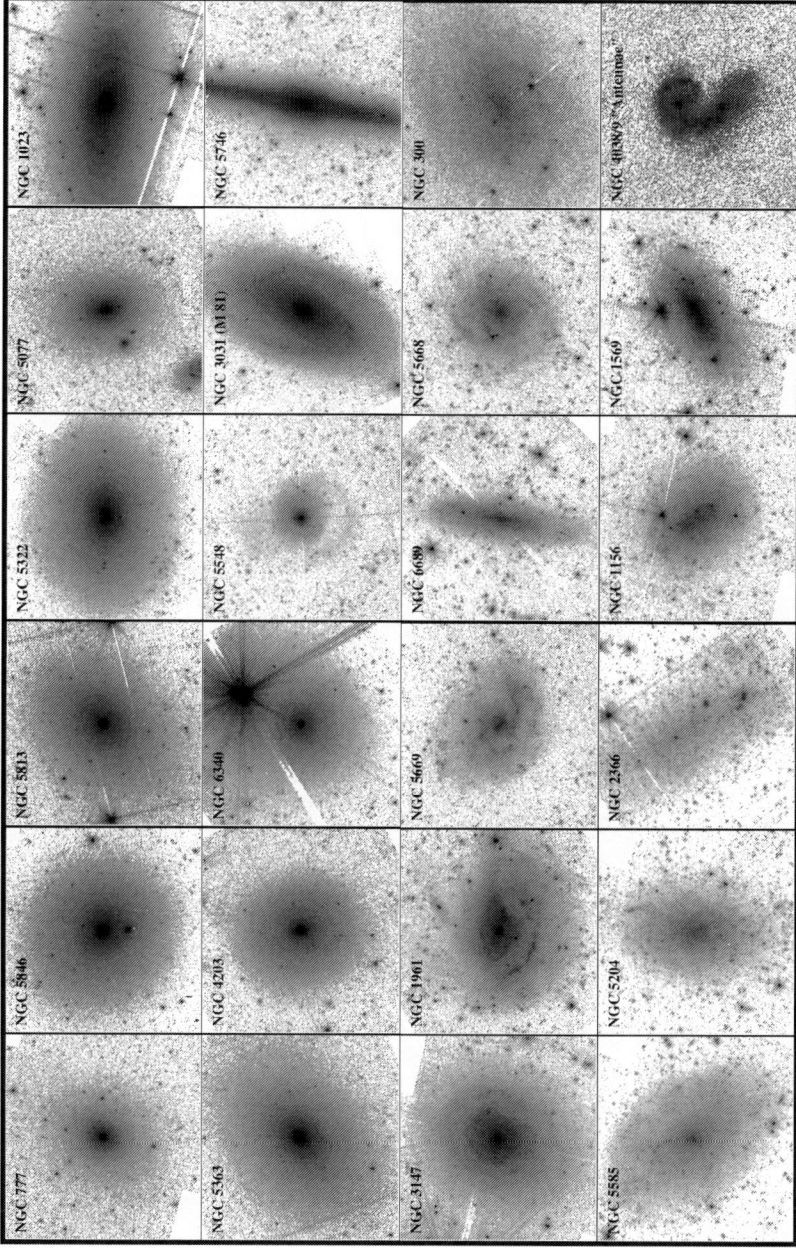

Figure 4.   Images of the galaxies at $\lambda = 3.6$ μm. The mosaic is organized from early- to late-types from left to right.

Other IR quantities which correlate with optical morphological type are shown in Figure 5. Two colors correlate well: the stellar $[3.6] - [4.5]$ color

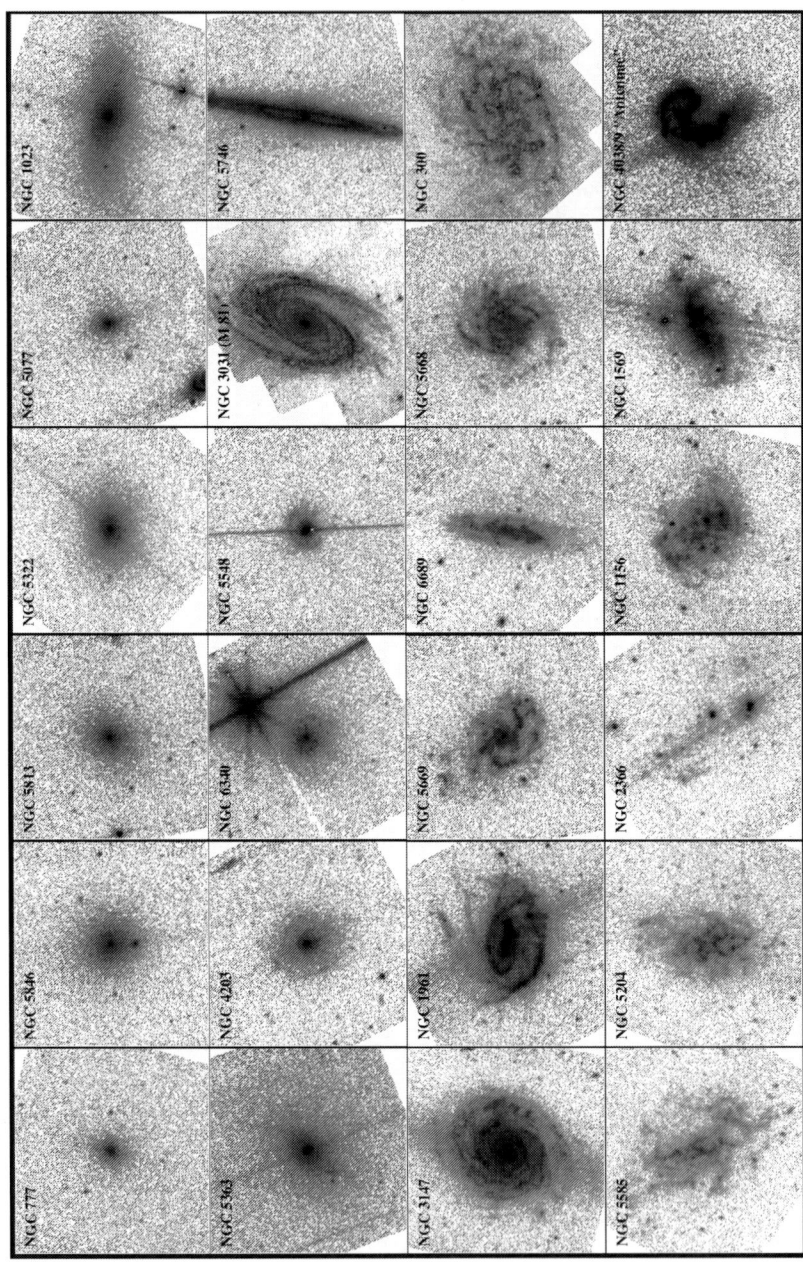

Figure 5.   Images of the galaxies at $\lambda = 8.0$ μm.

and the stellar vs. warm dust color $[3.6] - [8.0]$. The color $[3.6] - [4.5]$ is
redder for late-type galaxies because they have a young stellar population that

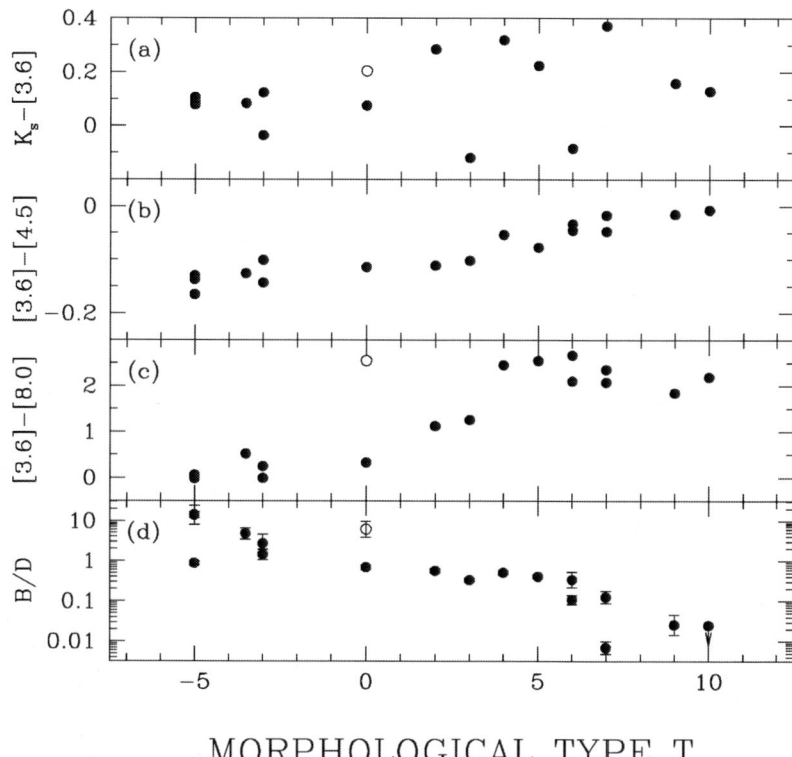

MORPHOLOGICAL TYPE T

Figure 6. Correlation between various colors and the 3.6 μm bulge-to-disk-ratio with optical morphological type. The $[3.6] - [4.5]$ color shows that the starlight changes from blue for early-types to red for late-types. A similar trend is found for $[3.6] - [8.0]$, which samples the ratio of starlight to warm dust emitting at the longer wavelengths.

partially masks the the CO absorption, characteristic of late-type giant stars, in the 4.5 μm bandpass. (See Figure 2). The color $[3.6] - [8.0]$ is redder for late-type galaxies due to the warm dust emission in the PAH lines in the 8.0 μm bandpass. The IR bulge-to-disk ratio also shows a good correlation. Any of these quantities, or all of them together, are suitable stand-ins for galaxy morphological type.

Why do infrared morphological classification and these other three quantities measured at infrared wavelengths work so well? The 3.6 and 4.5 μm light is a good tracer of stellar mass (for a wide range of metallicity and age) free of dust obscuration, hence B/D measured at these wavelengths samples a fundamental galaxy property. The dust emission is a high contrast tracer of the interstellar medium, thereby making the morphology easier to discern. In galaxies of various types, the 8.0 μm bandpass can exhibit anything from $\sim 100\%$ starlight (elliptical galaxies) to $\sim 95\%$ dust emission (NGC 1961,

3147, 4038/4039), and thus the $[3.6] - [8.0]$ color provides a direct measure of increasing ISM content along the galaxy sequence.

Spatial Distribution of Warm Dust in Lenticular Galaxies

The warm gas phase was detected in early-type galaxies via its dust emission at far-IR wavelengths with IRAS (Jura et al. 1987; Knapp et al. 1989) and at optical wavelengths from atomic gas emission (Caldwell 1984; Phillips et al. 1986). Later, ISO detected polycyclic aromatic hydrocarbon (PAH) lines in the mid-IR (Lu et al. 2003; Xilouris et al. 2004). Of order half or more early-type galaxies exhibit either far-IR dust or optical ionized gas emission, although the detection rate in the infrared could be lower if the effects of an AGN are removed (Bregman et al. 1998). Current knowledge of the distribution of warm dust emission within early-type galaxies indicates that it mostly follows the light distributions (e.g., Athey et al. 2002) and hence is thought to arise from processes other than star formation such as AGN mass loss. The exceptions to this statement are a few galaxies with somewhat extended structures in 15 μm ISO images that have been interpreted as dust lanes (Xilouris et al. 2004).

The mid-IR imaging capability of the Spitzer improves the spatial resolution over previous missions by a factor of a few and the sensitivity by more than an order of magnitude. These new images (Figure 7; Pahre et al. 2004a) show that several early-type galaxies have substantial structure in the distribution of their warm dust as traced by PAH emission bands at $6.2 - 8.6$ μm. In particular, two lenticular galaxies show warm dust organized into spiral arms, and a third shows an extended, smooth distribution. For NGC 4203, this warm dust emission is not seen as obscuration at optical wavelengths with the Hubble Space Telescope (Erwin & Sparke 2003), further demonstrating the power of the infrared to highlight new galaxy morphological features that are invisible at optical wavelengths.

NGC 5746 at Infrared Wavelengths

A good example of the power of infrared galaxy classification is NGC 5746 (Figure 8), an edge-on spiral that cannot be reliably classified at optical wavelengths. Even the near-infrared emission from 2MASS shows a little bit of obscuration, but the near-IR light is otherwise mostly smooth. At mid-IR wavelengths, however, a bright ring appears along with several arms outside of it. The galaxy's suggested morphological type is Sab(r).

Identifying the AGN in M 81

The presence of an AGN in M 81 can be inferred in two ways: (1) nuclear colors significantly redder than the bulge, and (2) point source residual after

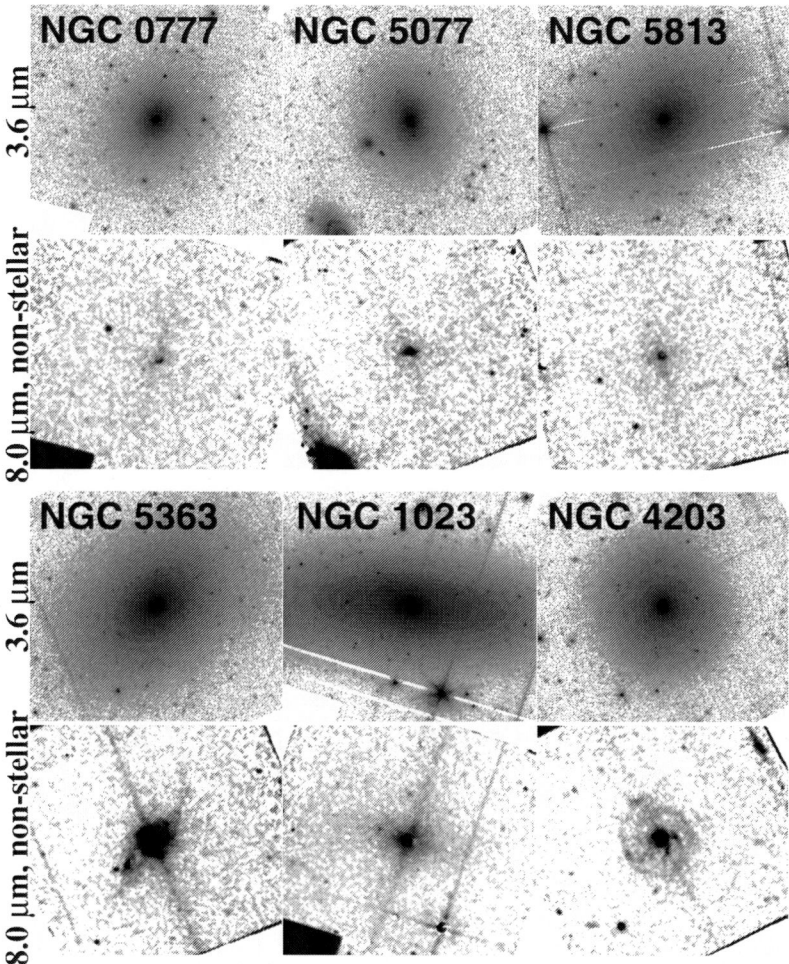

Figure 7. Warm dust emission in early-type galaxies. Six galaxies are shown, each with separate images in stellar emission (3.6 μm) and warm dust emission (8.0 μm "non-stellar"). The top three galaxies are classified as ellipticals and show no sign of warm dust emission at 8.0 μm. The bottom three galaxies are classified as lenticulars and show resolved warm dust. NGC 4203 appears face-on with several spiral arms emanating out from its nucleus, NGC 5363 appears somewhat inclined and shows two arms, and NGC 1023 shows a smooth distribution of dust.

fitting and subtracting two-dimensional models of the bulge and disk light. The first one is shown in Figure 9, and both are described by Willner et al. (2004). The nucleus of M 81 is more than 0.4 mag redder than the inner bulge, which cannot be explained by any reasonable stellar photosphere model.

NGC 5746

DSS   **IRAC 3.6 μm**   **IRAC 8.0 μm**

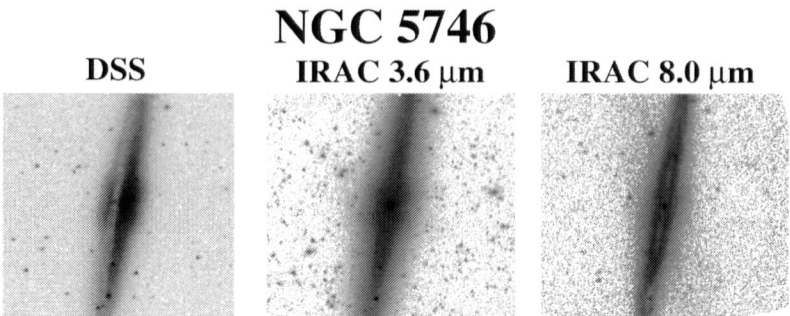

Figure 8.   Edge-on spiral galaxy NGC 5746. At optical wavelengths (left), the galaxy shows a prominent dust lane. The IRAC image at 3.6 μm (middle) is free of dust obscuration and shows a hint of a spiral arm. The 8.0 μm image (right) shows the ring prominently as well as several arms outside of it, demonstrating that this is a Sab(r) galaxy. The ring and arms are prominent in the PAH emission lines in the 8.0 μm band.

Summary of Spitzer Results on Galaxy Morphology and Classification

Images of two dozen galaxies taken at $3.2 < \lambda < 9.4$ μm with the IRAC instrument on the *Spitzer Space Telescope* show that:

1 The mid-IR clearly separates emission from interstellar matter and starlight without the effects of extinction.

2 The mid-IR dust emission, particularly the PAH feature at $\lambda = 7.7$ μm, is a clear tracer of the presence of interstellar matter. The emission shows high contrast against stellar emission at the same wavelength.

3 The mid-IR light provides an entirely new scheme by which to classify galaxies–primarily based on the ratio of their ISM to starlight emission.

4 The colors of stellar photospheres in the mid-IR vary only a small amount with population age or mass function, and hence the stellar emission is a direct tracer of stellar mass. The bulge-to-disk-ratios measured at 3.6 and 4.5 μm therefore sample the mass ratio of the stellar content, not a mixture of stellar content and recent massive star formation activity.

5 Two colors ($[3.6]-[4.5]$ and $[3.6]-[8.0]$), as well as the 3.6 μm bulge-to-disk-ratio, correlate well with traditional morphological type and hence can be regarded as a means of galaxy classification.

6 Three of six early-type galaxies observed exhibit dust emission that is organized into spiral arm or inner disk-like structures,

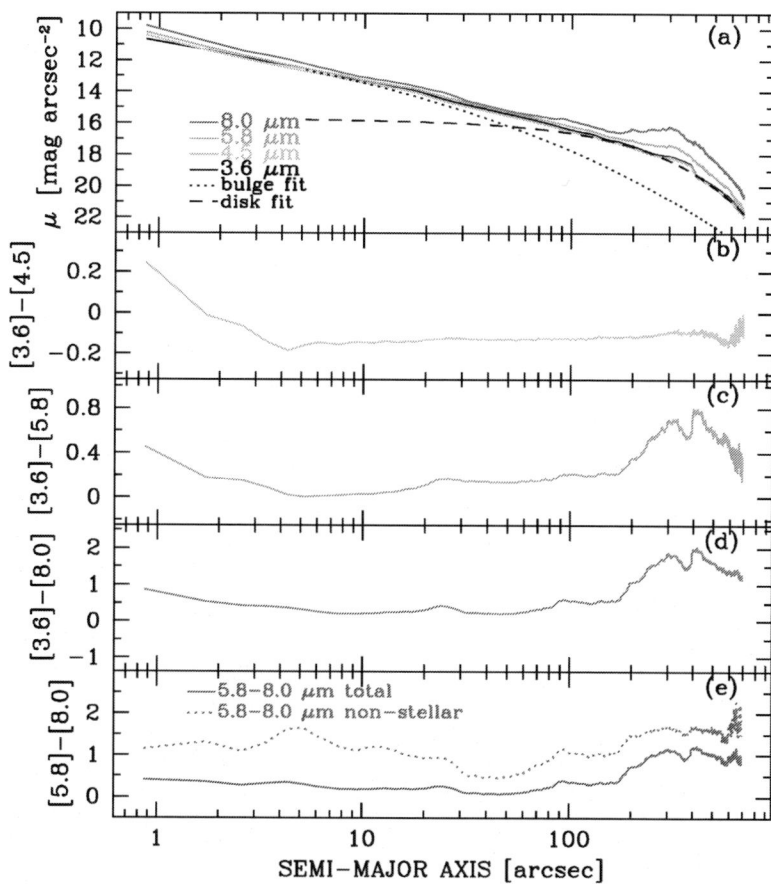

Figure 9. Surface photometry and color profiles for M 81. (a) Surface photometry in the IRAC bands along with bulge plus disk model fit to the 3.6 μm data. (b) Color profile in [3.6] − [4.5] which shows +0.4 mag redder colors in the nucleus, indicative of an AGN. (c) and (d) Color profile in [3.6] − [5.8] and [3.6] − [8.0], which are both redder than starlight alone due to PAH emission at the long wavelengths. (e) Color profiles in [5.8] − [8.0] for both the combined starlight and warm dust and the warm dust alone. The color of the warm dust is roughly consistent with PAH color [5.8] − [8.0] = 2.06 mag predicted by Li & Draine (2001).

7 The [5.8] − [8.0] color of the dust emission matches that for dust in actively star-forming galaxies and theoretical models of PAH emission, and

8 Active galactic nuclei can be readily identified both via their mid-infrared colors and as point source residuals in two-dimensional modelling.

Star Formation in Colliding Galaxies: The Antennae

Imaging of the Antennae galaxies (NGC 4038/4039) with IRAC has revealed large concentrations of star-forming activity away from both nuclei of the two merging galaxies (Wang et al. 2004) and confirm earlier ISO observations with lower resolution (Viguroux et al. 1996; Mirabel et al. 1998). IRAC images (Figure 10) have provided a new view of the total underlying star-forming activities unaffected by dust obscuration. The local star formation rate in the Antennae, using the flux ratio of non-stellar to stellar emission as measured by IRAC, can be mapped and compared to the measurements in starburst and ultraluminous galaxies. The rate in the active regions of the Antennae is found to be as high as those observed in starburst and ultraluminous galaxies on a "per unit mass" basis. This more complete picture of star formation in the Antennae will help us better understand the evolution of colliding galaxies, and the eventual fate of our own galaxy.

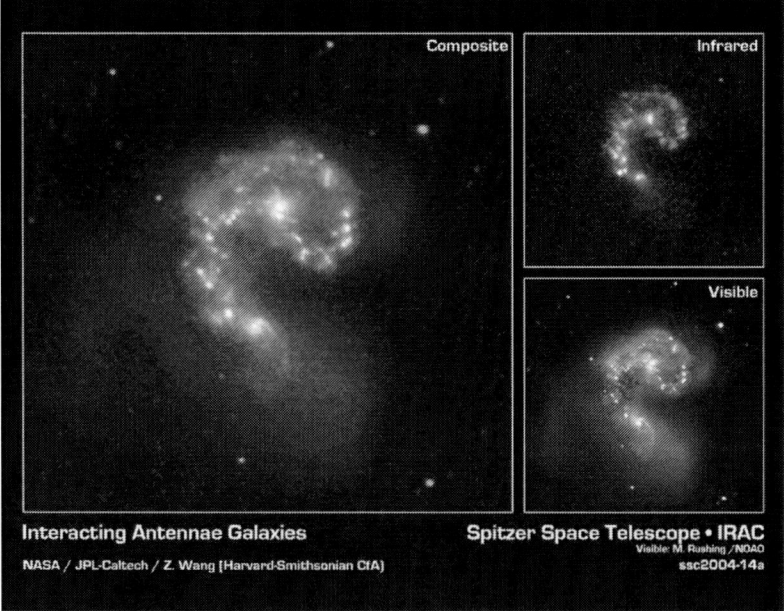

Figure 10.   Interacting Antennae Galaxies (NGC 4038/4039). The left image is a false-color composite of IRAC data and visible-light data from Kitt Peak National Observatory (M. Rushing). Visible light from stars in the galaxies (blue and green) is shown together with infrared light from warm dust clouds heated by newborn stars (red). The two nuclei of the merging system show up as yellow-white areas, one above the other. The upper right image shows the IRAC image by itself, with 3.6 μm wavelength in blue, 4.5 μm in green, 5.8 μm in orange and 8.0 μm in red. The lower right image shows the true-color, visible-light image by itself. Note in this image that the bright star-forming regions seen in the infrared image are blocked by dust clouds in the optical image.

Spitzer Observations of Young Stellar Clusters

The *Spitzer Space Telescope* promises to further revolutionize the study of star formation by providing the capability to image young stellar groups and clusters at mid-infrared wavelengths with the sensitivity to detect young stars down to the hydrogen burning limit and below. With this new capability, the legacies of IRAS, ISO and numerous ground-based near-infrared observations can be expanded by probing the spectral energy distributions of stars, brown dwarfs and protostars in young stellar clusters out to distances of 1 kpc or greater.

The IRAC team has imaged 31 young stellar groups and clusters with the MIPS and IRAC instruments. These have been selected from a catalog of 63 star forming regions within 1 kpc of the Sun containing 10 or more members (Porras et al. 2003). In parallel, 7 square degrees in the Orion molecular clouds are also being mapped. These surveys will sample the full continuum of multiple star forming regions in the nearest kiloparsec, from small groups of stars in Taurus to the rich Orion Nebula and Mon R2 clusters.

Initial results for four young stellar clusters in our sample with properties between those of Taurus and Orion: S140, S171, Cepheus C, and NGC 7129 have been reported by Megeath et al. (2004). These regions are at similar distances, but span a range of FIR luminosities, molecular gas masses, and cluster membership. In each cluster, color-color diagrams, based on IRAC photometry, were used to identify young stars with disks and protostars (Figure 11). Classification was based on the results of a paper by Allen et al. (2004) in which the IRAC colors of the observed young stars are compared to colors derived from models of stars with disks or infalling envelopes. In each of the four clusters, there are between 39 and 85 objects with colors inconsistent with reddened stellar photospheres.

The Distribution of Young Stars

We show IRAC mosaics of all four clusters in Figures 12 and Figure 13. In preparation for the *Spitzer* young stellar cluster survey, Ridge et al. (2003) mapped each of these regions in the ^{13}CO and C^{18}O ($J = 1 \rightarrow 0$) transition. overlay contours of the the C^{18}O emission, which is an excellent tracer of the structure of the molecular gas in star forming regions (Goldsmith, Bergin & Lis 1997). We find 72% of the class I and 56% of the class II sources fall within the detected C^{18}O emission, and 93% of the class I and 86% of the class II sources fall within the detected ^{13}CO emission. These percentages are qualitatively consistent with the class I sources being younger than the class II sources.

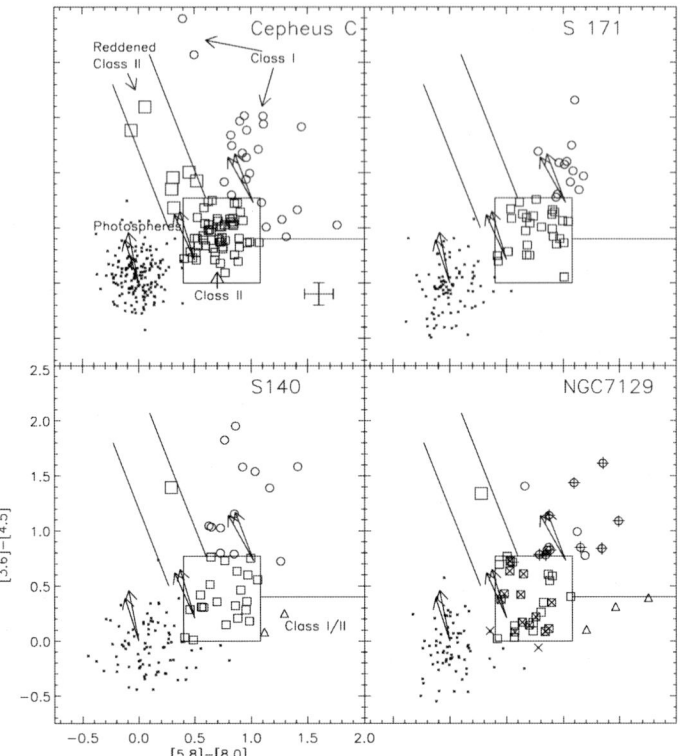

Figure 11.   The IRAC color-color diagram is displayed for all four clusters. Using a selection criteria outlined in the text and in Allen et al. 2004, the squares are identified as class II sources, the large square are reddened class I sources, and the circles are class I sources. The two parallel lines border the positions of sources we identify as reddened class II objects. In two of the regions, S140 and NGC 7129, we mark the class I/II sources (which share the characteristics of both class I and class II sources) with triangles; the horizontal line above the triangles shows the adopted division between class and class I/II sources. We show reddening vectors for $A_V = 30$ derived from the Draine & Lee (1984) and Mathis (1990) extinction laws, in each case the Draine & Lee vector points to the left of the Mathis vector. The vectors were calculated for Vega (at position 0,0), a young star with disks taken from the models of D'Allessio et al. (2004) (0.2,0.5), and a flat spectrum source (0.73,1). Twenty-four sources in NGC 7129 were classified using IRAC photometry combined with MIPS 24 μm photometry by Muzerrole et al. (2004); to display these classifications we overplot a plus sign for class I objects and an X for class II objects. The error bars in the Cepheus C plot show the median uncertainty in the colors for all the sources in all four clusters. A conservative 0.1 mag calibration uncertainty was added in quadrature to the median photometric uncertainty.

Summary of Spitzer Observations of Young Stellar Clusters

In all four clusters, the very young stars and protostars were identified by their excess emission in the mid-infrared and their distribution over multi-

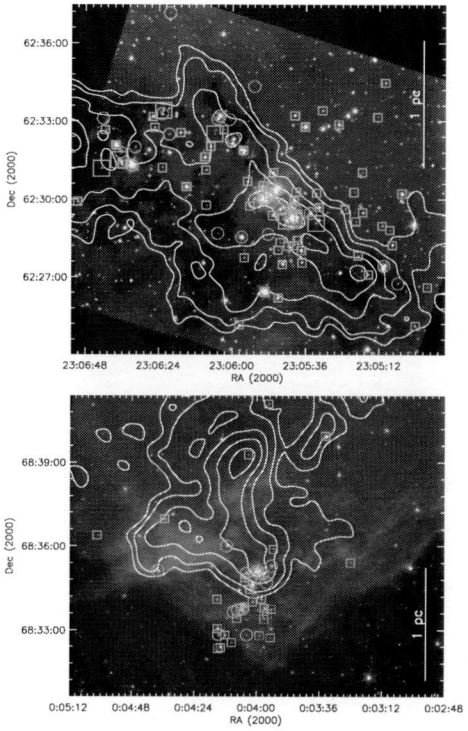

Figure 12. Images of Cepheus C (top) and S171 (bottom) constructed from the IRAC 3.6 μm (blue), 4.5 μm (green) and 8.0 μm (red) images. The contours are the maps of C$^{\cdot\cdot}$O ($1 \rightarrow 0$) emission from Ridge et al. (2003). The C$^{\cdot\cdot}$O observations have an angular resolution of $50''$. We show the position of each young star identified in the IRAC color-color diagram. We mark class II sources with squares, reddened class II objects with large squares, and class I objects with circles. The class I/II sources are marked by triangles.

parsec distances was determined. In contrast, the diameters of the clusters identified by near-IR star counts are typically 1 pc or less (Lada et al. 2003). This suggests that a significant fraction of stars in each star forming region form outside the dense clustered regions identified in star counts analyses. Gutermuth et al. (2004) find that half of the stars in NGC 7129 are located in a halo outside the cluster core. Furthermore, star formation in NGC 7129 is continuing in the halo, while the molecular gas has been dispersed toward the cluster core.

The distribution of sources in each region is strikingly different. In Cepheus C, the structure of the molecular cloud breaks up into distinct mid-IR dark cores, and several distinct concentrations of stars are also apparent to the eye. The observed distribution of gas and stars in this region is similar to the hi-

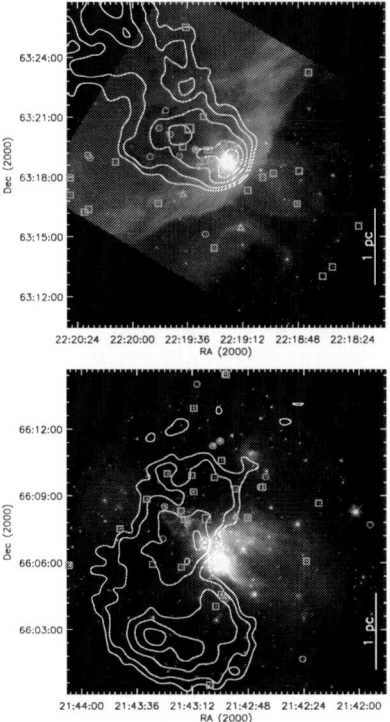

Figure 13.   Images of S140 (top) and NGC 7129 (bottom) using the the same scheme described in Figure 12.

erarchical morphologies generated in numerical models of star formation in turbulent clouds (Bonnell, Bate & Vine 2003). IRAC images of NGC 7129 show a dense cluster of primarily class II sources surrounded by a more extended halo of class I and II objects. Both S171 and S140 contain compact clusters at the edges of bright-rimmed clouds. These varied morphologies hint that environmental factors, such as the presence of external OB stars, may play a significant role in the formation of clusters.

Spitzer Reveals a Molecular Outflow from a Low Mass Protostar

Jets and outflows arising from low mass protostars can produce some of the most spectacular images of the star formation process. The HH 46/47 system is a striking example of a low mass protostar, surrounded by a circumstellar disk, ejecting a jet and creating a bipolar outflow perpendicular to the disk (Figure 14). The central protostar lies inside a dark cloud (known as a Bok globule) which is illuminated by the nearby Gum Nebula. At a distance of 450 pc in the constellation Vela, the protostar is hidden from view in the visible–light image (inset, Figure 14). When observed with IRAC the star and its dazzling jets of molecular gas appear with clarity (Noriega-Crespo et al. 2004). The clear definition of the counterlobe, similar to a large loop, which joins the bow shock with the source, is the most remarkable feature of the picture. For the first time the complete picture of the outflow at mid-infrared wavelengths is revealed. The jets arising from such protostars can reach sizes of trillions of kilometers and velocities of hundred of thousands of kilometers per hour.

A spectrum of HH 46/47, taken with the IRS's Short-Low and Long-Low Infrared Spectrometers over the wavelength range from 5.5 to 20 μm, is shown in Figure 15. Detected in the spectrum are water and carbon dioxide ices, silicates, and organic molecules, such as methyl alcohol and methane gas, which are thought to exist in the circumstellar disk surrounding the protostar. Observations of ices around low-mass protostars have been limited because of the unavailability of much of the 5-20 μm spectral region, where many of the molecular bending-mode transitions occur. For the first time, using *Spitzer*, high quality spectra can now be obtained in low-mass protostellar systems.

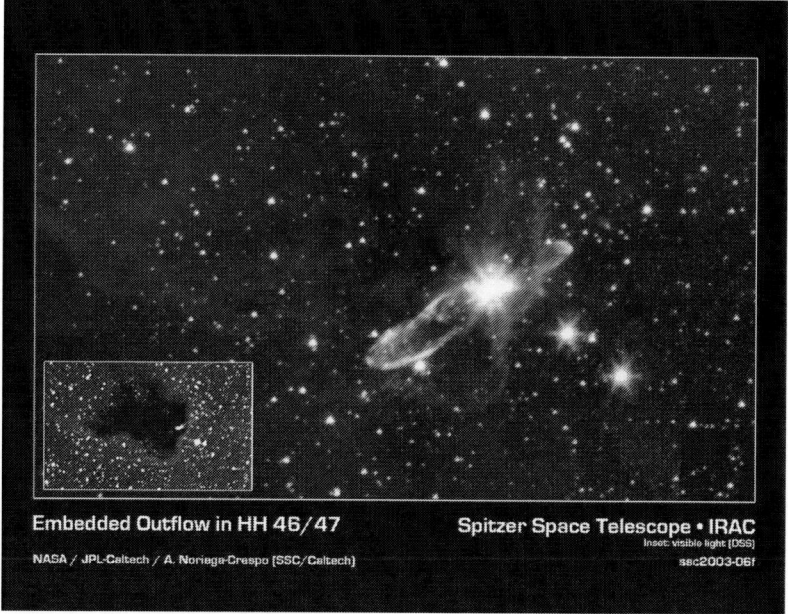

Embedded Outflow in HH 46/47 — Spitzer Space Telescope • IRAC

Inset: visible light [DSS]

NASA / JPL-Caltech / A. Noriega-Crespo [SSC/Caltech]

ssc2003-06f

Figure 14. Image of the HH 46/47 system arising from the Bok globule (ESO 210-6A) obtained with IRAC. The image covers a region of approximately 6.5 × 10.6 arcmin. Emission at 3.6, 4.5 and 5.8, and 8.0 μm is shown as blue, green, and red, respectively. The inset is a visible light comparison image.

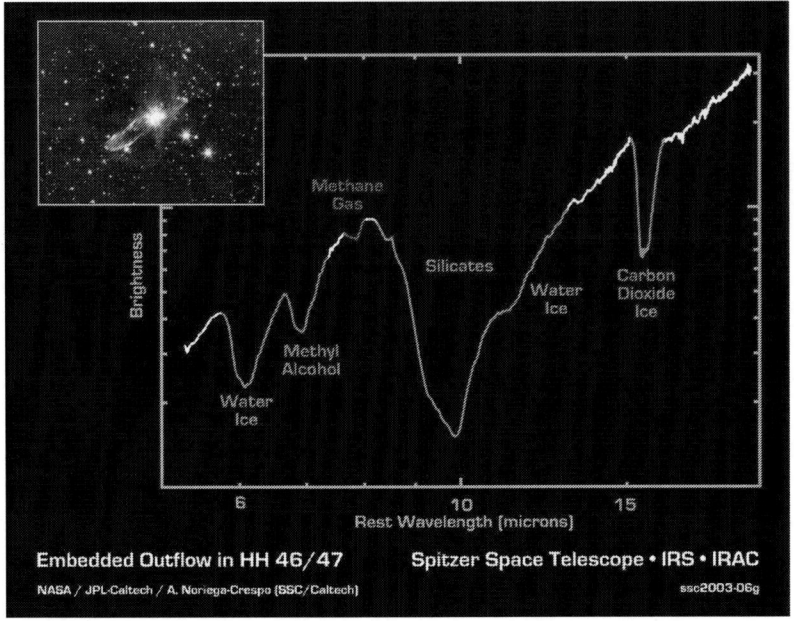

Figure 15.   A spectrum of HH 46/47 obtained with IRS from 5.5 to 20 μm.

Ring of Stellar Death

Shown in Figure 16 (Hora et al. 2004) is a false color image from *Spitzer* of the planetary nebula NGC 246, a dying star (center) surrounded by a cloud of glowing gas and dust which has been ejected from the star. Because of *Spitzer's* ability to penetrate the dust, this image highlights a never-before-seen feature: a giant ring of material slightly offset from the cloud's core. It is believed the ring consists of molecular hydrogen molecules that were ejected from the star in the form of atoms, then cooled to make hydrogen molecules. This new data will help explain how planetary nebulae take shape and eject matter into the interstellar medium.

A Portrait of Life and Death in the Universe: Henize 206

A wonderful example of the cycle of life and death that gives rise to stars throughout the universe is demonstrated in the *Spitzer* image of the nebula Henize 206 (He 206), which is located in the Large Magellanic Cloud (Figure 17). It is similar in size the the belt and sword of Orion, but is about 10 times as luminous in the infrared. Henize 206 was created when a supernova exploded, sending shock waves which impacted a molecular cloud of gas and dust, causing stars to be formed. The supernova remnant is seen as a ring of shocked gas, which is green in the figure. By mapping in the infrared, *Spitzer*

Planetary Nebula NGC 246   Spitzer Space Telescope • IRAC

NASA / JPL-Caltech / J. Hora [Harvard-Smithsonian CfA]   ssc2004-13a

Figure 16. The planetary nebula NGC 246.

was able to see through blankets of dust in the molecular cloud that block visible light and view the embedded young stars (bright white spots) that were formed as a result of the impact (V. Gorjian et al. 2004). Before the *Spitzer* observations there were only hints that newborn stars may be present.

Summary

The *Spitzer Space Telescope* has been operating very successfully since it started routine observations in December, 2003. The sample of new scientific results presented in this paper are indicative of its future potential for discovery over the next five years. The *Spitzer Space Telescope* is designed for the entire scientific community and all are urged to use it to expand our knowledge of the universe.

Figure 17.   Henize 206 in the Large Magellenic Cloud.

Acknowledgments

This work is based on observations made with the *Spitzer Space Telescope*, which is operated by the Jet Propulsion Laboratory, California Institute of Technology under NASA contract 1407. Support for the IRAC instrument was provided by NASA through Contract Number 960541 issued by JPL. The IRAC GTO program is supported by JPL Contract # 1256790. IRAF is distributed by the National Optical Astronomy Observatories, which are operated by the Association of Universities for Research in Astronomy, Inc., under cooperative agreement with the National Science Foundation. This publication makes use of data products from the Two Micron All Sky Survey, which is a joint project of the University of Massachusetts and the Infrared Processing and Analysis Center/California Institute of Technology, funded by the National Aeronautics and Space Administration and the National Science Foundation. This research has made use of the NASA/IPAC Extragalactic Database (NED) which is operated by the Jet Propulsion Laboratory, California Institute of Technology, under contract with the National Aeronautics and Space Administration. The Digitized Sky Surveys were produced at the Space Telescope Science Institute under U.S. Government grant NAG W-2166.

Notes

1. http://www.spitzer.caltech.edu

References

Allen, L. E., Calvet, N., D'Alessio, P., Merin, B., Megeath, S. T., Gutermuth, R. A., Pipher, J. L., Hartmann, L., Myers, P. C., & Fazio, G. G. 2004, ApJS, 154, 363.

Athey, A., Bregman, J., Temi, P., & Sauvage, M. 2002, ApJ, 571, 272.

Bonnell, I. A., Bate, M. R., & Vine, S. G. 2003, MNRAS 343, 413.

Bregman, J. N., Snider, B. A., Grego, L., & Cox, C. V. 1998, ApJ, 499, 670.

Buta, R. 1995, ApJS, 96, 39.

Caldwell, N. 1984, PASP, 96, 287.

D'Alessio, P., Merin, B., Calvet, N. & Hartmann, L. 2004, in preparation.

de Vaucouleurs, G. 1959, Handbuch der Physik, 53, 275.

Draine, B. T. & Lee, H. M. 1985, ApJ, 290, 211.

Erwin, P. & Sparke, L. S. 2003, ApJS, 146, 299.

Fazio, G. G., et al. 2004, ApJS, 154, 10.

Goldsmith, P. F., Bergin, E. A. & Lis, D. C. 1997, ApJ, 491, 615.

Gorjian, V., et al. 2004, ApJS, 154, 275.

Gutermuth, R. A., Megeath, S. T., Allen, L. E., Muzerolle, J., Pipher, J. L., Myers, P. C., & Fazio, G. G. 2004, ApJS, 154, 374.

Ho, L. C., Filippenko, A. V., & Sargent, W. L. W. 1997, ApJS, 112, 315.

Hora, J. L., Latter, W. B., Allen, L. E., Marengo, M., Deutsch, L. K., & Pipher, J. L. 2004, ApJS, 154, 296.

Houck, J. R., et al. 2004, ApJS, 154, 18.

Hubble, E. P. 1926, ApJ, 64, 321.

Hubble, E. P. 1936, Realm of the Nebulae (New Haven: Yale University Press).

Jura, M., Kim, D. W., Knapp, G. R., & Guhathakurta, P. 1987, ApJ, 312, L11.

Knapp, G. R., Guhathakurta, P., Kim, D., & Jura, M. A. 1989, ApJS, 70, 329.

Kormendy, J. 1979, ApJ, 227, 714.

Lada, C. J. & Lada. E. A. 2003, ARAA, 41, 57.

Li, A. & Draine, B. T. 2001, ApJ, 554, 778.

Lu, N., et al. 2003, ApJ, 588, 199.

Megeath, S. T., Allen, L. E., Gutermuth, R. A., Pipher, J. L., Myers, P. C., Calvet, N., Hartmann, L., Muzerolle, J. & Fazio, G. G. 2004, ApJS, 154, 367.

Mathis, J. S. 1990, ApJ, 28, 37.

Mirabel, I. F. et al. 1998, A&A, 333, L1.

Muzerolle, J., Megeath, S. T., Gutermuth, R. A., Allen, L. E., Pipher, J. L., Gordon, K. D., Morrison, J. E., Rieke, G. H., Myers, P. C., & Fazio, G. G. 2004, ApJs, 154, 379.

Noriega-Crespa, A. et al. 2004, ApJS, 154, 352.

Pahre, M. A., Ashby, M. L. N., Fazio, G. G., & Willner, S. P. 2004a, ApJS, 154, 229.

Pahre, M. A., Ashby, M. L. N., Fazio, G. G., & Willner, S. P. 2004b, ApJS, 154, 235.

Phillips, M. M., Jenkins, C. R., Dopita, M. A., Sadler, E. M., & Binette, L. 1986, AJ, 91, 1062.

Porras, A., Christopher, M., Allen, L., Di Francesco, J., Megeath, S. T., Myers, P. C., 2003, AJ, 126, 1916.

Ridge, N. A.. Wilson, T. L., Megeath, S. T., Allen, L. E., Myers, P. C. 2003, AJ, 126, 286.

Rieke, G. H., et al. 2004, ApJS, 154, 25.

Sandage, A. 1961, The Hubble Atlas of Galaxies (Washington: Carnegie Institution).

Viguroux, L. et al. 1996, A&A, 315, L93.

Wang, Z., et al. 2004, ApJS, 154, 193.

Werner, M. W., et al. 2004, ApJS, 154, 1.

Willner, S. P., et al. 2004, ApJS, 154, 222.

Xilouris, E. M., Madden, S. C., Galliano, F., Vigroux, L., & Sauvage, M. 2004, A&A, 416, 41.

Figure 18. Mosaic of galaxies at infrared wavelengths taken with the IRAC instrument on the Spitzer Space Telescope. Colors are coded as 3.6 (blue), 4.5 (green), and 8.0 μm (red). The galaxies are arranged according to traditional, optical morphological classifications. Starlight appears blue at these mid-infrared wavelengths, while warm dust (emitting in PAH emission lines) appears red. AGN emission is also red and point-like (e.g., NGC 5548). There is a clear transition from blue, stellar-dominated emission for the early-type galaxies to red, ISM-dominated emission for the late-type galaxies.

BEYOND THE STANDARD MODEL IN PARTICLE PHYSICS AND COSMOLOGY: CONVERGENCE OR DIVERGENCE?

Maxim Yu. Khlopov[1,2,3]

[1]*Center for Cosmoparticle physics "Cosmion",*
Keldysh Institute of Applied Mathematics,
Miusskaya Pl. 4, 125047, Moscow, Russia

[2]*Physics department, Universita' degli studi "La Sapienza",*
Piazzale Aldo Moro 5,CAP 00185 Roma, Italy

[3]*Moscow Engineering Physics Institute, MEPI , Moscow, Russia*

Abstract   The modern development of cosmology and particle physics naturally leads to analysis of physics beyond the standard model and of its cosmological impact. We have first positive experimental evidences for new physics. We need physics beyond the standard model for inflationary cosmology with baryosynthesis and nonbaryonic matter and energy. The principal ideas of cosmoparticle physics, studying mutual relationship between cosmology and particle physics are briefly reviewed. These ideas offer the way to study the theory of everything and the true history of the Universe, based on it, in the proper combination of their indirect physical, astrophysical and cosmological signatures.

Keywords: Cosmology, particle physics, cosmoparticle physics, inflation, baryosynthesis, dark matter, dark energy, mass of neutrino, cosmic microwave background, large scale structure, weakly interacting massive particles, neutralino, fermion families, primordial black holes, antimatter, positrons, antiprotons, cosmic gamma back ground, cosmic rays

Particles and Universe

Fundamental relationship between cosmology and particle physics originates from the well established links between microscopic and macroscopic descriptions in theoretical physics. Remind the links between statistical physics and thermodynamics, or between electrodynamics and theory of electron. To the end of the XX Century the new level of this relationship was realized. It followed both from the cosmological necessity to go beyond the world of known elementary particles in the physical grounds for inflationary cosmology with

M. M. Shapiro et al. (eds.), Neutrinos and Explosive Events in the Universe, 73–82.
© 2005 *Springer. Printed in the Netherlands.*

baryosynthesis and dark matter/energy as well as from the necessity for particle theory to use cosmological tests as the important and in many cases unique
way to probe its predictions.

The convergence of the extreme frontiers of our knowledge in micro- and
macro worlds leads to the wrong circle of problems, illustrated by the mystical
Uhroboros (self-eating-snake). The Uhroboros puzzle may be formulated as
follows: *The theory of the Universe is based on the predictions of particle
theory, that need cosmology for their test.* Cosmoparticle physics [1], [2] offers
the way out of this wrong circle. It studies the fundamental basis and mutual
relationship between micro-and macro-worlds in the proper combination of
physical, astrophysical and cosmological signatures. The divergence in model-
dependent realizations of this relationship is importnat for these studies.

Particles in the Universe

The role of particle content in the Einstein equations is reduced to its contribution into energy-momentum tensor. So, the set of relativistic species, dominating in the Universe, realizes the relativistic equation of state $p = \varepsilon/3$ and the
relativistic stage of expansion. The difference between relativistic bosons and
fermions or various bosonic (or fermionic) species is accounted by the statistic
weight of respective degree of freedom. The treatment of different species of
particles as equivalent degrees of freedom assumes strict symmetry between
them.

Such symmetry is not realized in Nature. There is no exact symmetry between bosons and fermions (e.g. supersymmetry). There is no exact symmetry
between various quarks and leptons. The symmetry breaking implies the difference in particle masses. The particle mass pattern reflects the hierarchy of
symmetry breaking.

Noether's theorem relates the exact symmetry to conservation of respective charge. The lightest particle, bearing the strictly conserved charge, is absolutely stable. So, electron is absolutely stable, what reflects the conservation
of electric charge. In the same manner the stability of proton is conditioned
by the conservation of baryon charge. The stability of ordinary matter is thus
protected by the conservation of electric and baryon charges, and its properties
reflect the fundamental physical scales of electroweak and strong interactions.
Indeed, the mass of electron is related to the scale of the electroweak symmetry
breaking, whereas the mass of proton reflects the scale of QCD confinement.

Most of the known particles are unstable. For a particle with the mass m the
particle physics time scale is $t \sim 1/m$, so in particle world we refer to particles
with lifetime $\tau \gg 1/m$ as to metastable. To be of cosmological significance
metastable particle should survive after the temperature of the Universe T fell
down below $T \sim m$, what means that the particle lifetime should exceed $t \sim$

$(m_{Pl}/m) \cdot (1/m)$. Such a long lifetime should find reason in the existence of an (approximate) symmetry. From this viewpoint, cosmology is sensitive to the most fundamental properties of microworld, to the conservation laws reflecting strict or nearly strict symmetries of particle theory.

However, the mechanism of particle symmetry breaking can also have the cosmological impact. Heating of condensed matter leads to restoration of its symmetry. When the heated matter cools down, phase transition to the phase of broken symmetry takes place. In the course of the phase transitions, corresponding to given type of symmetry breaking, topological defects can form. One can directly observe formation of such defects in liquid crystals or in superfluids. In the same manner the mechanism of spontaneous breaking of particle symmetry implies restoration of the underlying symmetry. When temperature decreases in the course of cosmological expansion, transitions to the phase of broken symmetry can lead, depending on the symmetry breaking pattern, to formation of topological defects in very early Universe. The defects can represent the new form of stable particles (as it is in the case of magnetic monopoles), or the form of extended structures, such as cosmic strings or cosmic walls.

Particle physics beyond the Standard model

Extensions of the standard model imply new symmetries and new particle states. The respective symmetry breaking induces new fundamental physical scales in particle theory. If the symmetry is strict, its existence implies new conserved charge. The lightest particle, bearing this charge, is stable. The set of new fundamental particles, corresponding to the new strict symmetry, is then reflected in the existence of new stable particles, which should be present in the Universe and taken into account in the total energy-momentum tensor.

New physics follows from the necessity to extend the Standard model. The white spots in the representations of symmetry groups, considered in the extensions of the Standard model, correspond to new unknown particles. The extension of the symmetry of gauge group puts into consideration new gauge fields, mediating new interactions. Breaking of global symmetries results in the existence of Goldstone boson fields.

For a long time the necessity to extend the Standard model had purely theoretical reasons. Aesthetically, because full unification is not achieved in the Standard model; practically, because it contains some internal inconsistencies. It does not seem complete for cosmology. One has to go beyond the Standard model to explain inflation, baryosynthesis and nonbaryonic dark matter. Recently there has appeared a set of experimental evidences for the existence of neutrino oscillations (see review in [3]), of cosmic WIMPs [4], and of double neutrinoless beta decay [5]. Whatever is the accepted status of these evidences, they indicate that the experimental searches may have already crossed the border of new physics.

In particle physics direct experimental probes for the predictions of particle theory are most attractive. The predictions of new charged particles, such as supersymmetric particles or quarks and leptons of new generation, are accessible to experimental search at accelerators of new generation, if their masses are in 100GeV-1TeV range. However, the predictions related to higher energy scale need non-accelerator or indirect means for their test.

The search for rare processes, such as proton decay, neutrino oscillations, neutrinoless beta decay, precise measurements of parameters of known particles, experimental searches for dark matter represent the widely known forms of such means.

New physics in the modern cosmology

In the old Big bang scenario the cosmological expansion and its initial conditions was given *a priori*. In the modern cosmology the expansion of the Universe and its initial conditions is related to the process of inflation. The global properties of the Universe as well as the origin of its large scale structure are the result of this process. The matter content of the modern Universe is also originated from the physical processes: the baryon density is the result of baryosynthesis and the nonbaryonic dark matter represents the relic species of the hidden sector of particle theory. Physics, underlying inflation, baryosynthesis and dark matter, is referred to the extensions of the standard model, and the variety of such extensions makes the whole picture in general ambiguous.

The nontrivial path of cosmological evolution, specific for each particular realization of inflational model with baryosynthesis and nonbaryonic dark matter, always contains some additional model dependent cosmologically viable predictions, which can be confronted with astrophysical data. The part of cosmoparticle physics, called cosmoarcheology, offers the set of methods and tools probing such predictions.

Cosmoarcheology considers the results of observational cosmology as the sample of the experimental data on the possible existence and features of hypothetical phenomena predicted by particle theory. To undertake the *Gedanken Experiment* with these phenomena some theoretical framework to treat their origin and evolution in the Universe should be assumed. As it was pointed out in [6] the choice of such framework is a nontrivial problem in the modern cosmology.

One can specify the new phenomena by their net contribution into the cosmological density and by forms of their possible influence on properties of matter and radiation. In the first aspect we can consider strong and weak phenomena. Strong phenomena can put dominant contribution into the density of the Universe, thus defining the dynamics of expansion in that period, whereas the contribution of weak phenomena into the total density is always subdominant. The phenomena are time dependent, being characterized by their timescale, so that permanent (stable) and temporary (unstable) phenomena can take place. They can have homogeneous and inhomogeneous distribution in space. The amplitude of density fluctuations $\delta \equiv \delta\varrho/\varrho$ measures the level of inhomogeneity relative to the total density, ϱ. The partial amplitude $\delta_i \equiv \delta\varrho_i/\varrho_i$ measures the level of fluctuations within a particular component with density ϱ_i, contributing into the total density $\varrho = \sum_i \varrho_i$. The case $\delta_i \geq 1$ within the considered i-th component corresponds to its strong inhomogeneity. Strong inhomogeneity is compatible with the smallness of total density fluctuations, if the contribution of inhomogeneous component into the total density is small: $\varrho_i \ll \varrho$, so that $\delta \ll 1$.

The phenomena can influence the properties of matter and radiation either indirectly, say, changing of the cosmological equation of state, or via direct interaction with matter and radiation. In the first case only strong phenomena are relevant, in the second case even weak phenomena are accessible to observational data. The detailed analysis of sensitivity of cosmological data to various phenomena of new physics are presented in [2].

Cosmophenomenology of new physics

To study the imprints of new physics in astrophysical data cosmoarcheology implies the forms and means in which new physics leaves such imprints. So, the important tool of cosmoarcheology in linking the cosmological predictions of particle theory to observational data is the *Cosmophenomenology* of new physics. It studies the possible hypothetical forms of new physics, which may appear as cosmological consequences of particle theory, and their properties, which can result in observable effects.

The simplest primordial form of new physics is the gas of new stable massive particles, originated from early Universe. For particles with the mass m, at high temperature $T > m$ the equilibrium condition, $n \cdot \sigma v \cdot t > 1$ is valid, if their annihilation cross section $\sigma > 1/(m m_{Pl})$ is sufficiently large to establish the equilibrium. At $T < m$ such particles go out of equilibrium and their relative concentration freezes out. More weakly interacting species decouple from plasma and radiation at $T > m$, when $n \cdot \sigma v \cdot t \sim 1$, i.e. at $T_{dec} \sim (\sigma m_{Pl})^{-1}$. The maximal temperature, which is reached in inflationary Universe, is the reheating temperature, T_r, after inflation. So, the super weakly interacting particles with the annihilation cross section $\sigma < 1/(T_r m_{Pl})$, as well as very heavy particles with the mass $m \gg T_r$ can not be in thermal equilibrium, and the detailed mechanism of their production should be considered to calculate their primordial abundance.

Decaying particles with the lifetime τ, exceeding the age of the Universe, t_U, $\tau > t_U$, can be treated as stable. By definition, primordial stable particles survive to the present time and should be present in the modern Universe. The net effect of their existence is given by their contribution into the total cosmological density. They can dominate in the total density being the dominant form of cosmological dark matter, or they can represent its subdominant fraction. Even in the latter case astrophysical data can be sensitive to their presence, as it is the case [7] for hypothetical stable neutrinos of the 4th generation with the mass about 50 GeV, contributing less than 0,1 % to the total density.

New particles with electric charge and/or strong interaction can form anomalous atoms and contain in the ordinary matter as anomalous isotopes. For example, if the lightest quark of 4th generation is stable, it can form stable charged hadrons, serving as nuclei of anomalous atoms of e.g. crazy helium [8].

Primordial unstable particles with the lifetime, less than the age of the Universe, $\tau < t_U$, can not survive to the present time. But, if their lifetime is sufficiently large to satisfy the condition $\tau \gg (m_{Pl}/m) \cdot (1/m)$, their existence in early Universe can lead to direct or indirect traces. Cosmological flux of decay products contributing into the cosmic and gamma ray backgrounds represents the direct trace of unstable particles. If the decay products do not survive to the present time their interaction with matter and radiation can cause indirect trace in the light element abundance or in the fluctuations of thermal radiation. If the particle lifetime is much less than 1s the multi-step indirect traces are possible, provided that particles dominate in the Universe before their decay. On the dust-like stage of their dominance black hole formation takes place, and the spectrum of such primordial black holes traces the particle properties (mass, frozen concentration, lifetime) [9]. The particle decay in the end of dust like stage influences the baryon asymmetry of the Universe. In any way cosmophenomenoLOGICAL chains link the predicted properties of even unstable new particles to the effects accessible in astronomical observations. Such effects may be important in the analysis of the observational data.

So, the only direct evidence for the accelerated expansion of the modern Universe comes from the distant SN I data. The data on the cosmic microwave background (CMB) radiation and large scale structure (LSS) evolution (see e.g. [10]) prove in fact the existence of homogeneously distributed dark energy and the slowing down of LSS evolution at $z \leq 3$. Homogeneous negative pressure medium (Λ-term or quintessence) leads to *relative* slowing down of LSS evolution due to acceleration of cosmological expansion. However, both homogeneous component of dark matter and slowing down of LSS evolution naturally follow from the models of Unstable Dark Matter (UDM) ([11], see [2] for review), in which the structure is formed by unstable weakly interacting particles. The weakly interacting decay products are distributed homogeneously. The loss of the most part of dark matter after decay slows down the LSS evolution. The dominantly invisible decay products can contain small ionizing component [11]. Thus, UDM effects will deserve attention, even if the accelerated expansion is proved.

The parameters of new stable and metastable particles are also determined by the pattern of particle symmetry breaking. This pattern is reflected in the succession of phase transitions in the early Universe. The phase transitions of the first order proceed through the bubble nucleation, which can result in black hole formation. The phase transitions of the second order can lead to formation of topological defects, such as walls, string or monopoles. The observational data put severe constraints on magnetic monopole and cosmic wall production, as well as on the parameters of cosmic strings. The succession of phase transitions can change the structure of cosmological defects. The more complicated forms, such as walls-surrounded-by-strings can appear. Such structures can be unstable, but their existence can lead the trace in the nonhomogeneous distribution of dark matter and in large scale correlations in the nonhomogeneous dark matter structures, such as *archioles* [12]. The large scale correlations in topological defects and their imprints in primordial inhomogeneities is the indirect effect of inflation, if phase transitions take place after reheating of the Universe. Inflation provides in this case the equal conditions of phase transition, taking place in causally disconnected regions.

If the phase transitions take place on inflational stage, they can provide different initial conditions at sufficiently large scales. Then new forms of primordial large scale correlations can appear. The example of global U(1) symmetry, broken spontaneously in the period of inflation and successively broken explicitly after reheating, was recently considered in [13]. In this model, closed walls of any size can be formed. After their size equals the horizon, closed walls can collapse into black holes. This mechanism can lead to formation of primordial black holes of a whatever large mass (up to the mass of AGNs [14]). Such black holes appear in the form of primordial black hole clusters, exhibiting fractal distribution in space [13]. It can shed new light on the problem of galaxy formation.

Primordial strong inhomogeneities can also appear in the baryon charge distribution. The appearance of antibaryon domains in the baryon asymmetrical Universe, reflecting the inhomogeneity of baryosynthesis, is the profound signature of such strong inhomogeneity [15]. On the example of the model of spontaneous baryosynthesis (see [16] for review) the possibility for existence of antimatter domains, surviving to the present time in inflationary Universe with inhomogeneous baryosynthesis was revealed in [17]. Evolution of sufficiently dense antimatter domains can lead to formation of antimatter globular clusters [18]. The existence of such cluster in the halo of our Galaxy should lead to the pollution of the galactic halo by antiprotons. Their annihilation can reproduce [19] the observed galactic gamma background in the range tens-hundreds MeV. The prediction of antihelium component of cosmic rays [20], accessible to future searches for cosmic ray antinuclei in PAMELA and AMS II experiments, as well as of antimatter meteorites [21] provides the direct experimental test for this hypothesis.

So the primordial strong inhomogeneities in the distribution of total, dark matter and baryon density in the Universe is the new important phenomenon of cosmological models, based on particle models with hierarchy of symmetry breaking.

Towards the unified theory of microworld and Universe

The theories of everything should provide the complete physical basis for cosmology. The problem is that the string theory is still in the form of "theoretical theory", for which the experimental probes are widely doubted to exist. The development of cosmoparticle physics can remove these doubts. In its framework there are two directions to approach the test of theories of everything.

One of them is related with the search for the experimentally accessible effects of heterotic string phenomenology. The mechanism of compactification and symmetry breaking leads to the prediction of homotopically stable objects and shadow matter [22], accessible to cosmoarcheological means of cosmoparticle physics. The condition to reproduce the Standard model naturally leads in the heterotic string phenomenology to the prediction of fourth generation of quarks and leptons [23] with a stable massive 4th neutrino [7], what can be the subject of complete experimental test in the near future. The comparison between the rank of the unifying group E_6 ($r = 6$) and the rank of the Standard model ($r = 4$) implies the existence of new conserved charges and new (possibly strict) gauge symmetries. New strict gauge U(1) symmetry (similar to U(1) symmetry of electrodynamics) is possible, if it is ascribed to the fermions of 4th generation. This hypothesis explains the difference between the three known types of neutrinos and neutrino of 4th generation. The latter possesses new gauge charge and, being Dirac particle, can not have small Majorana mass due to sea saw mechanism. If the 4th neutrino is the lightest particle of the 4th quark-lepton family, strict conservation of the new charge makes massive 4th neutrino to be absolutely stable. Following this hypothesis [23] quarks and leptons of 4th generation are the source of new long range interaction (y-electromagnetism), similar to the electromagnetic interaction of ordinary charged particles. New strictly conserved local U(1) gauge symmetries can also arise in the development of D-brane phenomenology [24]. If proved, the practical importance of this property could be hardly overestimated.

It is interesting, that heterotic string phenomenology embeds even in its simplest realisation both supersymmetric particles and the 4th family of quarks and leptons, in particular, the two types of WIMP candidates: neutralinos and massive stable 4th neutrinos. So in the framework of this phenomenology the multicomponent analysis of WIMP effects is favorable.

In the above approach some particular phenomenological features of simplest variants of string theory are studied. The other direction is to elaborate the extensive phenomenology of theories of everything by adding to the symmetry of the Standard model the (broken) symmetries, which have serious reasons to exist. The existence of (broken) symmetry between quark-lepton families, the necessity in the solution of strong CP-violation problem with the use of broken Peccei-Quinn symmetry, as well as the practical necessity in supersymmetry to eliminate the quadratic divergence of Higgs boson mass in electroweak theory is the example of appealing additions to the symmetry of the Standard model. The horizontal unification and its cosmology represent the first step on this way, illustrating the approach of cosmoparticle physics to the elaboration of the proper phenomenology for theories of everything [11].

Convergence and divergence

We can conclude that observational cosmology offers strong evidences favoring the existence of processes, determined by new physics, and the experimental physics approaches to their investigation. So modern cosmology and physics beyond the standard model converge in their mutual relationship.

However, there is a divergence in various extensions of standard model and in cosmological scenarious, based on them. Such divergence is important to realize all the necessary links between cosmology and particle physics. Cosmoparticle physics [1] offers the set of methods to arrange these links within the true unified picture of Universe and its particle world.

Acknowledgments

I am grateful to T.Stanev and J.Wefel for kind hospitality in Erice. The work was performed in the framework of the State Contract 40.022.1.1.1106 and was partially supported by the RFBR grant 02-02-17490 and grant UR.02.01.008.

References

[1] A.D.Sakharov: Vestnik AN SSSR **4**, 39 (1989); M.Yu.Khlopov: Vestnik of Russian Academy of Sciences **71**, 1133 (2001)

[2] M.Yu. Khlopov: *Cosmoparticle physics*, World Scientific, New York - London-Hong Kong - Singapore, 1999

[3] S. Mikheyev: This volume

[4] R. Bernabei et.al.: Riv. Nuovo. Cim. **26**, 1 (2003)

[5] H.V.Klapdor-Kleingrothaus: Phys. Atom. Nucl. **65**, 2198 (2002)

[6] M.Yu. Khlopov: In: *Cosmion-94*, Eds. M.Yu.Khlopov et al. Editions frontieres, 1996. PP. 67-76

[7] D. Fargion et al: JETP Letters **69**, 434 (1999); D. Fargion et al: Astropart. Phys. **12**, 307 (2000); K.M. Belotsky, M.Yu. Khlopov: Gravitation and Cosmology **8**, Suppl., 112 (2002)

[8] K.M. Belotsky, M.Yu. Khlopov: Gravitation and Cosmology **7**, 189 (2001)

[9] M.Yu. Khlopov, A.G. Polnarev: Phys. Lett. **B97**, 383 (1980)

[10] D.N. Spergel et. al.: Astrophys. J. Suppl. **148**, 175 (2003)

[11] Z. Berezhiani, M.Yu. Khlopov: Yadernaya Fizika **51**, 1157; 1479 (1990); **52**, 96 (1990); Z. Berezhiani, M.Yu. Khlopov, R.R. Khomeriki: Yadernaya Fizika **52**, 104, 538 (1990); A.S. Sakharov, M.Yu. Khlopov: Yadernaya Fizika **57**, 690 (1994)

[12] A.S. Sakharov, M.Yu. Khlopov: Yadernaya Fizika **57**, 514 (1994)

[13] M.Yu. Khlopov, S.G. Rubin, A.S. Sakharov: Gravitation and Cosmology **8**, Suppl., 57 (2002); astro/ph-0202505

[14] S.G. Rubin, A.S. Sakharov, M.Yu. Khlopov: JETP **92**, 921 (2001); hep/ph-0106187

[15] V.M. Chechetkin et al: Phys. Lett. **118B**, 329 (1982)

[16] A.D. Dolgov: Nucl. Phys. Proc. Suppl. **113**, 40 (2002)

[17] M.Yu. Khlopov, S.G. Rubin, A.S. Sakharov: Phys. Rev. **D62**, 0835051 (2000)

[18] M.Yu. Khlopov: Gravitation and Cosmology **4**, 69 (1998)

[19] Yu.A. Golubkov, M.Yu. Khlopov: Phys.Atom.Nucl. **64**, 1821 (2001)

[20] K.M. Belotsky et al: Phys. Atom.Nucl. **63**, 233 (2000)

[21] D. Fargion, M.Yu. Khlopov: Astropart. Phys. **19**, 441 (2003)

[22] Ia.I. Kogan, M.Yu. Khlopov: Yadernaya Fizika **46**, 314 (1987); Ia.I. Kogan, M.Yu. Khlopov: Yadernaya Fizika **44**, 1344 (1986)

[23] M.Yu. Khlopov, K.I. Shibaev: Gravitation and Cosmology **8**, Suppl., 45 (2002)

[24] G. Aldazabal et al: JHEP **0008**, 002 (2000); L.F. Alday, G. Aldazabal: JHEP **0205**, 022 (2002)

B-L GENESIS: GENERALITIES AND IMPACT ON NEUTRINO PHYSICS

M. Yoshimura
Department of Physics, Okayama University
Tsushima-naka 3-1-1 Okayama Japan 700-8530
yoshim@icrr.u-tokyo.ac.jp

Abstract   The theory of baryon asymmetry generation is reviewed, paying attention to leptogenesis scenarios, which do not require baryon number nonconservation in perturbation theory, and may link the problem to neutrino physics.

Introduction

The subject of generation of the baryon asymmetry in our universe is a still rapidly growing field and there are many interesting contributions in recent months. Here, basics of the theory of baryogenesis and leptogenesis are reviewed, focussing on topics considered most important in my view. After a brief introduction, I shall elaborate on the famous 3 conditions [1] taking specific examples, which starts from sources of baryon nonconservation. I shall discuss electroweak baryon nonconservation at high temperatures, explaining the difficulty of electroweak baryogenesis and the role of electroweak theory in redistributing both B and L numbers. Original GUT scenario [2] is used to explain important elements of the baryogenesis theory. Thermal leptogenesis scenario [3] is explained in view of interesting possible connection to neutrino masses recently discovered. In both GUT and leptogenesis the gravitino problem [4] presents important constraints, which make both scenarios fragile. I shall discuss both the problem and its proposed resolutions.

Before going into the main points, let me remind you of recent, experimental developments in particle physics and cosmology, because fascinating progress in recent years makes our topic today even more interesting than originally thought. In particle physics there are two important progresses relevant to our problem, hinting the unification. The unification at high energy of $\approx 10^{16} GeV$, of three fundamental coupling constants which appear separated in low energies is most importantly due to the nature of strong interaction; the asymptotic freedom which predicts that strong force becomes weaker and weaker as en-

M. M. Shapiro et al. (eds.), Neutrinos and Explosive Events in the Universe, 83–92.

ergy scale is increased. This has a far reaching consequence that the early universe is described by simple physical laws such as an equilibrium state of ideal gass and the first order perturbation theory. The coupling unification is achieved around $10^{16} GeV$, only when supersymmetry is assumed to work in intermediate energy regions. On the other hand, neutrino oscillation experiments such as SuperK [5], SNO [6], and KamLand [7] indicate that neutrino masses are of order $0.05 eV$ or smaller. This smallness is nicely explained by the seesaw mechanism [8], which then indicates the new physics scale roughly consistent with the scale of coupling unification. Thus, both of new discoveries in particle physics indicate roughly the same new physics beyond the standard theory we already know so well.

Yet another important hint comes from cosmological observations. Elaborate analysis of precision WMAP and the large structure data revealed the energy budget of the universe. Presence of the dark energy as a major component of the energy of the present universe is something unexpected to most of us, and is perhaps the most important experimental fact facing fundamental physics today. Furthermore, the universe is found to be a Euclidian space, excluding exotic non-Euclidean geometry as the law of nature. The fact that the dark matter component is about a quarter of the energy budget imposes some strong constraint on minimal supergravity theories, as well.

Despite all these impressive progress, we are still too far from the ultimate theory of everything. I listed some obvious avenues for future research in particle physics and cosmology. If I am allowed to say my personal prejudice, I would say that the flavor problem is beyond our reach for years to come, but our understanding of the law of force may further be advanced by a new discovery of violation of empirical conservation laws. The Majorana nature of neutrino masses and proton decay are just manifestation of violation of lepton and baryon numbers, and in my view there is no fundamental obstacle against these being discovered in future, however remote it might be.

A deep reason for this fanatic belief is the gauge principle. I would say that all empirical conservation laws not protected by the gauge principle are doomed to be violated at some level of strength. The only relevant question is at which energy scales these conservation laws are violated. Both theory and experiment should give definitive signatures for this energy scale. We already seem to have some hint on this.

Basics of baryogenesis

Thus we come to the problem of baryogenesis: how to generate the imbalance between matter and antimatter in the present universe. The question is directly related to existence of intelligent life, or more directly, presence of the matter world in our universe. The problem becomes urgent once one realizes almost symmetric microphysical laws and existence of high temperatures in the early universe, which makes presence of antimatter inevitable.

Need for the theory of baryogenesis rests with the almost certain fact that our present universe is made of matter, and lacks antimatter. Observational evidence against the symmetric universe, the universe which contains equal amount of matter and anitimatter globally, are many-fold. Most recently, baloon experiments such as BESS [9] that launched magnets showed that earlier indication of a large amount of antiproton is false, and the present precision data hint no evidence of exotic antiproton and antihelium. The data show that energy dependence of the low energy antiproton to the proton ratio is well explained by secondary antiproton production due to cosmic ray collision off interplanetary matter, taking into account the solar modulation. Thus, the evidence that our local part of the universe is matter dominant is very solid. One might think that in larger scales the universe might be globally matter-antimatter symmetric, but without lack of a reliable theory of matter-antimatter separation the amount of matter and antimatter left behind from the hot big bang is a function of microphysics parameters such as the annihilation cross section, and numerically is far too smaller than the observed baryon to the photon ratio..

The theory of baryogenesis requires new physics beyond the standard model. Fortunately, we already have nice hits; universality of three gauge couplings, the discovery of a finite neutrino mass, and recent results from WMAP, indicating convincing evidence for the dark matter, presumably some sort of still unknown elementary particles.

It is then both natural and even compelling to be able to explain the baryon asymmetry quantified by the baryon to the photon number density in the present universe. This ratio is the direct measure of the asymmetry prior to the cosmic disappearance of antimatter. This number is usually given in the form of the baryon to the entropy ratio n_B/n_s and is of order 10^{-10}, which seems too small at first; one excess of baryon over 10^{10} $B - \bar{B}$ pairs led to the present B-dominated universe. But actually this number is often too large to be explained in theoretical models. With this large number the standard electroweak theory fails as the microscopic theory for the baryogenesis, as explained below.

The necessary condition for baryogenesis is well documented since Sakharov, although explicit realization for any of these conditions is rather delicate. They are (1) baryon number nonconservation, (2) CP violation, (3) departure from equilibrium . The 3rd condition for the need of the arrow of time is due to presence of the inverse process, which should be possible if there is sufficient time for that to happen.

Let us elaborate some of these conditions. First, the baryon nonconservation. Needless to say, grand unified theories (GUT) predicts proton decay, along with more general baryon number violating processes. This is achieved at the Lagrangian level. Less known to physicists outside the field is that the standard electroweak theory itself violates the baryon number in non-perturbative ways, which I shall soon turn to.

The simplest GUT model [10], namely the SU(5) unification and its super-symmetric extension, is almost certainly ruled out from many reasons; lack of proton decay at the predicted lifetime level, inability to produce neutrino masses, and so on. On the other hand, the next target, SO(10) models, and their SUSY extensions in particular, are very promising. It has just the needed component of the right handed Majorana lepton for realization of the seesaw mechanism. The models also accommodates B - L violation which is needed for baryogenesis, as is explained later.

The next item is the electroweak baryon nonconservation. The effect is caused at zero temperature by quantum tunneling, thus suppressed by the exponential factor like $e^{-1/\alpha}$, first discovered by t'Hooft [11]. On the other hand, as pointed out by Dimopoulos and Susskind [12], and later elaborated by Kuzmin, Rubakov and Shaposhnikov [14], a similar effect at high TeV temperature scales is enhanced by barrier crossing over an unstable object, called the sphaleron configuration [13]. At even higher temperatures the effect is only suppressed by a power of the coupling [15]. The process precisely conserves B - L number, thus changing the baryon and lepton numbers simultanesouly. The physics behind this electroweak B violation is the change of fermion zero energy level caused by the gauge and Higgs fields, as clarified by Christ [16].

One might think that the electroweak theory can explain the baryon asymmetry. The 3rd out-of-equilibrium condition is met if the electroweak phase transition is of strongly 1st order. In this case the phase transition proceeds via bubble formation of the broken electroweak phase. Quarks propagating across the bubble wall may be reflected back, thereby producing the baryon asymmetry [17].

The electroweak scenario is now regarded very unlikely to be true [18]. Two main reasons are; (1) light Higgs mass needed for the 1st order phase transition is now experimentally excluded, (2) even if the Higgs mass bound is ignored, the estimated baryon to the photon ratio is too small by more than 10 orders.

Although the electroweak asymmetry generation is excluded, the electroweak baryon nonconservation poses a very important constraint on viable baryogenesis scenarios, because the process is effective to erase already existing baryon number and redistribute baryon and lepton numbers, while conserving B - L. Since the process may take place unhindered, baryon and lepton numbers generated by B - L nonconserving processes at higher temperatures are redistributed such that final B and L numbers are given by the thermal value;

$$B = a \times \Delta(B - L), \qquad a = \frac{8n_g + 4n_H}{22n_g + 13n_H} = \frac{28}{79}, \qquad (1)$$

for the minimum model of three generations and one Higgs doublet. Here $\Delta(B - L)$ is the preexisting asymmetry generated at a higher energy.

Three main contenders for baryogenesis are GUT genesis, leptogenesis, and SUSY scenarios. The GUT scenario is still alive, for instance in models based on SO(10) with B - L violation. The best candidate for B-genesis is the colored triplet of Higgs boson X_H, which has 2 types of channels with different baryon numbers, $q\,l$ and $\bar{q}\,\bar{q}$.

I shall spend some time on the GUT scenario, not because this is the most promising at present, but because in this case it is much easier to understand basic and common elements of the baryogenesis theory. First, about implication of the out-of-equilibrium condition. This requires to block the inverse Higgs decay at some level. The condition is expressed by an inequality $\Gamma_{IX} < H$ between the inverse decay rate Γ_{IX} and the Hubble expansion rate H at the typical time of the inverse decay, namely at around temperatures of the Higgs mass.. With this inequality met, the expansion time is shorter than the inverse decay time, which means that the inverse decay is difficult to occur within the cosmic time. The result gives a lower mass bound for the Higgs triplet, typically like $m_X > O[0.01]\alpha m_{pl}$, thus the Higgs mass m_X is close to the coupling unification scale. Moreover, the scenario is usually taken to imply a high temperature after inflation, because otherwise Higgs bosons may not be abundant to produce a large enough asymmetry.

Next, I would like to discuss intricacy of CP violation. In gauge theories one may compute the fundamental CP asymmetry, namely the difference of baryon numbers between particle and antiparticle decays using perturbation theory. It is thus given by an interference term of, for instance, the tree and the one-loop contributions. A convenient tool of computing the interference term is the Landau-Cutkovsky rule [19]. The result is like

$$\Delta \propto |g_1 f_1 + g_2 f_2 + \cdots|^2 - |g_1^* f_1 + g_2^* f_2 + \cdots|^2 = -4\Im(g_1 g_2^*)\Im(f_1 f_2^*) + \cdots.$$
(2)

A non-zero value for this difference immediately requires both CP violation $\Im(g_1 g_2^*) \neq 0$ and a non-trivial dynamical phase factor such as the rescattering phase in f_2 of a one-loop amplitude.

I shall next discuss some aspects of dynamical computation. The baryon to the entropy ratio in the present universe depends on a few factors. The CP asymmetry is computed as just explained, and directly reflects dependence on microscopic theoretical models. The other factors are more dynamical, resulting from integrating the set of Boltzmann equations or something more general. The result is summarized by dependence of the ratio, $\eta \equiv \Gamma_X/H$ where Γ_X is the Higgs decay rate and H is the Hubble rate at temperature of the Higgs boson mass. Numerical computation [20] suggests $n_B/n_s \propto \eta^{-1.2}$. The generated baryon to the photon ratio is also in proportion to the number density of agent Higgs X boson relative to its thermal density at the temperature of its mass.

Let us summarize constraints on GUT scenarios for a successful B-generation. We take the view that lepton number violating processes related to neutrino masses do occur at temperature scales lower than the GUT scale. Then the survival condition of the GUT generated baryon number places bounds on the magnitude of lepton number violation along with temperature scale of this violation. The result may be expressed by a formula like [21] $\langle m_\nu \rangle < 4eV/\sqrt{T_L/10^{10}GeV}$, where T_L is the temperature scale of L-violation. Some averaged neutrino mass $\langle m_\nu \rangle$, a measure of lepton violation, is constrained from above. This number has been very meaningful, indicating a relatively high L-violation temperature. The other grave problem is a possible overproduction of gravitinos, which will be discussed shortly,

One might say that with the GUT scenario, our existence is assured by the ultimate instability of matter that makes us, despite a very long lifetime. The baryon dominated universe is only a temporary existence in this picture.

Leptogenesis

Let us then turn to the leptogenesis. Generation of the lepton asymmetry at some high temperatures may lead to the baryon asymmetry, when coupled to the electroweak redistribution process. Relation between the final B-asymmetry and the initial L-asymmetry is given by the formula, $B = -\frac{28}{79}\Delta L$. Computation of the fundamental lepton asymmetry ΔL goes exactly as in the previous case of B-genesis, replacing the Higgs H_X by some other agent particle. Leptogenesis is interesting due to a possible connection to observed neutrino masses. Thus, it is a field studied very actively in recent years.

I shall only discuss some fundamentals on the original scenario of thermal leptogenesis [3], in which thermal abundance of the agent particle is assumed. The agent of the asymmetry generation is the heavy right-handed Majorana particle N_R, which is also responsible for the seesaw mass generation of small neutrino mass m_ν a la $m_\nu \approx m_D^2/M$, with M a typical N_R mass, assumed very large. Both N_R and the ordinary neutrino masses are of the Majorana type, and violate the lepton number by 2 units. The N_R decay into Higgs and leptons may lead to the lepton asymmetry. The CP-asymmetry is now a function of a combination of important parameters such as the lightest N_R mass and some average mass $\tilde{m}_\nu$ of ordinary neutrinos;

$$\propto \frac{M_1 \tilde{m}_\nu \delta}{v^2}, \tag{3}$$

where δ is a relevant CP violating phase and $v = 250GeV$. For simplicity, a mass hierarchy among three N_R masses is assumed for this formula, thus the mass M_1 of lightest right-handed Majorana particle appears in this formula.

The formula like this one has great impacts on neutrino physics. The most extensive analysis has been carried out recently by Giudice et al [22], and by [23]. They derive these mass bounds and constraint on the reheat temperature after inflation. The neutrino mass bound of order $0.1eV$ is even more stringent than the recent impressive WMAP result for the neutrino mass. The constraint on the reheat temperature $T_{RH} > 5 \times 10^8 GeV$ is also highly nontrivial.

Gravitino overproduction

The gravitino is a supersymmetric partner of gravitons, and hence it must exist in any supersymmetric theory including gravity. Its mass is around $1TeV$ and its decay rate is given by

$$\Gamma_{3/2} \approx \frac{m_{3/2}^3}{m_{pl}^2} = O[10^3 sec]^{-1} (\frac{m_{3/2}}{1TeV})^3 , \qquad (4)$$

in presently popular models of low energy SUSY breaking. The problem with this gravitino is that it decays very late, such that it might destroy light elements that are already processed at about 3 minutes after the big bang, ruining the great success of the big bang cosmology. Suppression of gravitino oevepro-duction is thus required. In the usual estimate the gravitino abundance $n_{3/2}/s$ relative to the entropy density is directly related to the reheat temperature after inflation, which then constrains

$$\frac{n_{3/2}}{s} = O[10^{-2}]\frac{T_{RH}}{m_{pl}} \Rightarrow T_{RH} < 10^6 - 10^8 GeV , \qquad (5)$$

where the numbers are taken from Kawasaki et al [4]. Thus, concordance with light element production requires a relatively low reheat temperature after inflation, which at its face value seems difficult to reconsile both B- and L-genesis.

A possible resolution of the gravitino problem may be provided by using the preheating mechanism after inflation. The mechanism is an application of the parametric resonance [24] which occurs in dynamical systems. Since the gravitino problem presents severe constraint on both leptogenesis and baryoge-nesis, I shall digress here on the theory of preheating, which in my view must be considered seriously in cosmological scenarios based on inflation.

To simplify discussion, I shall take the simplest kind of inflation model, called chaotic inflation [25]. The inflaton potential is just a mass term, $m_\xi^2 \xi^2/2$, its mass m_ξ taken to be $\approx 10^{13} GeV$ to explain the right order of magnitude of the observed density perturbation $\delta\rho/\rho \approx 10^{-6}$. This potential is then very flat at the Planck epoch of inflation. A notable feature of this inflaton oscillation is that the initial dimensionless amplitude is very large. The inflaton amplitude is gradually damped until the Hubble rate becomes comparable to the inflaton mass $H = m_\xi$, and after $O[10 - 100]$ oscillations explosive ξ particle production takes place from the oscillating background field..

As to the model of particle production, let us take the matter-inflaton coupling of the form, $g\xi\varphi^2$ with g a dimensioful coupling, which is simplest. Since the background classical inflaton field ξ is spatially homogeneous, one may Fourier-decompose the quantum field φ that couples to ξ. The mode equation for the field φ_k of momentum k is then equivalent to the harmonic oscillator equation,

$$\ddot{\varphi}_k + 3\frac{\dot{a}}{a}\dot{\varphi}_k + (k^2 + m_\varphi^2 + g\xi_0 \cos(m_\xi t))\varphi_k = 0\,, \tag{6}$$

where $a(t)$ is the cosmic scale factor. Presence of the frequency term varying in time periodically $\cos(m_\xi t)$ is what dominates the particle production.

The old view of reheating is based on the independent particle decay given by the simple perturbative formula for $\xi \rightarrow \varphi\varphi$. It is further assumed that all decays occur simultaneously and inflaton energy is immediately converted into created φ particle energy. This leads to the reheat temperature given by $\approx \sqrt{\Gamma_\xi m_{pl}}$, with Γ_ξ the decay rate in perturbation theory.

However, this is not what may actually happen. The inflaton periodic oscillation gives rise to stability-instability bands in the two parameter $(k^2 + m_\varphi^2, g\xi_0)$ space of the mode equation for produced particles. The instability bands occupy a large region of the parameter space, and if the initial data lie in these regions, particle production occurs at an exponentially rate. The further deep in the region, the more enhanced the instability, and particle creation is enormously enhanced. A simple diagramatic understanding of the n-th instability band is possible [26] if one extrapolates into the small amplitude region; in terms of Feyman diagrams the decaying ξ field has n quantum involved in the perturbative language, thus the process like $n\xi \rightarrow \varphi\varphi$ occurs. This is quite different from the simplest interpretation of particle production based on the two-body decay $\xi \rightarrow \varphi\varphi$, in which only one ξ quantum is involved. Of course, in the small amplitude oscillation limit, the perturbative interpretation is a valid description. But for large amplitudes a drastic change occurs. A notable feature of the non-perturbative effect is the possibility of producing massive particles, if the produced φ mass $m_\varphi < nm_\xi/2$. Namely, n times larger energetic particle may be produced.

A simple example of the parametric resonance is how one swings. One changes the center of one's body periodically, which enhances the swinging amplitude. Large swing corresponds to copious particle production in our problem.

For details we refer to some result of numerical computation of particle production [27], taking into account both the back-reaction due to particle production and effect of cosmic expansion. Typically, after ≈ 10 oscillations, violent particle production occurs, resulting in sudden transformation of the inflaton oscillation energy $\dot{\xi}^2$ into produced φ particle energy.

The new effect of preheating has great impactd on high energy particles. It becomes possible to produce either H_X or N_R by a large amount, which then generates a large baryon or a lepton asymmetry in the preheating stage. What may happen at later stages is that both the baryon asymmetry and the gravitino abundance is diluted in the late stage of perturbative decay, to an acceptable level [28]. Even with this dilution I believe that the right amount of both baryons and gravitinos may be produced, although detailed analysis based on particular models have to be worked out in specific cases.

There are other interesting recent proposals to suppress gravitino production [29], which also affects the LSP dark matter scenario, but I have no time to discuss this.

Let me conclude with a few obvious statements.

1. B - L genesis links the micro and the macro worlds and is a great hint on physics beyond the standard model.

2. The old baryogenesis idea is still alive, waiting for discovery of nucleon decay.

3. The new leptogenesis scenario is very interesting due to its possible connection to the neutrino sector and lepton number violation.

4. One must watch out the gravitino overproduction in both scenarios.

5. One needs some new, fresh idea to establish a link with other low energy CP violation such as observed K and B decays.

References

[1] A.D. Sakharov, *Pis'ma Zh. Eksp. Teor. Fiz.* **5**, 32(1967); [JETP Lett. **5**, 24(1967)].

[2] M. Yoshimura, *Phys. Rev. Lett.* **41**, 281(1978);**42**,746(E) (1979); *Phys. Lett.***88B**, 294(1979); S. Dimopoulos and L. Susskind, *Phys. Rev.***D18**, 4500(1978); D. Toussaint, S.B. Treiman, F. Wilczek and A. Zee, *Phys. Rev.***D19**, 1036(1979); S. Weinberg, *Phys. Rev. Lett.***42**,850(1979).

[3] M. Fukugita and T. Yanagida, *Phys. Lett.***174B**, 45(1986).

[4] M. Kawasaki, K. Khori, and T. Moroi, astro-ph/0408426 (2004).

[5] Y. Fukuda et al., *Phys. Rev. Lett.***81**,1562(1998).

[6] Q.R. Ahmad et al., *Phys. Rev. Lett.***87**, 071301(2001); *Phys. Rev. Lett.***89**,011301(2002).

[7] K. Eguchi et al., *Phys. Rev. Lett.* **90**,021802(2003).

[8] T. Yanagida, in *Proceedings of the Workshop on the Unified Theory and the Baryon Number in the Universe*, Tsukuba, Ibaraki, Japan, edited by A. Sawada and A.Sugamoto (KEK Report No.KEK-79-18,1979); M. GellMann, P. Ramond, and R. Slansky, in *Supergravity*, edited by D.Z. Freedman and P.Van Niewenhuizen (North-Holland, Amsterdam, 1979).

[9] Y. Asaoka et al., *Phys. Rev. Lett.***84**, 1078(2000).

[10] P. Langacker, *Phys. Rep.***72**, 185(1981).

[11] t' Hooft, *Phys. Rev. Lett.***37**, 8(1976).

[12] S. Dimopoulos and L. Susskind, *Phys. Rev.***D18**, 4500(1978).

[13] F.R. Klinkhamer and N.S. Manton, *Phys. Rev.***D30**, 2212(1984).

[14] V.A. Kuzmin, V.A. Rubakov and M.E. Shaposhnikov, *Phys. Lett.***155B**, 36(1985).

[15] V.A. Rubakov and M.E. Shaposhnikov, hep-ph/9603208 (1996).

[16] N. Christ, *Phys. Rev.***D21**, 1591(1980).

[17] G.R. Farrar and M.E. Shaposhnikov, *Phys. Rev.***D50**, 774(1994).

[18] M.B. Gavela, et al, *Nucl. Phys.***B430**, 345(1994);**B430**, 382(1994); P. Huet and E. Sather, *Phys. Rev.***D51**, 379(1995).

[19] M. Yoshimura, *Cosmological Baryon Production and Related Topics* , in Proceedings of the Fourth Kyoto Summer Institute on Grand Unified Theories and Related Topics, ed. by M. Konuma and T. Maskawa, World Scientific Pub., Singapore (1981).

[20] I. Joichi, S. Matsumoto, and M. Yoshimura, *Phys. Rev.***D58**, 043507(1998); J. N. Fry, K. A. Olive, and M. S. Turner, *Phys. Rev.* **D 22**, 2953, 2977 (1980).

[21] J.A. Harvey and M.S. Turner, *Phys.Rev.***D42**, 3344(1990).

[22] G. F. Giudice, A. Notari, M. Raidal, A. Riotto, and A. Strumia, *Nucl. Phys.***B685**, 89(2004) and hep-ph/0310123(2003).

[23] W. Buchmuller and M. Plumacher, hep-ph/9904310 (1999).

[24] L. Kofman, A. Linde, and A.A. Starobinsky, *Phys. Rev. Lett.* **73**, 3195(1994); M. Yoshimura, *Prog. Theor. Phys.* **94**, 873(1995); H. Fujisaki, K. Kumekawa, M. Yamaguchi, and M. Yoshimura, *Phys. Rev.* **D53**, 6805(1996) and hep-ph/9508378.

[25] A.D. Linde, *Phys. Lett.* **129B**, 177(1983).

[26] M. Yoshimura, hep-ph/9603356 (1996).

[27] H. Fujisaki, K. Kumekawa, M. Yamaguchi, and M. Yoshimura, *Phys. Rev.* **D54**, 2494(1996) and hep-ph/9511381.

[28] M. Yoshimura, J.Korean Phys.Soc. 29 (1996) S236 and hep-ph/9605246 (1996).

[29] J.L. Feng et al., hep-ph/0410178 (2004).

II

EXPLOSIVE EVENTS AND THEIR AFTERMATH

STELLAR EXPLOSION: FROM SUPERNOVAE TO GAMMA-RAY BURSTS

Konstantin Postnov[1,2]

[1]*Sternberg Astronomical Institute, Moscow, Russia*

[2]*University of Oulu, Finland*
pk@sai.msu.su

Abstract   Current understanding of core collapse and thermonuclear supernovae is reviewed. Recent progress in unveiling the nature of cosmic gamma-ray bursts
(GRB) is discussed, with the focus on the apparent link of several GRBs with
an energetic subclass of stellar explosions, type Ib/c core-collapse supernovae.
This relation provides the strong case that the GRB phenomenon is connected
with the final stages of massive star evolution and possibly with the formation
of neutron stars and black holes.

Keywords: Supernovae, core collapse, thermonuclear explosions, gamma-ray bursts

Introduction

Seventy years ago W. Baade and F. Zwicky (1934) [8] were the first to
point out that one of the brightest astronomical phenomenon, supernovae stars
(SNe), can be due to explosions of massive stars at the end of their evolution.
The formation of a dense neutron core (neutron star) results in a sudden energy release of order of the gravitational binding energy of the neutron star
which amounts to $E_g \sim -GM_{NS}^2/R_{NS} \approx 10^{53}$ ergs for the canonical values of the NS mass $M_{NS} \approx 1 M_\odot$ and radius $R_{NS} \approx 10$ km. It was soon
recognized by Gamow and Schoenberg (1941) [49] that most of this energy
comes into neutrino emission. Twenty years after, Hoyle and Fowler (1960)
[59] showed that energy released in type Ia supernovae are connected to thermonuclear burning of a degenerate stellar core. Here the available energy is
$\sim 0.007 M c^2 \approx 2 \times 10^{51}$ ergs for the Chandrasekhar mass of the white dwarf.

In the end of 1960s cosmic gamma-ray bursts (GRBs) were discovered by
gamma-ray satellites [79, 98]. Largely due to the inability of precise localization of the GRB position on the sky using gamma-ray detectors only, the
origin of GRBs was as enigmatic as that of SNe before Baade and Zwicky's

M. M. Shapiro et al. (eds.), Neutrinos and Explosive Events in the Universe, 95–117.

suggestion until late 1990s, when the first successful localizations of GRBs using their afterglow emission were made by BeppoSAX satellite in X-rays [35]. The detection of X-ray afterglows several hours after GRB allowed dedicated follow-up observations of the GRB error boxes to be carried out using powerful optical (e.g. [118]) and radio telescopes [40], in which rapidly decaying afterglow emissions were also detected. Quite soon after that, in the spring 1998, a bright peculiar nearby supernova 1998bw was found within the error box of GRB 980425 [45, 80], suggesting the link between GRBs and SNe. Presently we have several unequivocal associations of GRBs with very energetic type Ibc supernovae called "hypernovae" (see below, Section II).

Being connected to the evolution of stars, SN studies overlap with practically all fields of the modern astronomy, from physics of tiny interstellar medium (e.g. [90]) to the formation of superdense neutron stars [82]. As was also pioneered by Baade and Zwicky, they are sources of astrophysical shocks in which cosmic ray particles are accelerated [17].

Here we focus on some recent highlights in both core collapse and thermonuclear supernova studies, which became possible mainly due to increasingly accurate radiation hydrodynamic calculations with a detailed treatment of neutrino processes. We also briefly describe recent success of asymmetric SN simulations (2D magneto-rotational collapse). Next we focus on recently established link between GRB explosions and energetic type Ibc supernovae (hypernovae) and discuss recent ideas on the GRB progenitors. We hypothesize that different core collapse outcomes may lead to the formation of different classes of GRBs.

Core collapse supernovae

An extensive discussion of basic physics of core collapse supernovae can be found in [12]; the evolution of massive stars, core collapse, formation of stellar remnants and supernova nucleosynthesis are reviewed in [163]; a recent concise discussion of problems and prospects for core collapse supernovae can be found in [105].

In the end of thermonuclear evolution, the core of a massive star can lose mechanical stability for various reasons. In the stellar mass range $8M_\odot < M < 20M_\odot$ a partially degenerate core with mass close to the Chandrasekhar limit $M_{core} \sim M_{Ch}$ and high density ($\rho \sim 10^9 - 10^{10}$ g/cm^3) appears. Under these physical conditions, the chemical potential of degenerate electrons becomes so high that neutronisation reactions $e^- + (A, Z) \to (A, Z - 1) + \nu_e$ become effective even at zero temperature. At densities $> 10^6$ g/cm^3 degenerate electrons becomes relativistic so the adiabatic index of matter $\gamma = d\log P/d\log \rho \to 4/3$, the critical value for loss of mechanical stability. Neutronisation of matter means its deleptonisation (decrease in the lepton number

$Y_e = N_e/N_b$), so the pressure at some moment increases slower than $\rho^{4/3}$ though γ is formally above $4/3$ calculated at constant Y_e, and a catastrophic collapse begins. Temperatures are higher at larger stellar masses $M > 20M_\odot$ so the collapse is initiated by photodissociation of nuclei (here γ really becomes $< 4/3$). For most massive stars with $M > 60M_\odot$ pair creation makes $\gamma < 4/3$.

The collapse occurs on the dynamical (free-fall) time scale $t_{ff} \sim 1/\sqrt{G\rho} \sim$ a fraction of second. Adiabaticity holds so the entropy per baryon $s = S/k_B \approx const \sim 1$ and even can increase due to non-equilibrium beta-processes. Low specific entropy (compared to H-burning phase at the main sequence, $s \approx 10-15$) prevents dissociation of nuclei until they "touch" each other at densities of the order of the nuclear density, $\rho_n \sim 2 \times 10^{14}$ g/cm^3. The collapse stops (if the core mass is below some M_{max}), and bounce of the shock occurs at ~ 50 km from the center.

The bounce shock heats up deleptonized matter and rapidly spends most of its kinetic energy to destroy nuclei and produce plenty of free nucleons (n, p). Modified URCA-processes [38] becomes important: $e^- + p \rightarrow n + \nu_e$, $e^+ + n \rightarrow p + \bar{\nu}_e$ and pair-neutrino annihilation takes place: $e^+ + e^- \rightarrow \nu_e + \bar{\nu}_e$. At the typical collapse temperatures $T \sim 10$ MeV a lot of ν's is produced [168] However, at densities $\rho \sim 10^{12}$ g/cm^3 the mean free path of 10 MeV neutrinos is by 5-6 orders of magnitude smaller than the size of the proto neutron star ($R \sim 50$ km) so the opaque "neutrinosphere" forms. Most of the core collapse neutrinos diffuse out of the neutrinosphere on a time scale ~ 10 seconds. First calculations of ν spectra in core collapse SN were performed by D.K. Nadyozhin [108, 109] We should note that subsequent detailed calculations (e.g. [101] and references therein) did not change much these spectra. Thus, the modest $\sim 10\%$ fraction of the total neutrino energy released in the core collapse ($\sim 10^{53}$ ergs) would be sufficient to unbind the overlying stellar envelope and produce the phenomenon of type II supernova explosion.

Neutrino-driven explosions

Thermal SN explosion mechanism was proposed by Colgate and White [32]. In this picture, part of the neutrino flux liberated in the core collapse is deposited to the stellar mantle to make it unbound ($\sim 10^{51}$ ergs is needed). Specific mechanisms include neutrino-driven fluid instabilities, for example convection both above neutrinosphere and inside the proto-NS. Neutrino-driven convection, however, may not be as important as thought before, as follows from recent detailed 2D studies of convection [24]. Instead, other fluid instabilities such as newly found double-diffusive instability (the so-called "lepto-entropy fingers") [23], may effectively increase neutrino luminosity to help successful explosion. Note here that process $\nu_e + \bar{\nu}_e \rightarrow e^+ + e^- \rightarrow \gamma + \gamma$

may also be important above ν-sphere [9]. This process was proposed as the energy source for GRB fireballs.

It is now realized that even detailed 1D-calculations of the core collapse supernovae fail to produce explosion (see e.g. reviews by [87, 105] and references therein). The main reasons are that the bounce shock rapidly stalls (turns into a pure accretion front) over neutrinosphere at $R_s \sim 200$ km because of nuclear dissociations, and net neutrino heating of the freshly accreted material immediately after the front is insufficient. Burrows [25] notes that 1D-models are about to produces a successful explosion, only 25% to 50% increase in the energy deposition rate inside the "gain region" is required. The stalled shock is not revived because the fresh fall-back matter is advected more rapidly than heated up by neutrinos. An additional heating by the viscous dissipation in differentially rotating models is considered in [142]. Differential rotation of collapsing core seems quite plausible (see evolutionary calculations of rotating stars [54, 55]). The energy stored in shear can amount to $\sim 10^{52}$ ergs for reasonably short 6 ms spin periods of proto neutron stars. The magneto-rotational instability (MRI) which arises in differentially rotating magnetized fluids with negative angular velocity gradients $d\Omega/dr < 0$ [149, 27]. MRI is thought to be responsible for turbulence in accretion disks and may operate in stars as well. This viscosity was found [142] to be comparable with neutrino-driven convective viscosity and much larger than the neutrino viscosity. The account for additional energy dissipation due to the MRI viscosity in differentially rotating stellar cores allowed [142] to obtain a successful explosion in their 1D-calculations.

Transition from 1D to multi-dimensional hydrodynamic calculations has enabled a more accurate treatment of fluid instabilities. Presently, it is recognized that even with detailed account of Boltzmann neutrino transport, additional physical properties of the collapsing core, e.g. rotation, need to be coupled with multi-dimensional calculations to produce successful neutrino-driven explosions (see e.g. [72, 105]).

Asymmetric explosions

There are increasing observational evidence that supernova explosions are generically asymmetric. The indications are as follows.

Spectroscopic observations.   The asphericity of the explosion is directly inferred from spectropolarimetric observations [86]. The smaller the hydrogen mass of the presupernova envelope, the higher the degree of linear polarization is observed. In the canonical type II-plateau supernovae with heavy hydrogen envelopes (SN1999em, 2003gd etc.) polarization is small (less than 0.5%), indicating low degree of asphericity. In contrast, in the stripped-envelope progenitors giving rise to type IIb (SN1993J), type IIn (SN1998S), type Ic

(SN2003gf) or peculiar Ic (SN2002ap, maybe hypernova, see below) super-novae, th polarization degree is large, of order of 1-2%. The degree of polarization tends to increase with time (i.e. as more and more deeper layers of the ejecta become transparent). This suggests the relation of asymmetric supernova explosion to pulsar kicks and, in extreme cases, to GRB explosions (see below).

Pulsar kick velocities. Based on measurements of proper motions of $\sim$ 100 radio pulsars, it is well established now [91, 7] that young neutron stars demonstrate space velocities up to several 100 km/s which cannot be explained without introducing additional natal kicks.

Shape of young galactic SNR. The asphericity of SNRs 1987a, Cas A, N132D etc. is directly seen in multiwavelength observations [116, 61]. Recent *Chandra* X-ray observations of galactic SNR W49B [77] suggest this SNR may be the remnant of an asymmetric hypernova explosion in our Galaxy.

Strong mixing of Ni-56. Strong mixing of $Ni - 56$ in the outer layers of the ejecta is required to explain the observed SN light curves (especially SN1987a, [107]).

Different mechanisms for the SN asymmetry have been proposed.

Hydrodynamic mechanism. Large-scale neutrino-driven overturn between the shock and neutrinosphere can cause the SN anisotropy explosion [26]. Pulsar kicks up to 500 km/s from neutrino convection were obtained in recent 2D hydrodynamic calculations with Boltzmann neutrino transport [24], while in 3D-calculations (but with less rigorous treatment of neutrino transport) [43] high initial velocities of proto-NS were found to be damped by neutrino emission generally to less than 200 km/s. However, the postbounce stellar core flow is found (both in 2D and 3D) to be subjected to generic non-spherical perturbations of the accretion shock leading to the development of the so-called "stationary accretion shock instability" [22]. This large-scale ($l = 1, 2$ modes) fluid instability was also found in multi-dimensional simulations in Ref. [73], and may be responsible for observed bimodal pulsar velocity distribution [7]. It is important that the developed accretion-shock instability can result in bipolar explosions even in the absence of rotation.

Magnetorotational mechanism. Rotation of the pre-supernova core and magnetic field increase during the collapse, and was proposed in the beginning of 1970s [13] as an alternative mechanism for core-collapse SNe. The magnetorotational mechanism by its nature is asymmetric and can launch antiparallel jets during the explosion. First simulations were made in Ref. [84, 141]. Results of recent 2D MHD calculations [4, 5] are very encouraging: strong differential rotation in the presence of (maybe initially weak) magnetic field was shown to increase the magnetic pressure and form a MHD shock. As a result, rotational energy of neutron star is converted to the energy of the radial expansion of the envelope. The exponential growth of magnetic field in differentially rotating collapsing stellar cores due to MRI [1] appears to help the magnetorotational explosion as well. The magnetorotational explosion was found in [5] to be essentially divided into three stages: linear growth of the toroidal component of the magnetic field due to twisting of the magnetic filed lines, the exponential growth of both toroidal and poloidal field components due to development of MHD instabilities, and the formation of a MHD shock leading to the explosion. These 2D MHD simulations [5] have shown that the obtained energy of the magnetorotational explosion $\sim 0.6 \times 10^{51}$ ergs is sufficient to explain type II and type Ib core-collapse supernovae.

Formation and subsequent collision of a close binary NS+NS system inside the stellar interiors.   V.S. Imshennik [64] suggested the following scenario: a rapidly rotating core collapse results in the core fusion due to rotational instability (the original idea goes back to von Weizsaecker [158]), then the binary NS system coalesces due to gravitational radiation losses. In this scenario, the lighter NS (with larger radius) first fills its Roche lobe and start losing mass as the orbit shrinks due to gravitational radiation. The NS gets unstable when $M < M_{min} \approx 0.1 M_\odot$ providing an energy release of $\sim 10^{50}$ erg in neutrinos [18, 33, 34]. This mechanism has been further elaborated in Ref. [65, 2]. It was suggested for SN1987a explosion [66] to explain double LSD neutrino signal separated by ~ 5 hours observed from SN1987a. A point of concern here may be a rapidly rotating pre-supernova core required for this mechanism to operate. Recent calculations by the Geneva group of stellar evolution with rotation but without magnetic field [57] indicate that there are enough angular momentum in the pre-supernova core, while calculations with (even approximate) account for magnetic field [55] show a severe (30-50 times) reducing of the final rotation of the collapsing iron core in massive stars. The core fragmentation during collapse with rotation was also not found in 3D calculations in Ref. [44].

Non-standard neutrino physics.   To explain high pulsar kicks, neutrino asymmetries in high magnetic fields pertinent to young pulsars have been invoked [29, 14, 83]. The neutrino asymmetry is a natural consequence of the asymmetry of basic weak interactions in the presence of a strong magnetic field. Neutrino oscillations requiring a sterile neutrino with mass 2-20 keV and a small mixing with active neutrinos was proposed in Ref. [81]. However, a critical analysis in Ref. [73] suggests that neither rapid rotation nor strong magnetic field of the young neutron star can be unequivocally inferred from observations of core-collapse SNR 1987A and Cas A, so the solution to the problem of high pulsar kicks probably should be looked for within the frame of the conventional hydrodynamic SN explosion mechanism (see above).

Thermonuclear supernovae

Thermonuclear supernovae constitute a separate very important class of stellar explosions. In contrast to core-collapse SNe they demonstrate very similar light curves and which allowed their using as "standard candles" in modern cosmology (e.g. [128] and references therein). SN Ia are due to thermonuclear burning of a (C+O) white dwarf with $M \sim M_{Ch}$ [59]. A recent deep review of the SN 1a explosion models can be found in Ref. [56]. In the modern picture (e.g. [165]), thermal instabilities in degenerate matter with $\rho \sim 2 - 9 \times 10^9$ g/cm^3, $T \sim 7 \times 10^8$ K, after ~ 1000 years of core convection initiate flame ignition within a typical ignition radius $\sim 150 - 200$ km and could result in the complete destruction of the white dwarf. In principle, at higher density a neutron star could be formed due to electron-capture processes (via accretion-induced collapse of a O+N+Mg core or accretion onto a O-Ne-Mg white dwarf in a binary system [133, 62]), but higher densities are disfavored from evolutionary viewpoint.

Possible scenarios for the formation of type Ia SNe include double degenerate white dwarf mergings or accretion onto a massive white dwarf in a symbiotic binary (see discussion of different formation channels for SN Ia progenitors in Refs. [166, 131, 167]). Sub-Chandrasekhar mass models for SN1a due to an external trigger (e.g. detonation of the accreted He layer) may result in subluminous SN Ia (like 1991bg). A SN 1.5 model in which the degenerate core explodes at the late asymptotic giant phase of the evolution of an intermediate mass star was suggested in Ref. [63] SN 2002ic and SN 1997cy in whose spectra hydrogen absorption lines were discovered probably belong to this class [30]. In this model the hydrogen envelope ejection is synchronized (within ~ 600 years) with the explosion of a contracting white dwarf as its mass approaches the Chandrasekhar limit.

After the initial thermonuclear ignition of the degenerate core interior, a strong temperature dependence of the nuclear reaction rates ($\propto T^{12}$ at $\sim$ MeV temperatures) leads to the formation of very thin burning layers propagating conductively with subsonic speeds (deflagration, i.e. a flame) or burning due to shock compression (supersonic detonation). Both propagation modes are linearly unstable to spatial perturbations and presently are treated using multi-dimensional calculations. The prompt detonation of the degenerate core as SN 1a mechanism was first studied by D. Arnett [6]. However, the mechanism is inconsistent with observations as too little intermediate-mass elements are born in detonation, and was found not to operate because the core at ignition is insufficiently isothermal [160]. Deflagration to detonation (delayed detonation) burning regime was suggested in Ref. [67] and further elaborated in [78]. Pure carbon deflagration with convective heat transport was proposed in Ref. [110].

The main problem encountered in studies of the SN Ia explosion mechanisms can be formulated as follows: the prompt detonation produces enough energy for explosion (thermonuclear runway energy a white dwarf $\sim 0.007 M c^2 \sim 1.5 \times 10^{51}$ ergs) but gives incorrect nucleosynthesis yields (mainly iron peak elements, in contrast with observations showing significant abundance of the intermediate mass elements), the pure deflagration is too slow and must be accelerated. The acceleration of the deflagration front is achieved by involving different flame instabilities: Landau-Darrieus flame instability, which leads to the flame front fractalization - wrinkles and folds, so the surface area of the flame effectively increases (see analytical treatment in [19] and numerical simulations in [129]); (2) Rayleigh-Taylor (RT) instability resulting from the buoyancy of hot, burned material in the dense, unburned surroundings. The RT instability generically develops after the initial ignition in degenerate cores turbulent deflagration and leads to the appearance of the hot bubbles ("mushrooms") floating upwards while spikes of cold fluid falls down. Secondary hydrodynamic instabilities due to shear along the bubble surface rapidly leads to the turbulence. The key role of the RT instability in the presently concurrent turbulent deflagration and delayed-detonation models for SN Ia explosions is fully confirmed in 3D numerical simulations [127, 47].

Recent progress in the physics of the SN Ia explosions has been done mainly due to multi-dimensional numerical simulations of the flame propagation in the degenerate star. For example, multidimensional Chandrasekhar mass deflagration simulations [127] indicate acceleration of the turbulent combustion front up to 30% of the speed of sound, which is enough to produce an explosion without transition to detonation. Generally, 3D-models are found to be more energetic. Multi-dimensional calculations of nucleosynthesis in the pure deflagration Chandrasekhar model [144] shows that turbulent flame converts about 50% of carbon and oxygen to ash with different composition depending on the density of the unburned material. To burn most of the material in the center (so that to avoid unobserved low-velocity carbon, oxygen and intermediate-mass elements in the spectra), the model requires a large number of ignition spots. In contrast, 3D simulations of the delayed-detonation SN Ia explosion [48] indicate that there is no such problem in this model and the explosion energy is higher than that obtained in the pure deflagration burning. It remains to be seen from future observations which model is more adequate.

Note that uncertainties in the ignition conditions of the degenerate star leads to some irreducible diversity of the explosion kinetic energy, peak luminosity, nickel production for the same initial configuration. Modeling of light curves of SN Ia turns out to be a powerful tool to check the SN Ia explosion models (see recent calculations by multi-group radiation hydrocode STELLA [21]).

Gamma-ray bursts

GRBs have remained in the focus of modern astrophysical studies for more than 30 years. After the discovery of GRB afterglows in 1997 [35], the model of GRB as being due to a strong explosion with isotropic energy release of 10^{53} ergs in the interstellar medium became widely recognized. Various aspects of GRB phenomenology are discussed in many reviews: observational and theoretical studies are summarized in [60], first observations of afterglows are specially reviewed in [119], GRB theory is extensively discussed in [102, 169].

A widely used paradigm for GRBs is the so-called fireball model (e.g. reviews by Piran [120, 121] and references therein). In this model, the energy is released in the form of thermal energy (its initial form is usually not specified) near the compact central source (at distances and is mostly converted into leptons and photons (the fireball itself). The relativistic outflow (wind) is formed driven by the high photon-lepton pressure (generically in the form of two oppositely directed narrow collimated jets) [113, 135]. The fireball internal energy is converted to the bulk motion of ions so that relativistic speed with high Lorentz-factors (typically, $\Gamma > 100$) is achieved during the initial stage of the expansion; the ultrarelativistic motion is in fact dictated by the need to solve the fireball compactness problem (see [20] for a detailed discussion and references). The kinetic motion of ions is reconverted back into heat in strong collisionless relativistic shocks at typical distances of 10^{12} cm. Assuming the appropriate turbulence magnetic field generation and particle acceleration in the shocks, energy thermalized in the shocks is emitted via synchrotron and inverse-Compton radiation of shock-accelerated electrons [125] (see [156] for a review), which is identified with the GRB emission. A shell of ultrarelativistically moving cold protons produces a blast wave in the surrounding medium, forming an external shock propagating outward and reverse shock that propagates inward and decelerates the explosion debris. Most energy of the explosion is now carried by the external shock which decelerates in the surrounding medium. Assuming magnetic field generation and particle acceleration in the external shock, the afterglow synchrotron emission of GRB is produced in radio [115], optical [75, 104] and X-rays [151, 104]. Note that at this stage the memory of the initial explosion conditions is cleaned, and the dynamical evolution of the external shock is well described by the Blandford-McKee self-similar solution [15], a relativistic analog of the Sedov-von-Neumann-Taylor solution for strong point-like explosion. This explains the apparent success in modeling the GRB afterglow spectral and temporal behavior in the framework of the simple synchrotron model [159], irrespective of the actual nature of the GRB explosion.

There is no consensus thus far about the origin of the GRB emission itself. Within the fireball model, the GRB can be produced by internal fireball dissipation (the internal shock wave model, e.g. [126]), or in the external blast wave decelerating in the ambient (inhomogeneous) medium [103, 37]. The fireball model is known to face some important problems (for example, baryon contamination of the fireball, the microphysics of magnetic field generation and particle acceleration in collisionless ultrarelativistic shocks etc., see a critical review in [93, 36]). In Ref. [93] an alternative to the fireball model is analyzed in which large-scale magnetic fields are dynamically important. Whether the GRB jets are hot (fireball model) or cold (electromagnetic model) remains to be determined from future observations. Here crucial may be spotting the very early GRB afterglows and measuring polarization of prompt GRB emission (see [92] for the short-list of the electromagnetic model predictions). Note that irrespective of the mechanism mediating the energy transfer from the central source to the baryon-free region, many essential features of the observed non-thermal GRB spectra can be reproduced in some general physical models of prompt gamma-ray emission of GRBs, e.g. by synchrotron self-Compton emission of plasma with continuously heated nearly monoenergetic electrons [140].

An important open issue is whether GRBs can be the sources of ultrahigh energy cosmic rays (UHECR). This association was suggested in Refs. [155, 106, 150] based on similarity of the energy release in GRBs $\sim 10^{44}$ erg per year per cubic Megaparsec with what is observed in UHECRs and assuming effective proton acceleration in relativistic collisionless shocks. The mechanism of UHECR production in GRBs is still uncertain, but basic requirements for proton acceleration to high energies in mildly-relativistic shocks (pertinent to the internal shock model of GRBs) appear to be satisfied. See Ref. [157] for more detail and discussion. See also lectures by M. Teshima and M. Ostrowski on this School.

Below we focus on the observed association of GRBs with an energetic subclass of core-collapse supernovae, type Ibc SNe, which with each new finding provides an increasing evidence that the GRB phenomenon is related to the evolution of most massive stars and formation of relativistic compact objects (neutron stars and black holes).

Supernova - GRB connection

Theoretical grounds: the collapsar model

The connection of GRBs with stellar explosions was first proposed theoretically. Woosley (1993) [161] considered a model of accretion onto a newly formed rotating black hole to power the GRB fireball. The progenitor to GRB in this model is a rapidly rotating Wolf-Rayet (WR) star deprived of its hydrogen and even helium envelop due to powerful stellar wind or mass transfer in a binary system. Dubbed by Woosley himself as "failed type Ib supernovae", this model is now called the collapsar model [94]. In this model, a massive ($\geq 25 M_\odot$) rotating star with a helium core $\geq 10 M_\odot$ collapses to form a rapidly rotating BH with mass $\geq 2 - 3 M_\odot$. The accretion disk from the presupernova debris around the BH is assumed to be the energy source for GRB and is shown to be capable of providing the prerequisite $10^{51} - 10^{52}$ ergs via viscous dissipation into neutrino-antineutrino fireball. The energy released is assumed to be canalized in two thin antiparallel jets penetrating the stellar envelope.

Another possible energy source in the collapsar model could be the electromagnetic (Poynting-dominated) beamed outflow created via MHD processes, much alike what happens in the active galactic nuclei powered by accretion onto a supermassive BH. The estimates show that the Blandford-Znajek (BZ) process [16] in the collapsar model (e.g. [85]) can be a viable candidate for the central engine mechanism for GRBs, provided somewhat extreme values for BH spin (the Kerr parameter $a = Jc/GM^2 \sim 1$) and magnetic field strength in the inner accretion disk around the BH ($B \sim 10^{14} - 10^{15}$ G). In that case the rotating energy of BH (up to $0.29 M_{bh} c^2$ for $a = 1$) is transformed to the Poynting-dominated jet with energy sufficient to subsequently produce a GRB.

Yet another source of energy in the collapsar model could be the rotation energy of a rapidly spinning neutron star with high magnetic field (magnetar), as originally proposed by Usov [146, 147]. As in the BZ-based models, the GRB jets are Poynting-dominated. Lyutikov and Blandford [93] develop the electromagnetic model, which postulates that the rotating energy of the GRB central engine is transformed into the electromagnetic energy (for example, in a way similar to the Goldreich-Julian pulsar model) and is stored in a thin electromagnetically-dominated "bubble" inside the star. The bubble expands most rapidly along the rotational axis, breaks out of the stellar envelopes and drives the ultrarelativistic shock in the circumstellar material. In contrast to the synchrotron GRB model, here GRB is produced directly by the magnetic field dissipation due to current-driven instabilities in this shell after the breakout. The energy transfer to GRB is mediated all the way by electromagnetic field and not by the ion bulk kinetic energy. It remains to be checked by observations whether the EM or fireball model for GRB emission is correct.

A different scenario is the so-called "supranova" model for GRB proposed in [152] involves a delayed collapse to BH through the formation of a super-massive rotationally supported neutron star (see also [153] for another vari-ant of this model). In the supranova scenario GRB is associated with the BH formation and occurs in baryon-clean surrounding after the initial supernova explosion. The original scenario predicts several weeks - month time delay between the SN explosion and GRB. The critical comparison of the supranova mechanism with observations (showing its inconsistency) can be found in [36]. In this paper an interesting extension of the supranova scenario is considered in which the second collapse occurs minutes or hours after the primary SN explosion. In this model, a rotationally-supported magnetized neutron star is formed in a SN type Ic explosion and rapidly loses energy via powerful pulsar wind along the rotational axis. The wind drills the baryon-clean polar cones through which a newly formed rapidly spinning BH generates relativistic jets producing a GRB (for example, by BZ mechanism). Though no rapid rotation of magnetized neutron star has been obtained in stellar evolution calculations (nor can it definitely be inferred from the existing observations, cf. [73]), the spin increase with the remnant's mass found in Ref. [55] may provide credence to this mechanism. Some signatures from the pulsar-wind heated supernova shell can be tested in future observations.

Observational evidence: GRB-supernova associations

First hint on the association of GRBs with SNe came from the apparent time coincidence (to within about a day) of GRB 980425 with a peculiar supernova SN 1998bw [45]. SN 1998bw occurred in a spiral arm of nearby (redshift $z = 0.0085$, distance ~ 40 Mpc) spiral galaxy ESO 184- G82. Such a close location of GRB 980425 rendered it a significant outliers by (isotropic) energy release $\Delta E_{iso} \approx 10^{48}$ erg from the bulk of other GRBs with known energy release, and even from a beaming-corrected mean value of GRB energies of $\sim 10^{51}$ erg [41].

Now the most convincing evidence for GRB-SN association is provided by spectroscopic observations of expanding photosphere features in late GRB af-terglows. Especially strong is the case of a bright GRB 030329 associated with SN 2003dh [58, 138, 97, 99, 76]. Spectral observations of the optical afterglow of this GRB revealed the presence of thermal excess above non-thermal power-law continuum typical for GRB afterglows. Broad absorption troughs which became more and more pronounced as the afterglow faded indicated the pres-ence of high-velocity ejecta similar to those found in spectra of SN 1998bw. Despite these strong evidences, there are some facts which cannot be explained by simple combination of the typical SN Ibc spectrum and non-thermal power-law continuum. For example, the earliest spectroscopic observations of GRB

030329 of optical spectra taken on the 6-m telescope SAO RAS 10-12 hours after the burst [136] showed the presence of broad spectral features which could
not be produced by a SN at such an early stage. The complicated shape of the
optical light curve of this GRB with many rebrightenings [88] and polarization
observations made by VLT [52] suggest a clumpy circumburst medium and require additional refreshening of shocks (if one applies the synchrotron model,
e.g. [51]).

Another interesting example of GRB-SN connection is provided by GRB
031203. This GRB is one of the closest ($z = 0.105$) known GRBs and is found
to be intrinsically faint, $\Delta E_{iso} \sim 10^{50}$ ergs ([154, 132] [1]. The low energy
release in gamma-rays is confirmed by the afterglow calorimetry derived from
the follow-up radio observations [137] and allows this GRB to be considered as
an analog to GRB 980425. It is important that the low energy release in these
bursts can not be ascribed to the off-axis observations of a "standard" GRB
jet (unless one assumes a special broken power-law shape of GRB luminosity
function, see [53]). However, a bright type Ib/c supernova SN 2003lw was
associated with GRB 031203 as suggested by the rebrightening of the R light
curve peaking 18 days after the burst and broad features in the optical spectra
taken close to the maximum of the rebrightening [31, 143, 96, 46].

The comparison of radio properties of 33 SNe type Ib/c with those of measured radio GRB afterglows allowed Berger et al. [10] to conclude that not
more than few per cents of SNe type Ib/c could be associated with GRBs,
which explains the observed small galactic rate of GRBs. However, it still remains to be studied how much intrinsically faint GRBs like 980425 and 031203
can contribute to the total GRB rate.

Hypernovae

Core-collapse supernovae with kinetic energy of the ejecta $\sim 10 - 30$ times
as high as the standard 1 foe (1foe $= 10^{51}$ erg) are now collectively called
"hypernovae". The term was introduced by B. Paczynski shortly after the discovery of first GRB afterglows in 1997 by the Beppo-SAX satellite [114] based
on qualitative analysis of possible evolutionary ways leading to cosmic GRB
explosions.

SN 1998bw was exceptionally bright compared to other Ib/c SNe (the peak
bolometric luminosity of order 10^{43} erg/s, comparable to the SN Ia peak luminosities). This points to the presence of a substantial amount of ^{56}Ni isotope,
the radioactive decay thereof being thought to power the early SN light curves.
The spectra and light curve of SN 1998bw was modeled by the explosion of
a bare C+O of a very massive star that has lost its hydrogen and helium envelopes with a kinetic energy more than ten times typical SNe energies[68],
and they called SN 1998bw a hypernova.

Since then several other SNe were classified as SN 1998bw-like hypernovae by their spectral features and light curves: SN 1997ef, SN 2002ap, SN 2003dh/GRB030329, SN 2003lw/031203. Recently, SN 1997dq was dubbed a hypernova by its similarity with SN 1997ef [100].

Extensive numerical modeling of light curves and spectra of hypernovae (see [112] for a recent review) confirmed the need of atypically high for corecollapse SNe mass of nickel ($\sim 0.1 - 0.5M_\odot$) to be present in the ejecta in order to explain the observed hypernova properties. The rapid rise in of the observed light curves of the "canonical" SN 1998bw requires a substantial amount of ^{56}Ni to be present near the surface. This strongly indicates the important role of mixing during the explosion as nickel is synthesized in deep layers during a spherical explosion. This fact can serve as an additional evidence for non-spherical type Ic explosions. As we already stressed above, the asphericity appears to be a ubiquitous feature of core-collapse supernovae in general, culminating in bipolar hypernova explosions associated with GRBs.

Spectral modeling suggests [112] that the broad-band spectral features generally seen in early and maximum light of hypernovae signal very rapid photospheric expansion. In particular, authors of Ref. [112] notice the very unusual for other SNe fact that OI ($\lambda = 7774A$) and CaII IR (at $\lambda \sim 8000A$) absorption lines merge into a single broad absorption in early spectra of SN 1998bw, which indicates a very large velocity of the ejecta (the line separation ~ 30000 km/s).

In general, varying (a) the progenitor C+O core mass from 2 to ~ 14 solar masses, choosing (2) the appropriate mass cut (corresponding to the mass of the compact remnant, a neutron star or black hole $M_c = 1.2 - 4M_\odot$), and (3) mass of ^{56}Ni isotope ($\sim 0.1 - 0.5M_\odot$) and its mixing allow [112] to reproduce the observed spectra and light curves of hypernovae.

The analysis of nucleosynthesis in hypernovae suggests a possible classification scheme of supernova explosions [111]. In this scheme, core collapse in stars with initial main sequence masses $M_{ms} < 25 - 30M_\odot$ leads to the formation of neutron stars, while more massive stars end up with the formation of black holes. Whether or not the collapse of such massive stars is associated with powerful hypernovae ("Hypernova branch") or faint supernovae ("Faint SN branch") can depend on additional ("hidden") physical parameters, such as the presupernova rotation, magnetic fields. [39], or the GRB progenitor being a massive binary system component [145, 117]. The need for other parameters determining the outcome of the core collapse also follows from the continuous distribution of C+O cores of massive stars before the collapse, as inferred from observations, and strong discontinuity between masses of compact remnants (the "mass gap" between neutron stars and black holes) [28][2].

The mass of ^{56}Ni synthesized in core collapse also appears to correlate with M_{ms}. In ordinary SNe (like 1987a, 1993j, 1994i), $M_{Ni} = 0.08 \pm 0.03M_\odot$, but for hypernovae this mass increases up to $\sim 0.5 M_\odot$ for the most energetic events. Large amount of ^{56}Ni in hypernovae suggests a different nucleosynthesis event. It was shown that unlike conventional core-collapse shock nucleosynthesis [162], nucleosynthesis in bipolar supernova explosions [95] or in relativistic modest-entropy massive wind from accretion disk around a BH [124] can in principle give the observed high amount of nickel in hypernovae. Another important consequence of hypernova nucleosynthesis can be larger abundances (relative to the solar one) of Zn, Co, V and smaller abundance of Mn, Cr, the enhanced ratios of α-elements, and large ratio of Si, S relative to oxygen (see [111] for further detail).

It is also necessary to note that energy requirements for hypernova explosions ($\Delta E > 2 \times 10^{51}$ ergs) can hardly be provided by the most elaborated delayed neutrino explosion mechanism. Indeed, the net explosion energy in this mechanism comes mostly from nuclear recombination of matter inside the gain radius. Analytical [70] and numerical calculations [71] suggest this mass to fall within $0.01 - 0.1 M_\odot$ range, so assuming the recombination energy release 8 MeV per nucleon results in $\sim 10^{51}$ ergs of the explosion energy (see also calculations of the SN explosion energy as a function of the progenitor mass in Ref. [42]). The magnetorotational mechanism [5] fails to produce the hypernova energies either (see above). Unless some non-standard physics does operate, the hypernova energies can be recovered from accretion of the rotating collapse debris onto BH ($\Delta E \sim 0.06 - 0.42 \Delta M_a c^2$ depending on the BH spin) or by BZ mechanism ($\Delta E \leq 0.29 M_{bh} c^2$). In both cases, rapid rotation (and extremely large magnetic field for BZ to operate) of the presupernova is required. Here realistic self-consistent calculations still have to be done, which is a very difficult task.

Progenitors of GRBs

The GRB-SN connection leads to the generally accepted concept that massive stars that lost their envelopes are progenitors of long GRBs (this limitation is due to the fact that predominantly long GRBs with duration ¿ 2 s can be well localized on the sky and provide rapid alerts for follow-up multiwavelength observations). For short single-pulsed GRBs (a quarter of all bursts, see e.g. catalog by Stern et al. [139]) the binary NS+NS/NS+BH merging hypothesis [18, 130, 69] remains viable (see also recent general relativistic hydrodynamic models of the launch and propagation of relativistic jets due to thermal energy deposition near the center of binary mergers [3]).

As we already noted, the emerging empirical evidence is that there exist intrinsically faint, single-pulsed, apparently spherically-symmetric GRBs (980425, 031203) associated with strong hypernovae. These hypernovae require maximal amount of nickel to be synthesized in explosion and large kinetic energies. On the other hand, another unequivocal hypernova SN 2003dh, associated with the "classical" GRB 030329, can be modeled with exceptionally high kinetic energy (4×10^{52} ergs) but smaller amount of nickel ($\sim 0.35 M_\odot$) and smaller mass of the ejecta ($8 - 10 M_\odot$) [99]. These parameters were obtained assuming spherical symmetry, which is of course not the case for GRB 030329[3]. But if this tendency is real and will be confirmed by later observations, we can return to our hypothesis [123] that there should be distinct classes of GRBs according to what is the final outcome of collapse of the CO-core of a massive star. If the collapse ends up with the formation of a neutron star, an intrinsically faint smooth GRB could be produced and a heavy envelope is ejected in the associated SN Ib/c explosion. The GRB energy in this case can be essentially the rotation energy of the neutron star $\sim 10^{49} - 10^{50}$ ergs, as in the electromagnetic model [146]. If a BH is formed, a lighter envelope is ejected with accordingly smaller amount of nickel and possibly with higher kinetic energy of the ejecta, and more energetic, highly variable GRB with a "universal" jet structure [122] emerges fed by non-stationary accretion onto the BH.

The GRB energy can be also interpreted in more exotic way requiring a new physics. The possible relation of GRBs to mirror dark matter was discussed in Ref. [20]. The conversion of light axions from SNe to photons as the source of the GRB fireballs was considered in [89, 11]. Recently it was suggested [50] that ultramassive axions in the mirror world with the Peccei-Quinn scale $f_a \sim 10^4 - 10^6$ GeV and mass $m_a \sim 1$ MeV can be produced in the gravitational collapse or in the merging of two compact stars. The axions tap most of the released energy and can decay ~ 1000 km away mostly into visible electron-positron pairs (with 100% conversion efficiency) thus creating the primary GRB fireball. The estimates show that successful short GRBs can be obtained in compact binary coalescences, while long GRBs can be created in collapsars. In extended SN II progenitors, this energy may help the mantle ejection. In compact CO-progenitors for SN Ib/c axions decay inside the star, so depending on the stellar radius weaker or stronger GRBs associated with SNe type Ib/c explosions can be observed. In this picture again the collapse with the formation of a neutron star or BH may have different signatures.

Conclusions

Cosmic explosions including various types of supernovae and GRBs are natural consequence of stellar evolution. The big efforts of different groups to theoretically understand the physical mechanism(s) of core collapse SNe appear to be approaching the final phase. The thermal (neutrino-driven) mechanism for core collapse SNe is mostly elaborated at present but still fails to produce a strong explosion. There is understanding why stars do not explode by this mechanism, and the road map is designed how to obtain the explosion. Recent multidimensional simulations with accurate incorporation of Boltzmann neutrino transport indicate that effectively a modest boost in the neutrino luminosity is still required. This can be due to inclusion of new fluid instabilities and more accurate treatment of neutrino processes and microphysics (the neutron star equation of state) into the full 3D hydrodynamic calculations. The rotation and magnetic fields is not yet fully included into calculations, which is a challenging task. First results of 2D MHD calculations of the magnetorotational supernova explosion seem to be encouraging. The most energetic (hypernova) explosions, however, require an additional to neutrino source of energy, and the rotation and magnetic fields can be the principal ingredients.

An impressive progress has been done in multidimensional calculations of thermonuclear explosions of degenerate dwarfs for type Ia supernovae. It is still however unclear whether pure deflagration or delayed detonation is at work in SN Ia. The important problem is to more precisely determine the initial ignition conditions. Detailed radiation hydrodynamic modeling revealed that SN Ia light curves proved to be very sensitive to the explosion models and thus can be used to check the models.

In contrast to SNe, the nature of cosmic GRBs remains unclear. The most important recent progress in understanding GRBs was in establishing the link of at least part of them to unusually energetic SN Ibc explosions (hypernovae). At present several unequivocal GRB-SN associations are known. The two closest GRBs discovered so far (GRB 989425 and GRB 031203) proved to be intrinsically weak compared to the bulk of other GRBs with measured redshifts. They both show a single-peak smooth gamma-ray light curve with no signs of jet-induced breaks in the afterglows. In the third (most strong) case of the GRB-SN association, GRB 030329/SN 2003dh, the GRB light curve is two-peaked, the afterglows show evidence for jet. Modeling of the underlaid hypernovae light curve and spectra revealed the first two cases to require smaller kinetic energies but higher mass of the ejecta and the amount of the synthesized nickel than SN 2003dh. We tentatively propose that the tendency "weaker, more spherically symmetric GRB - stronger hypernova" may indicate the formation of a NS in the case of weak GRBs and of a BH in the case of strong variable GRBs as the final outcome of the core collapse. In the NS case the GRB energy comes from the rotational energy of neutron star and is possibly mediated by the electromagnetic field. When BH is formed the GRB energy source is the gravitational energy released during non-stationary accretion onto the black hole or the black hole rotation. It is not still excluded that the GRB phenomenon signals some new physics underlying the formation of compact stars.

Observations of various types of supernovae in other galaxies and especially of a (long awaited!) galactic event by all available means (including neutrino and gravitational wave detectors) should undoubtedly be crucial for further understanding physical mechanisms of cosmic explosions. We are sure that the increasing statistics of GRB/SNe in the nearest future obtained with new GRB-dedicated space missions like SWIFT will tell us much more on the nature of GRBs and their progenitors.

Acknowledgments The author acknowledges the support through RFBR grants 02-02-16500, 03-02-17174 and 04-02-16720.

Notes

1. A bright soft X-ray flux was inferred from XMM observations of evolving X-ray halo for this burst [148], making it an X-ray rich GRB [154]; this point of view was questioned by [132].

2. Recent radial velocity measurements of the companion star in low-mass X-ray binary 2S 0921-630 limits the mass of the compact object within the range $1.9 \pm 0.25 M_\odot < M_x < 2.9 \pm 0.4 M_\odot$, making it the plausible high-mass neutron star or low-mass BH [74]. A slightly higher mass interval $2.0 M_\odot < M_x < 4.3 M_\odot$ (1-sigma) was obtained in [134].

3. For example, two-component modeling of SN 2003dh as a slowly moving high-mass equatorial ejection and almost discontinuous low-mass polar outflow [164] requires the high mass of synthesized Ni-56 $\sim 0.5 M_\odot$, as in other hypernovae

References

[1] S. Akiyama, C. Wheeler, D.L. Meier, I. Lichtenstadt: ApJ **584**, 954 (2003)

[2] A.G. Aksenov, E.A. Zavrodina, V.S. Imshennik, D.K. Nadezhin: Astron. Lett. **23**, 677 (1997)

[3] M.A. Aloy, H.-T. Janka, E. Müller: Astron. Astrophys. in press. Preprint astro-ph/0408291

[4] N.V. Ardeljan, G.S. Bisnovatyi-Kogan, K.V. Kosmachevskii, S.G. Moiseenko: Astrophysics **47**, 37 (2004)

[5] N.V. Ardeljan, G.S. Bisnovatyi-Kogan, S.G. Moiseenko: MNRAS submitted (2004). Prerint astro-ph/0410234

[6] D. Arnett: ApSS **5**, 180 (1969)

[7] Z. Arzoumaian, D.F. Chernoff, J.M. Cordes: ApJ **584**, 943 (2002)

[8] W. Baade, F. Zwicky: Proc. Natl. Acad. Sci. USA **20**, 254 (1934); W. Baade, F. Zwicky: Phys. Rev. **46**, 76 (1934)

[9] V.S. Berezinsky, O.F. Prilutsky O.F.: Astron. Astrophys. **175**, 309 (1987)

[10] E. Berger, S.R. Kulkarni, D.A. Frail, A.M. Soderberg: ApJ **599**, 408 (2003)

[11] O. Bertolami: Astropart. Phys. **11**, 357 (1999)

[12] H. Bethe: Phys. Rep. **62**, 801 (1990)

[13] G.S. Bisnovatyi-Kogan: AZh **47**, 813 (1970)

[14] G.S. Bisnovatyi-Kogan: Asymmetric neutrino emission and fromation of rapidly moving pulsars. In *Proc. 8th Workshop on Nuclear Astrophysics*. Ed. W.Hillebrandt, E. Müller (MPA: Garching, 1996), p.41

[15] R. Blandford, C.F. McKee: Phys. Fluids **19**, 1130 (1976)

[16] R. Blandford, R.L. Znajek: MNRAS **179**, 433 (1977)

[17] R. Blandford, D. Eichler: Phys. Rep. **154**, 1 (1987)

[18] S.I. Blinnikov, T. Perevodchikova, I.D. Novikov, A.G. Polnarev: SvA Letters **10**, 177 (1984)

[19] S.I. Blinnikov, P.V. Sasorov: Phys. Rev. E **53**, 4827 (1996)

[20] S.I. Blinnikov: Surveys High Energy Phys. **15**, 37 (2000). Preprint astro-ph/9911138

[21] S.I. Blinnikov, E.I. Sorokina: ApSS **290**, 13 (2004)

[22] J.M. Blondin, A. Mezzacappa, C. DeMarino: ApJ **584**, 971 (2003)

[23] S.W. Bruenn, E.A. Raley, A. Mezzacappa: preprint astro-ph/0404099 (2004)

[24] R. Buras, M. Rampp, H.-Th. Janka, K. Kifonidis: Phys. Rev. Lett. **90**, 241101-1 (2003)

[25] A. Burrows. Understanding core-collapse supernovae. In *Proc. 12th Workshop on Nuclear Astrophysics*. Ed. E. Müller, H.-Th. Jamka (2004). Preprint astro-ph/0405427

[26] A. Burrows, J. Hayes, B.A. Fryxell: ApJ **450**, 830 (1995)

[27] S. Chandrasekhar: Proc. Nat. Acad. Sci. **46**, 253 (1960)

[28] A.M. Cherepashchuk: Astron. Rep. **45**, 120 (2001)

[29] N.N. Chugaj: SvAL **10**, 87 (1984)

[30] N.N. Chugai, Yungelson L.R.: Astron. Lett. **30**, 65 (2004)

[31] B.E. Cobb, C.D. Bailyn, P.G. van Dokkum, et al.: ApJ **608**, L93 (2004)

[32] S.A. Colgate, R.H. White: ApJ **143**, 626 (1966)

[33] M. Colpi, S.L. Shapiro, S.A. Teukolsky: ApJ **339**, 318 (1989)

[34] . M. Colpi, S.L. Shapiro, S.A. Teukolsky: ApJ **369**, 422 (1991)

[35] E. Costa, F. Frontera, J. Heise, et al.: Nature **387**, 783 (1997)

[36] C.D. Dermer: In *Proc. 10th Marcel Grossman Meeting on General Relativity*, Rio de Janeiro, Brasil (July 20-26, 2003), in press. Preprint astro-ph/0404608

[37] C. Dermer, K.E. Mitman: ApJ **513**, L5 (1999)

[38] D.A. Dicus: Phys. Rev. D **6**, 941 (1972)

[39] E. Ergma, E.P.J. van den Heuvel: Astron. Astrophys. **331**, L29 (1998)

[40] D. Frail, S.R. Kulkarni, L. Nicastro, et al.: Nature **389**, 261 (1997)

[41] D.A. Frail, et al., 2001, ApJ, **562**, L55 (2001)

[42] C.L. Fryer: ApJ **522**, 412 (1999)

[43] C.L. Fryer: ApJ **601**, L175 (2004)

[44] C.L. Fryer, M.S. Warren: ApJ **601**, 391 (2004)

[45] T.J. Galama, P.M. Vreeswijk, J. van Paradijs, et al.: Nature **395**, 670 (1998)

[46] A. Gal-Yam, D.-S. Moon, D.B. Fox, et al.: 2004, ApJ, **609**, L59 (2004)

[47] V.N. Gamezo, A.M. Khokhlov, E.S. Oran, et al.: Science **299**, 77 (2003)

[48] V.N. Gamezo, A.M. Khokhlov, E.S. Oran: Phys. Rev. Let. **92**, 1102 (2004); submitted to ApJ (2004). Preprint astro-ph/0409598

[49] G. Gamow, M. Schoenberg: Phys. Rev. **59**, 539 (1941)

[50] L. Gianfanga, et al., 2004, preprint hep-ph/0409185

[51] J. Granot, E. Nakar, T. Piran: Nature, **426**, 138 (2003)

[52] J. Greiner, S. Klose, K. Reinsch, et al.: Nature, **426**, 157 (2003)

[53] D. Guetta, R. Perna, L. Stella, M. Vietri: pretint astro-ph/0409715 (2004)

[54] A. Heger, N. Langer, S. Woosley: ApJ **528**, 368 (2000)

[55] A. Heger, S. Woosley, H. Spruit: ApJ submitted (2004); preprint astro-ph/0409422

[56] W. Hillebrandt, J.C. Niemeyer: ARRA **38**, 191 (2000)

[57] R. Hirschi, G. Meynet, A. Maeder: Astron. Astrophys. **425**, 649 (2004)

[58] J. Hjorth, J. Sollerman, P. Moller, et al.: Nature **423**, 847 (2003)

[59] F. Hoyle, W.A. Fowler: ApJ **132**, 565 (1960)

[60] K. Hurley, R. Sari, S.G. Djorgovki S.G. In *Compact Stellar X-ray Sources*. Eds. W. Lewin and M. van der Klis (Cambridge Univ. Press, 2003). Preprint astro-ph/0211620

[61] U. Hwang, J.M. Laming, C. Badenes, et al.: ApJ submitted (2004). Preprint astro-ph/0409760

[62] I.Jr. Iben, J. Whelan: ApJ **186**, 1007 (1973)

[63] I.Jr. Iben, A. Renzini: ARAA **21**, 271 (1983)

[64] V.S. Imshennik: SvA Lett. **18**, 194 (1992)

[65] V.S. Imshennik, D.V. Popov: Astron. Let. **20**, 529 (1994)

[66] V.S. Imshennik, O.G. Ryazhskaya: Astron. Let. **30**, 14 (2004)

[67] L.N. Ivanova, V.S. Imshennik, V.M. Chechetkin: ApSS **31**, 497

[68] K. Iwamoto, P.A. Mazzali, K. Nomoto, et al.: Nature **395**, 672 (1998)

[69] H.-Th. Janka, et al.: ApJ **527**, L39 (1999)

[70] H.-Th. Janka: Astron. Astrophys. **368**, 527 (2001)

[71] H.-Th. Janka, R. Buras, K. Kifonidis, et al.: In *Stellar Collapse*. Ed. C.L. Fryer (Dordrecht: Kluwer, 2004). Preprint astro-ph/0212314

[72] H.-Th. Janka, R. Buras, F.S. Kitaura Joyanes, et al.: Core-Collapse Supernovae: Modeling between Pragmatism and Perfectionism. In *Proc. 12th Workshop on Nuclear Astrophysics*. Ed. E. Müller, H.-Th. Janka (2004). Preprint astro-ph/0405289.

[73] H.-Th. Janka, L. Scheck, K. Kifonidis, et al.: Supernova Asymmetries and Pulsar Kicks – Views on Controversial Issues. In *The Fate of the Most Massive Stars, Proc. Eta Carinae Science Symposium (Jackson Hole, May 2004)* (2004), in press. Preprint astro-ph/0408439

[74] P.G. Jonker, D. Sreeghs, G. Nelemans, M. van der Klis: MNRAS submitted (2004). Preprint astro-ph/0410151

[75] J.I. Katz: ApJ **432**, L107 (1994)

[76] K.S. Kawabata, J. Deng, L. Wang, et al.: ApJ **593**, L19 (2003)

[77] J. Keohane et al.: $http : //chandra.harvard.edu/press/04_releases/press.060204.html$ (2004)

[78] A.M. Khokhlov: Astron. Astrophys. **245**, 114 (1991)

[79] R. Klebesadel, I. Strong, R. Olson: ApJ **182**, L85 (1973)

[80] S.R. Kulkarni, D.A. Frail, M.H. Wieringa, et al.: Nature **395**, 663

[81] S. Kusenko: Int. J. Mod. Phys. D (2004) in press. Preprint astro-ph/0409521

[82] J.M. Lattimer, M. Prakash: Science **304**, 536 (2004)

[83] D. Lai, D.F. Chernoff, J.M. Cordes: ApJ **549**, 1111 (2001)

[84] J.M. LeBlanc, J.R. Wilson: ApJ **161**, 541

[85] H.K. Lee, R.A.M.J. Wijers, H.A. Brown: Phys. Rep. **325**, 83 (2000)

[86] D.C. Leonard, A.V. Filippenko. Spectropolarimetry of Core-Collapse Supernovae. In *Supernovae as Cosmological Lighthouses*. Ed. M.Turrato et al. (AIP Conf. ser., 2004), in press. Preprint astro-ph/0409518

[87] M. Liebendörfer: Fifty-Nin Reasons for a Supernova to not Explode. In *Proc. 12th Workshop on Nuclear Astrophysics*. Ed. E. Müller, H.-Th. Janka (2004), in press. Preprint astro-ph/0405429

[88] Y.M. Lipkin, E.O. Ofek, A. Gal-Yam, et al. 2004: ApJ **606**, 381 (2004)

[89] A. Loeb: Phys. Rev. D **48**, R3419 (1993)

[90] T.A. Lozinskaya: *Supernovae and Stellar Wind in the Interstellar Medium*, (Americam Inst. of Physics, New York 1992)

[91] A.G. Lyne, D.R. Lorimer: Nature **369**, 127 (1994)

[92] M. Lyutikov: preprint astro-ph/0409489 (2004)

[93] M. Lyutikov, R. Blandford: preprint astro-ph/0312347 (2003)

[94] A. MacFadyen, S. Woosley: ApJ **524**, 262 (1999)

[95] K. Maeda, K. Nomoto: ApJ **598**, 1163 (2003)

[96] D. Malesani, G. Tagliaferri, G. Chincarini, et al.: ApJ **609**, L5 (2004)

[97] T. Matheson, P.M. Garnavich, K.Z. Stanek K.Z., et al.: ApJ **599**, 394 (2003)

[98] E.P. Mazets, S.V. Golenetskii, V.N. Ilinskii: JETP Lett. **19**, 77 (1974)

[99] P.A. Mazzali, J. Deng, N. Tominaga, et al.: ApJ **599**, L95 (2003)

[100] P.A. Mazzali, J. Deng, K. Maeda, et al.: ApJ in press (2004). Preprint astro-ph/0409575

[101] O.E.B. Messer, A. Mezzacappa, S.W. Bruenn, M.W. Guidry M.W.: ApJ **507**, 353 (1998)

[102] P. Meszaros: ARAA **40**, 137 (2002)

[103] P. Meszaros, M.J. Rees: ApJ **405**, 278 (1993)

[104] P. Meszaros, M.J. Rees: ApJ **476**, 232 (1997)

[105] A. Mezzacappa: preprint astro-ph/0410085 (2004)

[106] M. Milgrom, V.Usov: ApJ **449**, L37 (1995)

[107] R.C. Mitchell, E. Baron, D. Branch, et al.: ApJ **556**, 979 (2001)

[108] D.K. Nadyozhin: ApSS **53**, 131 (1978)

[109] D.K. Nadyozhin, I.V. Otroshenko: Astr. Zhurn. **57**, 78 (1980)

[110] K. Nomoto, D. Sugimoto, S. Neo: ApSS**39**, L37 (1976)

[111] K. Nomoto, K, Maeda, H. Umeda, et al.: In *A Massive Star Odyssey, from Main Sequence to Supernovae*. Proc. IAU Symp. 212. Ed. K.A. van der Hucht and C. Esteban. (San Francisco: Astron. Soc. of Pacific, 2003), p. 395

[112] K. Nomoto, K. Maeda, P.A. Mazzali,et al.: In *Stellar Collapse*. Astrophysics and Space Science Library **302**. Ed. C.L.Fryer. (Kluwer Acad. Publ., Dordrecht, 2004). Preprint astro-ph/0308136

[113] B. Paczynski: ApJ **363**, 218 (1990)

[114] B. Paczynski: ApJ **494**, L45 (1998)

[115] B. Paczynski, J.E. Rhoads: ApJ **418**, L5 (1993)

[116] S. Park, D.N. Burrows, G.P. Garmire, et al.: ApJ **567**, 314 (2002)

[117] Ph. Podsiadlowski, P.A. Mazzali, K. Nomoto, et al.: ApJ **607**, L17 (2004)

[118] J. van Paradijs, P. Groot, T. Galama, et al.: Nature **386**, 686 (1997)

[119] J. van Paradijs, C. Kouveliotou, R.A.M.J. Wijers: ARAA **38**, 379 (2000)

[120] T. Piran: Phys. Rep. **333**, 529 (2000)

[121] T. Piran: Rev. Mod. Phys. (2004), in press (astro-ph/0405503)

[122] K.A. Postnov, M.E. Prokhorov, V.M. Lipunov: Astron. Rep. **45**, 236 (2001) (astro-ph/9908136)

[123] K.A. Postnov, A.M. Cherepashchuk: Astron. Rep. **45**, 517 (2001)

[124] J. Pruet, R. Surman, G.C. McLaughlin: ApJ **602**, L101 (2004)

[125] M.J. Rees, P. Meszaros: MNRAS, **258**, 41P (1992)

[126] M.J. Rees, P. Meszaros: ApJ, **430**, L93 (1994)

[127] M. Reinecke, W. Hillebrandt, J.C. Niemeyer: Astron. Astrophys. **391**, 1167 (2002)

[128] A. Riess, L.-G. Strolger, J. Tonry, et al.: ApJ **607**, 665

[129] F.K. Röpke, W. Hillebrandt, J.C. Niemeyer: Astron. Astrophys. **420**, 411 (2004); **421**, 783 (2004)

[130] M. Ruffert, H.-Th. Janka: Astron. Astrophys. **344**, 573 (1999)

[131] P. Ruiz-Lapuente, S. Blinnikov, R. Canal, et al.: Mem. Soc. Astron. Ital. **71**, 435 (2000)

[132] S.Yu. Sazonov, A.A. Lutovinov, R.A. Sunyaev: Nature **430**, 646 (2004)

[133] E. Schatzman. In *Star Evolution*. (Academic Press, New York and London, 1962), p. 389

[134] T. Shahbaz, J. Casares, C. Watson, et al.: ApJ Lett. in press (2004). Preprint astro-ph/0409752

[135] A. Shemi, T. Piran: ApJ **365**, L55 (1990)

[136] V.V Sokolov, T.A. Fatkhullin, V.N. Komarova, et al.: Bull. Special Astrophys. Obs. RAS **56**, 5 (2004). Preprint astro-ph/0312359

[137] A.M. Soderberg, S.R. Kulkarni, E. Berger, et al.: Nature **430**, 648 (2004)

[138] K.Z. Stanek, T. Matheson, P.M. Garnavich, et al.: ApJ **591**, L17

[139] B.E. Stern, Ya. Tikhomirova, D. Kompaneets, et al: ApJ **563**, 80 (2001)

[140] B.E. Stern, Ju. Poutanen: MNRAS **352**, L35 (2004)

[141] E.M.D. Symbalisty: ApJ **285**, 729 (1984)

[142] T.A. Thompson, E. Quataert, A. Burrows: preprint astro-ph/0403224 (2004)

[143] B. Thomsen, J. Hjorth, D. Watson, et al.: Astron. Astrophys. **419**, L21 (2004)

[144] C. Travaglio, W. Hillebrandt, M. Reinecke, F.-K. Thielemann: Astron. Astrophys. **425**, 1029 (2004)

[145] A.V. Tutukov, A.M. Cherepashchuk: Astron. Rep. **47**, 386 (2003)

[146] V.V. Usov: Nature **357**, 472 (1992)

[147] V.V. Usov: MNRAS **267**, 1035 (1994)

[148] S. Vaughan, R. Willingale, P.T. O'Brien, et al.: ApJ, **603**, L5 (2004)

[149] E.P. Velikhov: Sov.Phys. JETP **36**, 1398 (1959)

[150] M. Vietri: ApJ **453**, 883 (1995)

[151] M. Vietri: ApJ **478**, L9 (1997)

[152] M. Vietri, L. Stella: ApJ **507**, L45 (1998)

[153] M. Vietri, L. Stella: ApJ **507**, L45 (1998)

[154] D. Watson, J. Hjorth, A. Levan, et al.: ApJ, **605**, L101 (2004)

[155] E. Waxman: Phys. Rev. Lett. **75**, 386 (1995)

[156] E. Waxman: In *Supernovae and Gamma-Ray Bursts*. Ed. K.W.Weiler. Lecture Nores in Physics **598** (Springer-Verlag 2004), p. 393

[157] E. Waxman: ApJ **606**, 988 (2004)

[158] C.F. von Weizsaecker: Zeit. für Astrophys. **24**, 181 (1947)

[159] R.A.M.J. Wijers, M.J. Rees, P. Mészáros: MNRAS **288**, L51 (1997)

[160] S.E. Woosley: In *Supernovae*. Ed. A.G. Petschek (Springer-Verlag: Berlin, 1990), p. 182

[161] S.E. Woosley: ApJ **405**, 273 (1993)

[162] S.E. Woosley, T.A. Weaver: ApJ Suppl. **101**, 181 (1995)

[163] S.E. Woosley, A. Heger, T.A. Weaver: Rev. Mod. Phys. **74**, 1015 (2002)

[164] S.E. Woosley, A. Heger: ApJ in press (2004), preprint astro-ph/0309165

[165] S.E. Woosley, S. Wunsch, M. Kuhlen M.: ApJ **607**, 921 (2004)

[166] L.R. Yungelson, M. livio: ApJ **497**, 168 (1998)

[167] L.R. Yungelson: In *White Dwarfs: Galactic and Cosmological Probes*. Eds. E.M. Sion, H.L. Shipman and S. Vennes (Kluwer, 2004), in press. Preprint astro-ph/0409677

[168] Ya. B. Zeldovich, O.H. Gusseinov: DAN SSSR **162**, 791 (1965)

[169] B. Zhang, P. Meszaros: Int. J. Mod. Phys. A **19**, 2385 (2004)

CLOSE BY COMPACT OBJECTS AND RECENT SUPERNOVAE IN THE SOLAR VICINITY

Sergei Popov[1,2]

[1]*University of Padova*
via Marzolo 8, 35131, Padova, Italy

[2]*Sternberg Astronomical Institute*
Universitetski pr. 13, 119992, Moscow, Russia
polar@sai.msu.ru

Abstract   I discuss young close-by compact objects, recent supernovae in the solar neighbourhood, and point to their connection with cosmic ray studies. Especially the
role of the Gould Belt is underlined.

Keywords: neutron stars, evolution, supernovae, cosmic rays

Introduction

In this short paper at first I shall try to show links between studies of cosmic
rays (CRs) and studies of close-by compact objects (in particular neutron stars
– NSs). The reason for an existence of such relation is obvious: both phenomena (CRs and NSs) have the same origin – supernova (SN) explosions. Then I
discuss in more details the population synthesis of close-by NSs. The analysis
of the population of these sources makes us to conclude that the solar neighbourhood (by that I mean a region about few hundred of parsecs around the
Sun) in enriched with young NSs. It is a natural consequence of the existence
of the Gould Belt – local structure formed by massive stars.

The main message I want to deliver in this contribution is the following.
We are living in a region of the Milky way enriched with massive stars in
comparison with a typical place at 8 kpc from the center of the Galaxy. Due
to this fact SN rate in ~ 600 pc around the Sun during the last few tens of
Myrs is enhanced. It results also in the ebhanced number of near-by young
compact objects. Now we know about two tens of young (age < 4 Myrs) close
(distance < 1 kpc) NSs. This local (in space and time) increase of SN rate can
be important for modeling of CR flux on the Earth.

M. M. Shapiro et al. (eds.), Neutrinos and Explosive Events in the Universe, 119–130.
© 2005 *Springer. Printed in the Netherlands.*

Cosmic rays and recent close-by supernovae

Galactic CRs are genetically connected with SN: CRs are accelerated in SN remnants (see other materials in this volume, especially contributions by Michal Ostrowski, Vladimir Ptuskin and John Wefel). The local CR density is inevitably dependent on the rate of SN in the Galaxy in general and in the solar proximity in particular. CR flux can be increased if the Sun appears in a region of the Galaxy with enhanced SN rate, or if by chance several SN explosions in a row happen in the vicinity of the Sun.

For example Shaviv (2004) suggested, that passages of the Sun through galactic spiral arms are accompanied by increase of CR flux on Earth (at this conference some results were presented in the talk by Smadar Levi). As CRs can have influence on Earth climate (in particular on cloud formation) such passages can be responsible for the (quasi)periodic appearence of ice age epochs on our planet. However, I would like to note, that even local quasi-periodic fluctuations of starformation rate (and correspondent changes in SN rate) can give a similar effect. Such fluctuations were found by Vanbeveren and De Donder (2003) in their modeling of star formation in the solar vicinity. Another possibility to enhance SN rate is connected with the formation of structures like the Gould Belt (see description of the Belt below). [1] All these effects can work together, and therefore modeling of the CR flux history becomes a very complicated business because of a "turbulent history" of star formation in $\sim$ 1 kpc around the Sun.

For a long time local SN were considered as one of candidates for a catastrophic event responsible for mass extinction of species. Still SN potentially can play a "positive" role in the history of life on Earth. An interesting possibility of SN influence on bioworld of our planet can be connected with the hypothesis discussed by Tsarev (1999; it is also described in an easier available paper by Chernavskii 2000). In the original version the author suggested that chirality symmetry breaking existing in the bioorganic world can be connected with strong neutrino burst due to a close SN which happened in the early stages of the history of the Earth. However, a close-by SN (close enough to povide a strong neutrino flux) can just destroy terrestrial bioorganics. In that sense it is possible to suggest a more promising hypothesis which includes so called "choked gamma-ray bursts" (Meszaros, Waxman 2001). If a jet is unable to penetrate through the stellar envelope, then only a neutrino beam will break-out. In that case it is possible both: to increase the neutrino flux, and to decrease negative influence of hard radiation of a SN on the biosphere.

Recent local SN (and related close-by NS) can strongly influence not only the flux of CR but the spectrum as well. Erlykin and Wolfendale (1997) proposed that a knee in the spectrum is due to a single source. According to their estimates there should exist a near-by relatively young SN remnant. An estimated age of such a hypothetical object was found by these authors to be equal to 80-100 kyr, and its distance – to 230-350 pc. They suggested that the Monogem nebula (and related to it PSR 0656+14) can be such a source which produce the knee in the spectrum (see also the contribution by Jorg Hoerandel in this volume for a more detailed discussion of the knee).

How probable an appearence of such an object is? If the Sun is sutuated in a typical region of the Galaxy at 8 kpc from its center then the probability of appearence of a SN remnant of necessary age at necessary distance is very low. However, we are living in a region of enhanced SN rate, and there are many young compact objects around us. In the following section I am going to discuss this population and it origin in more details.

Young close-by NSs and the Gould Belt

The solar proximity is rich in young NSs. Some of them represent a new class of sources: radioquiet X-ray dim isolated NSs (see a review in Treves et al. 2000). The reason of such enrichment is an existence of a structure formed by massive stars – the Gould Belt. In the following subsection we discuss it in some details. Then we present new results on a population synthesis of young NSs, and finally we discuss a convenient way to represent this population.

Population of near-by young NSs is presented in the table (reproduced from Popov, Turolla 2004). We included there 20 objects. Thirteen of them are detected as thermal emitters, i.e. as *coolers*. They are seven radioquiet ROSAT NSs (aka the Magnificent seven – M7), Geminga and geminga-like source 3EG J1835+5918, and four normal radiopulsars (among which the famous Vela pulsar, and PSR B0656+14 which is probably connected with the single source, see above). Seven close-by NSs which are not detected as thermal emitters (shown at the bottom of the table) are normal radiopulsars with ages 2-4 Myrs.

The Gould Belt

The Gould Belt is a structure consisting of clusters of massive stars. About 2/3 of massive stars inside 600 pc around the Sun belong to the Belt (see a recent paper by Grenier 2004 for main references). The Gould Belt radius is about 300-500 pc. It is inclined at $\sim 18-22°$ respect to the galactic plane. The Sun is situated not far from the center of that disc-like structure. The age of the Belt is not well estimated, 30-60 Myrs can be a good guess. The origin of this system is unclear. One of the possibility is an impact of a high velocity cloud with the galactic disk. It is important to note, that the Sun and the Belt are not genetically related to each other.

Due to the presence of the Belt the rate of SN around us (say in a few hundred parsecs) during last several tens of million years is higher than it is at an average place at a solar distance from the galactic center. The enhancement is about 3-4 times over the Galactic average. Grenier (2003) estimates a SN rate in the Belt as 20-27 per Myr during the last few Myrs. Because of that there should be a local overabundance of young NSs which can appear as hot cooling objects, as gamma-ray sources etc. (in priciple in addition to NSs there should be young BHs in the solar neighbourhood, but these objects are much more elusive; it would be very interesting to detect them, see Prokhorov, Popov 2002). For example, such young NSs can form a subpopulation among EGRET unidentified sources. The existence of the Local Bubble can be also linked with the Belt (Berghofer, Breitwerdt 2002).

Log N - Log S distribution of close-by NSs

One of the most powerful methods of investigating properties of some population of sources is a Log N – Log S distribution. It is an integral distribution, which shows the number of sources with a flux larger than a given value. It is useful to remind two limiting cases. If sources are distributited homogeniously and isotropicaly then the slope of Log N – Log S is equal to -3/2: the number of sources grows as the cube of a distance, but the observed flux decreases as the square. In the case of a disk-like system the slope is -1 (the number of objects is growing only as a square of a distance). If sources are located at the same distance (for example, sources in a given far-away galaxy) then Log N – Log S shows distribution of sources in luminosity.

Popov et al. (2003) calculated Log N – Log S distribution of cooling NSs in the solar vicinity, which can be observed by ROSAT and other X-ray missions. The results are shown in the fig. 1.

The main output of these calculations can be formulated as follows: observed cooling isolated NSs originated in the Gould Belt. This result does not depend on the uncertainties of the population synthesis model. I want to note, that according to our calculations there should be about several tens more unidetified isolated NSs with fluxes detectable by ROSAT. These objects are hiding in crowded regions close to the galactic plane (and to the plane of the Gould Belt). In addition there should be many NSs with ages above $\sim$ 1 Myr, which now are too cold now to be detected. A total expected number of NSs depending on their ages and distances can be easily shown on the age-distance diagram.

If all uncertainties like initial spatial distribution of NSs, their kick velocity distribution, characteristics of emission (atmospheric effects etc.), mass spectrum of NSs are fixed, then Log N - Log S distribution of close-by NSs can be a powerful tool to put constraints on models of thermal evolution of NSs.

Table 1.   Local ($D < 1$ kpc) population of young (age < 4.25 Myrs) isolated neutron stars

Source name	Period	CR[a]	$\dot{P}$	D	Age[b]	Refs
	s	cts/s	$10^{-..}$ s/s	kpc	Myrs	
Magnificent seven						
RX J1856.5-3754	—	3.64	—	0.117^e	~ 0.5	[1,2]
RX J0720.4-3125	8.37	1.69	$\sim 30-60$	—	—	[1,3]
RX J1308.6+2127	10.3	0.29	—	—	—	[1.4]
RX J1605.3+3249	—	0.88	—	—	—	[1]
RX J0806.4-4123	11.37	0.38	—	—	—	[1,5]
RX J0420.0-5022	22.7	0.11	—	—	—	[1]
RX J2143.7+0654	—	0.18	—	—	—	[6]
Geminga type						
PSR B0633+17	0.237	0.54^d	10.97	0.16^e	0.34	[7]
3EG J1835+5918	—	0.015	—	—	—	[8]
Thermal. emit. PSRs						
PSR B0833-45	0.089	3.4^d	124.88	0.294^e	0.01	[7,9,10]
PSR B0656+14	0.385	1.92^d	55.01	0.762^f	0.11	[7,10]
PSR B1055-52	0.197	0.35^d	5.83	$\sim 1^c$	0.54	[7,10]
PSR B1929+10	0.227	0.012^d	1.16	0.33^e	3.1	[7,10]
Other PSRs						
PSR J0056+4756	0.472	—	3.57	0.998^f	2.1	[10]
PSR J0454+5543	0.341	—	2.37	0.793^f	2.3	[10]
PSR J1918+1541	0.371	—	2.54	0.684^f	2.3	[10]
PSR J2048-1616	1.962	—	10.96	0.639^f	2.8	[10]
PSR J1848-1952	4.308	—	23.31	0.956^f	2.9	[10]
PSR J0837+0610	1.274	—	6.8	0.722^f	3.0	[10]
PSR J1908+0734	0.212	—	0.82	0.584^f	4.1	[10]

[a]) ROSAT PSPC count rate
[b]) Ages for pulsars are estimated as $P/(2\dot{P})$,
for RX J1856 the estimate of its age comes from kinematical considerations.
[c]) Distance to PSR B1055-52 is uncertain (~ 0.9-1.5 kpc)
[d]) Total count rate (blackbody + non-thermal)
[e]) Distances determined through parallactic measurements
[f]) Distances determined with dispersion measure

[1] Treves et al. (2000) ; [2] Kaplan et al. (2002); [3] Zane et al. (2002);
[4] Hambaryan et al. (2001); [5] Haberl, Zavlin (2002); [6] Zampieri et al. (2001);
[7] Becker, Trumper (1997); [8] Mirabal, Halpern (2001); [9] Pavlov et al. 2001;
[10] ATNF Pulsar Catalogue (see Hobbs et al. 2003)

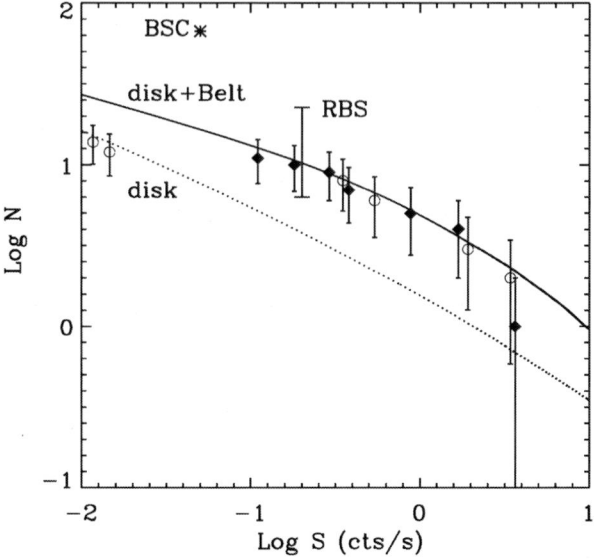

Figure 1.   Log N – Log S distribution for close-by cooling isolated NSs. Black diamonds are plotted if the dimmest source at specified flux is one of the "Magnificent seven". Otherwise we plot an empty circle. Two lines represent results of calculation. Dotted line — only stars from the galactic disc contribute to the Log N – Log S distribution. Solid line — contribution of the Gould Belt is added. "RBS" and "BSC" are two observational limits, obtained from the ROSAT data (RBS: Schwope et al. 1999; BSC: Rutledge et al. 2003). From Popov et al. (2003).

Age-distance diagram

To illustrate the population of young close-by NSs it is convenient to use the age-distance diagram (Popov 2004).

If we are speaking about observations of thermal emission of young cooling NSs then most of their properties depend on their age, and their detectability in X-rays obviously strongly depends on the distance (not only because of the flux dilution, but also because of the strong interstellar absorption of soft X-rays). In that sense it is useful to plot an age-distance diagram (ADD) for these objects.

There are several reasons to introduce ADDs. First an ADD easily illustrates distributions of sources in age and distance. Then it is possible to plot "expectation lines" for an abundance of sources of different types and compare them with data (i.e. line which show how many sources of smaller than a given value we can expect to find at distances smaller than a given one). Finally since the observability of sources depends mainly on their ages and distances, it is possible to illustrate observational limits.

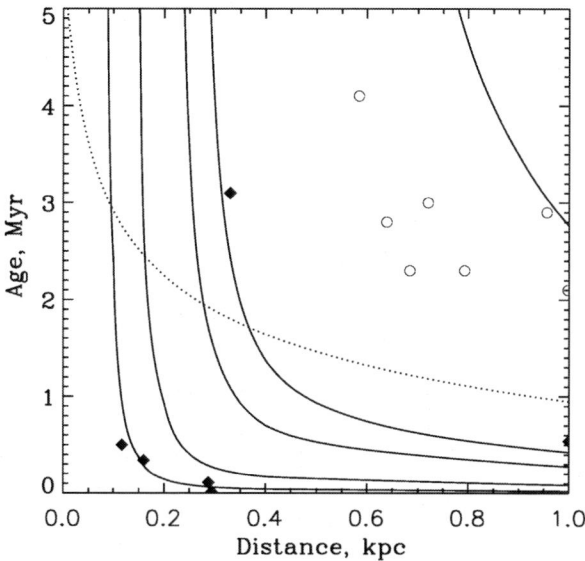

Figure 2.   Age-distance diagram with dynamical effects. Solid lines (from bottom to the top) corresponds to 1, 4, 13, 20 and 100 sources. The dotted line is the 'visibility" line (see details in Popov 2004).

In the fig. 2 an ADD for close-by young NSs is presented. These objects can be observed in soft X-rays due to their thermal emission. The thermal evolution strongly depends on the mass. According to different models (see Kaminker et al. 2002, Blaschke et al. 2004 and references there in) a NS with a mass ~ 1.3–$1.4\,M_\odot$ remains hot ($T \sim (0.5$–$1)\,10^6$ K, or equivalently ~ 50–100 eV) up to ~ 0.5–1 Myr. It corresponds to a luminosity $\sim 10^{32}$ erg s^{-1}. Usually NSs with low masses (1–$1.3\,M_\odot$) cool down slower, while with higher masses faster.

To plot an ADD it is necessary to fix maximal values for age and distance for selection of sources. I choose 4.25 Myrs as a limiting age for selection of observed sources. This is the time after which even a low-mass NSs cools down to $\sim 10^5$ K and becomes nearly undetectable in X-rays (see Kaminker et al. 2002, Popov et al. 2004). A limiting distance is taken to be equal to 1 kpc (because of the absorption it is difficult to detect in X-rays an isolated NS at larger distance unless the star is very hot). There are 20 sources of different nature which are supposed to be young close-by NSs (age < 4.25 Myr, distance < 1 kpc) (see Popov et al. 2004 and the table above): the M7, Geminga and the geminga-like source RX J1836.2+5925, four PSRs with detected thermal emission (Vela, PSR 0656+14, PSR 1055-52, PSR 1929+10[2]), and seven PSRs without detection of thermal emission. Not for all of these sources there are good estimates of ages and/or distances (especially for the M7). In the figure there are data points for 13 objects for which such estimates exist (distance to PSR B1055-52 is uncertain, and we accept it to be 1 kpc).

Note, that there are also two PSRs with distance <1 kpc and with ages in between 4.25 and 5 Myr (PSR B0823+26, PSR B0943+10) and PSR J0834-60 with distance 0.49 kpc and unknown age. These three objects are not included into the figures. Also PSR B1822-09 with age 0.23 Myr and distance 1 kpc is not plotted.

Two types of objects are distinguished on the graph: detected ones (shown as black diamonds) and undetected due to the thermal X-ray emission (remember, there are at least seven additional sources – six from the M7 and one geminga-like object – for which there are no definite determinations of age or/and distance).

In the picture the "expectation" lines are shown with an inclusion of the dynamical evolution. I.e. here NSs' movements in the Galactic potential are accurately calculated. To plot them we use the same model of progenitor distribution and SN rate as in (Popov et al. 2003). At birth NSs obtain kick velocities in accordance with Arzoumanian et al. (2002) model, and move in the galactic potential.

Five "expectation" lines (solid) are plotted for 1, 4, 13, 20 and 100 sources. For example each point on the line for 20 sources shows us values of age (t_{20}) and distance (R_{20}) such that according to our model we expect to see 20 sources with ages less than t_{20} at distances less than R_{20}.

The dotted line represents the "visibility" line. The idea of adding such a line is to show the maximal distance for a given age (or vice versa a maximal age for a given distance) at which a hot (i.e. low-mass) NS can be detected. To calculate this line I take the cooling curve from (Kaminker et al. 2002) for $M = 1–1.3\,M_\odot$ (in the model of these authors cooling of NSs with $M < 1.35\,M_\odot$ is nearly mass-independent). Such curves were used for example in (Popov et al. 2003). As soon as a cooling curve is fixed then the age determines the luminosity of the object. The limiting unabsorbed flux is assumed to be 10^{-12} erg cm^{-2} s^{-1}. According to WebPIMMS[3] it corresponds to ~ 0.01 ROSAT PSPS counts per second for $N_H = 10^{21}$ cm^{-2} and a blackbody spectrum with $T = 90$ eV, or to ~ 0.1 ROSAT PSPS counts per second for $N_H = 10^{20}$ cm^{-2} and a blackbody spectrum with $T = 50$ eV. The latter values corresponds to the dimmest source among the M7 – RX J0420.0-5022; the former to possibly detectable hot far away objects. Without any doubt such a simplified approach underestimates the absorption at large distances. So the age at which a NS is still observable is overestimated, but for distances < 1 kpc and ages < 1 Myr it should not be a dramatic effect.

The fact that the line for 20 sources lies below ∼1/2 of the observed points tells us that according to our model not all NSs are observed. For example, at R=1 kpc we expect to see 20 sources with ages ∼0.5 Myr, but in reality we observe 20 sources with age <4.25 Myr. It means that for a constant NS formation rate we see only about 10% of them. One has to bear in mind that for seven sources ages or/and distances are unknown, but they are suspected to be young and close, this is why here we discuss 20 and not 13 sources. Their inclusion or exclusion changes our conclusions quantitatively, but not qualitatively. Six of the M7 sources which are not plotted on the graphs due to lack of good distance measurement should populate the left bottom corner of the graphs, since according to models of NS thermal evolution (see for example Kaminker et al. 2002, Blaschke et al. 2004) their ages are expected to be < 1 Myr, and their distances should be < 500 pc. NSs can easily escape detection as *coolers* simply because they cooled down in ∼ 1 Myr. A fraction of PSRs can be undetected due to beaming (however, they may be observed as EGRET sources).

PSR B1929+10 lies above the "visibility line". However it is unclear if X-ray emission of this faint (0.012 ROSAT cts s^{-1}) source is due to non-thermal mechanism or not (Becker & Trumper 1997). Now this object is not widely accepted as a *cooler* and usually it is not plotted on the $T - t$ (temperature-age) graphs with cooling curves (see for example Kaminker et al. 2002 and Blaschke et al. 2004).

Surprisingly fluctuations in the part of the plot with small number of NSs (bottom left) are not large: four detected *coolers* lie just below the line for four sources. Probably we see nearly all low-massive NSs with ages < 1 Myr at $R < 300$–400 pc. It should be noted that we are very lucky to have such a young and close object as the Vela pulsar. If one believes in the mass spectrum with small fraction of NSs with $M > 1.5\,M_\odot$ then inevitably one has to conclude that Vela cannot be a massive NS. It can be important in selection of cooling models (see Blaschke et al. 2004). If in a model Vela is explained only by a cooling curve for $M > 1.5\,M_\odot$ then the model may be questionable. In fact, having such a young and massive NS so close is very improbable. A similar conclusion can be made for Geminga and PSR 0656+14.

There is a deficit of sources below the "visibility" line. These sources could be already detected as dim X-ray sources by ROSAT but were not identified as isolated NSs. The limit for the number of isolated NSs from the BSC (Bright Source Catalogue) is about 100 sources at ROSAT count rate >0.05 cts s^{-1} (see Rutledge et al. 2003). However we do not expect to see so many young isolated NSs due to their thermal emission. An expected number of *coolers* observable by ROSAT is about 40 objects with ages < 1 Myr inside $R < 0.5$–1 kpc (see the "visibility" line in the fig.2 in comparison with the line for 20 sources, for example). As it was shown by Popov et al. (2003) most of unidentified *coolers* in ROSAT data are expected to be located at low galactic latitudes in crowded regions.

According to the classical picture all NSs were assumed to be born as PSRs more or less similar to Crab. Depending on the initial parameters they were assumed to be active for $\sim 10^7$ yrs. In the last years this picture was significantly changed. Without any doubts we see a deficit of young PSRs in the solar vicinity in comparison with an expected value of young NSs from the Gould Belt and the beaming factor cannot be the only reason for this deficit. This deficit can be explained if one assumes that a significant part of NSs do not pass through the radio pulsar stage or that this stage is extremely short for them (long periods of NSs from the Magnificent seven prove it). Of course fluctuations (in time and space) of NS production rate can be important as far as statistics is not that high.

I conclude this subsection with a note that an ADD can be the useful tool for an illustration of the properties of close-by NSs. Its modifications can be applied to other types of sources. For example an addition of the third axis (for p or $\dot{p}$ for example) can be useful in discussing the population of radiopulsars.

Conclusions

I want to stress once again that the region of $\sim$ few hundred parsecs around the Sun is characterized by an enhanced SN rate. This is due to the Gould Belt. The Belt (as a structure formed by bright stars) was well known since 19th century. In addition now we observe young close-by NSs which, as it was shown by population synthesis calculations, originated in the Belt. Some of EGRET unidentified sources also are suspected to be young NSs born in the Belt.

The existence of the Gould Belt should inevitably influence properties of CRs detected on Earth.

Acknowledgments I thank the Organizers for hospitality and partial financial support. Special thanks to many participants for stimulating discussions. I thank Mirian Tsulaia for comments on the text.

Notes

1. Effects of the Belt on CR characteristics were discussed in Pohl et al. (2003) and in Dermer (2004).

2. It should be mentioned, that soft X-ray emission of PSR 1929+10 can be due to polar caps or due to non-thermal mechanism.

3. http://heasarc.gsfc.nasa.gov/docs/corp/tools.html

References

[1] Arzoumanian, Z., Chernoff, D.F., Cordes, J.M. 2002). "The velocity distribution of isolated radio pulsars", ApJ 568, 289.

[2] Becker, W., Trumper, J. (1997). "The X-ray luminosity of rotation-powered neutron stars", A&A 326, 682.

[3] Berghofer, T.W., Breitschwerdt, D. (2002). "The origin of the young stellar population in the solar neighborhood – A link to the formation of the Local Bubble?", A&A 390, 299.

[4] Blaschke, D., Grigorian, H., Voskresensky, D. (2004). "Cooling of neutron stars. Hadronic model.", A&A (in press) [astro-ph/0403170].

[5] Chernavskii, D.S. (2000), "The origin of life and thinking from the modern physics point of view", Physics Uspekhi 170, 157.

[6] Dermer, C.D. (2004). "Gamma ray bursts, sepernovae, and cosmic ray origin", in: Proc. of ISCRA 13th Course, Eds. M. Shapiro, T. Stanev and J. Wefel, World Scientific Publ., p.189.

[7] Erlykin, A.D., Wolfendeil, A.W. (1997). "A single source of cosmic rays", J. Phys. G 23, 979.

[8] Grenier, I.A. (2003). "Unidentified EGRET sources in the Galaxy", astro-ph/0303498.

[9] Grenier, I.A. (2004). "The Gould Belt, star formation, and the local interstellar medium", astro-ph/0409096.

[10] Haberl F., Zavlin V. (2002). "XMM-Newton observations of the isolated neutron star RX J0806.4-4123 ", A&A 391, 571.

[11] Hambaryan, V., Hasinger, G., Schwope, A. D., Schulz, N. S. (2001). "Discovery of 5.16 s pulsations from the isolated neutron star RBS 1223", A&A 381, 98.

[12] Hobbs, G. Manchester, R. Teoh, A. Hobbs, M. (2003). "The ATNF Pulsar Catalogue", in: Proc. of IAU Symp. 218. "Young neutron stars and their environment" [astro-ph/0309219].

[13] Kaminker, A.D., Yakovlev, D.G., Gnedin, O.Y. (2002). "Three types of cooling superfluid neutron stars: theory and observations", A&A 383, 1076.

[14] Kaplan, D.L., van Kerkwijk, M.H., Anderson, J. (2002). "The parallax and proper motion of RX J1856.5-3754 revisited", ApJ, 571, 447.

[15] Meszaros, P., Waxman, E. (2001). "TeV Neutrinos from Successful and Choked Gamma-Ray Bursts", Phys. Rev. Lett. 87, 171102.

[16] Mirabal, N., Halpern, J.P. (2001). "A neutron star identification for the high-energy gamma-ray source 3EG J1835+5918 detected in the ROSAT All-Sky Survey", ApJ 547, L137.

[17] Pavlov, G. et al. (2001). "The X-ray spectrum of the Vela pulsar resolved with the Chandra X-ray observatory", ApJ 552, L129.

[18] Pohl, M., Perrot, C., Grenier, I., Digel, S. (2003). "The imprint of Gould's Belt on the local cosmic-ray electron spectrum", A&A 409, 581.

[19] Popov, S.B., Turolla, R., Prokhorov, M.E., Colpi, M., Treves, A. (2003). "Young close-by neutron stars: the Gould Belt vs. the galactic disc", astro-ph/0305599.

[20] Popov, S.B., Turolla, R., Prokhorov, M.E., Colpi, M., Treves, A. (2004). "Young compact objects in the solar vicinity", in: Proc. of ISCRA 13th Course, Eds. M. Shapiro, T. Stanev and J. Wefel, World Scientific Publ., p.101.

[21] Popov, S.B., Turolla, R. (2004). "Isolated neutron stars: An astrophysical perspective", in: Proc. of the NATO Advanced Research Workshop On Superdense QCD Matter and Compact Stars (in press) [astro-ph/0312369].

[22] Popov, S.B. (2004). "Age-Distance diagram for close-by young neutron stars", astro-ph/0407370.

[23] Prokhorov, M.E., Popov, S.B. (2002). "Close young isolated black holes", Astronomy Letters 28. 609.

[24] Rutledge, R.E., Fox, D.W., Bogosavljevic, M., Mahabal, A. (2003). "A limit on the number of isolated neutron stars detected in the ROSAT Bright Source Catalogue", astro-ph/0302107.

[25] Schwope, A.D., Hasinger, G., Schwarz, R., Haberl, F., Schmidt, M. (1999). "The isolated neutron star candidate RBS 1223 (1RXS J130848.6+212708)", A&A 341, L51.

[26] Shaviv, N. (2004). "Cosmic ray diffusion in the Milky Way: model, measurement and terrestrial effects", in: Proc. of ISCRA 13th Course, Eds. M. Shapiro, T. Stanev and J. Wefel, World Scientific Publ., p.113.

[27] Treves, A., Turolla, R., Zane, S., Colpi, M. (2000). "Isolated neutron stars: accretors and coolers", PASP 112, 297.

[28] Tsarev, V.A. (1999). "Chiral effects of neutrino after supernova explosions", Bulletin of the Lebedev Physics Institute 2, 22.

[29] Vanbeveren, D., De Donder, E. (2003). "The effect of the process of star formation on the temporal evolution of the WR and O-type star populations in the Solar neighbourhood", astro-ph/0309104.

[30] Zampieri, L. et al. (2001). "1RXS J214303.7+065419/RBS 1774: a new isolated neutron star candidate", A&A 378, L5.

[31] Zane, S. et al. (2002). "Timing analysis of the isolated neutron star RX J0720.4-3125", MNRAS 334, 345.

SUPERNOVA REMNANTS AS COSMIC ACCELERATORS

Vladimir Ptuskin
Institute for Terrestrial Magnetism, Ionosphere and Radiowave Propagation of the Russian Academy of Science (IZMIRAN), Troitsk, Moscow Region 142190, Russia
vptuskin@izmiran.rssi.ru

Abstract   Most cosmic rays are thought to be accelerated by the shocks of supernova explosions. The data on cosmic rays at the Earth and the observations of nonthermal radiation from supernova remnants testify that the particles are accelerated with high efficiency and in a wide range of energies. We discuss the acceleration of galactic cosmic rays in supernova remnants in a general context of the problem of cosmic-ray origin. The scenario of cosmic ray acceleration will be probably verified in full details when the data from the new generation of ground-based and space gamma-ray experiments in conjunction with new X-ray satellites will be available.

Keywords: cosmic rays – cosmic-ray sources – supernova remnants – acceleration of particles – shock waves – turbulence – gamma rays

Introduction

To a first approximation, the all-particle spectrum of cosmic rays can be described by a power law on more than 11 decades on particle energy, so that the dependence of cosmic ray intensity on particle energy is close to $E^{-2.7}$ at energy more than about 10 GeV. Closer examination reveals some structure in the galactic cosmic ray spectrum that includes the knee at 4×10^{15} eV, the second knee at about 10^{18} eV, and the ankle at 10^{19} eV. The steady-state spectrum is shaped by two principle processes - the acceleration at the sources and the subsequent propagation in the Galaxy.

As early as in 1934, Baade and Zwicky [1] linked the appearance of supernovae with neutron-star formation and cosmic ray generation. Fermi (1949) [2] regarded cosmic rays as a gas of relativistic charged particles moving in interstellar magnetic fields. Once a close link between radioastronomy and cosmic rays has been conclusively proved in late fifties, the basic model of the origin of galactic cosmic rays was worked out in 1964 by Ginzburg and Syrovatskii [3]. It was established that released into the interstellar medium from

M. M. Shapiro et al. (eds.), Neutrinos and Explosive Events in the Universe, 131–142.

their galactic sources, relativistic particles spend tens millions years moving in magnetic fields before exit from the galactic cosmic ray halo (if interactions with interstellar gas and radiation do not expel them earlier).

The propagation of cosmic rays in the interstellar medium is usually described as diffusion. The modeling of cosmic ray diffusion in the Galaxy includes the solution of transport equation with a given source distribution and boundary conditions for all cosmic ray species. The transport equation describes diffusion, convection by the hypothetical galactic wind, and changes of energy (energy losses and possible distributed acceleration). In addition, nuclear collisions with interstellar gas atoms resulting in the production of secondary energetic particles and the decay of radioactive isotopes should be taken into account when considering proton-nucleus component. The basic features of the galactic diffusion model seem to be well established, see e.g. [4], [5], [6]. This model explains the data on particle energy spectra, composition and anisotropy and provides a basis for the interpretation of radio astronomical, X-ray and gamma ray measurements. The measured up to ~ 100 GeV/nucleon ratio of fluxes of secondary to primary nuclei in cosmic rays is decreasing with energy at $E > 1$ GeV/nucleon. It allows to determine the energy dependence of cosmic ray diffusion coefficient $D \sim v(p/Z)^a$, where the index $a = 0.3...0.6$ depending on the specific model of cosmic ray propagation in the Galaxy, see e.g. [7] (here v is the particle velocity, p is the particle momentum, Z is the particle charge). Thus the cosmic-ray source spectrum is more flat than the observed steady state spectrum and it is close to $E^{-2.1}...E^{-2.4}$ at energies below the knee.

The diffusion approximation works for particles with energies not much larger than about $10^{17}Z$ eV (Z is the particle charge). The ultra high energy cosmic rays experience only weak scattering in magnetic fields and their propagation in the Galaxy is studied with the use of direct trajectory calculations. It is usually assumed that cosmic rays with the highest detected energies, $E > 10^{19}$ eV, are of extragalactic origin. They might be generated in active galactic nuclei, relativistic jets, interacting galaxies, or result from the decays of hypothetical heavy relic particles that were produced during the early cosmological epochs.

Requirements on galactic sources of cosmic rays

If one uses the cosmic ray energy requirements and the nonthermal radiation as a guideline, then the most powerful accelerators of relativistic particles in the Galaxy should be supernovae and supernova remnants, pulsars, neutron stars in close binary systems, and winds of young massive stars. The total power L_{cr} needed to maintain the observed energy density of cosmic rays is estimated as 10^{41} erg/s. For the acceleration by a supernovae, this estimate implies the release of energy in the form of cosmic rays of approximately 10^{50} erg per supernova if the supernova rate in the Galaxy is 1 every 30 years. The found value comes to about 10% of the kinetic energy of the ejecta that is in agreement with the prediction of the theory of diffusive shock acceleration discussed below. It assumes the acceleration of cosmic rays by the outward propagating shock, which results from the supernova explosion and propagates in the interstellar medium or in the wind of the progenitor star.

The rotational energy of a young pulsar with period P that remains after the supernova explosion is estimated to be $2 \times 10^{50}(10 \text{ ms}/P)^2$ erg and may provide an additional energy reservoir for the acceleration of cosmic rays.

The data on the nonthermal radio emission of supernova remnants reinforce the above energy estimates. Assuming the equipartition between cosmic-ray and magnetic field energy densities and assuming an electron to proton ratio of 10^{-2}, one finds the total energy of relativistic electrons to be $3 \times 10^{48} - 10^{49}$ erg in each of the remnants Cas A, IC 443 and Cygnus Loop [8]. Shell-type supernova remnants exhibit a broad range of spectral indices, centered roughly on $\alpha = -0.5$. This can be related to the electron power law index through the equation $\gamma = 2\alpha + 1 \approx 2.0$ but the range is from about 1.4 to 2.6. The analysis of synchrotron emission in Cas A [9] showed the presence of electrons with energies up to 200 GeV at the strength of magnetic field about 500 μG in this young supernova remnant. The interpretation of nonthermal radio emission from external galaxies confirms that the supernova remnants are the sites of acceleration of relativistic electrons with the same efficiency which is needed to provide the observed intensity of galactic cosmic-ray electrons.

Gamma-ray emission associated with few bright supernova remnants has been found using the EGRET catalogue of gamma-ray sources at $E > 30$ MeV [10]. The gamma-ray fluxes from the two most prominent sources gamma-Cygni and IC443 indicate an energy of about 3×10^{49} erg for relativistic protons and nuclei confined in each envelope, assuming that gamma rays are generated through $\pi^0 \rightarrow 2\gamma$ decay.

The non-thermal X-rays radiation with a characteristic power law tail at energies more than few KeV from the bright rims in supernova remnants including SN1006, RX J1713.7-3946, IC443, RCW 86, and Cas A was found in the experiments fulfilled by the X-ray observatories ASCA, RXTE, ROSAT, Chandra, XMM/Newton. It was interpreted as synchrotron emission by electrons accelerated up to energies as high as 100 TeV, see [11], [12] for review. The inverse Compton scattering of background photons by these electrons and/or the gamma rays generated via π^0 decays are the most probable mechanisms of the emission of TeV gamma rays detected from RX J1713.7-3946 [13], and Cas A [14]. (The TeV gamma-ray flux detected from SN1006 [15] was not confirmed by recent HESS observations [16].

The correction of observed at the Earth cosmic ray composition for nuclear fragmentation in the interstellar medium makes it possible to determine the initial elemental and isotopic composition of accelerated particles, to clarify the process of cosmic ray acceleration and the nature of cosmic ray sources.

The relative abundance of chemical elements in cosmic ray sources is in general similar to the solar and to the local galactic abundance but show some significant deviations. The elements that appear underabundant by a factor of about 5 are those elements that are difficult to ionize. The critical ionization potential is approximately 10 eV that corresponds to ionization at an equilibrium temperature 10^4 K (characteristic e.g. of the solar photosphere). The correlation of abundance with the first ionization potential is also known for solar energetic particles. Thus, it is possible that the outer layers of relatively cool stars serve as injectors of the seed particles required for the subsequent acceleration [17]. For most elements the volatility is correlated with the first ionization potential, so that volatility is also considered as a possible selection factor for the cosmic ray population. Then predominant acceleration and breakup of grains that is natural for the diffusive shock acceleration could explain the situation [18].

The popular at present scenarios which explain the cosmic ray source composition include the acceleration of grains together with relatively less abundant volatile ions by shocks in the interstellar medium [19], the acceleration of freshly formed material particularly grains in young supernova remnants [20], and the acceleration of material in hot superbubbles with multiple supernova remnants [21].

Corrected on the first ionization potential, the elemental composition at cosmic ray sources as compared to the solar and local galactic composition exhibits the underabundance of H and He by a factor of 10 (if normalized to Fe). It also demonstrates higher Pt/Pb ratio by a factor of 5 and higher actinides abundance, the ratio $(Z \geq 88)/(74 \leq Z \leq 87)$ by a factor of 3. These anomalies may be an indication that cosmic rays arise from supernova material (synthesized in the r-process) mixed with the interstellar gas.

The isotopic composition of cosmic rays is now measured for all stable isotopes for elements from H through Ni. The isotopic composition of cosmic ray source material is strikingly similar to the composition found in the solar system. Thus, the solar-like mix of isotopes of Fe and Ni suggests that Type II and Type Ia supernovae contribute to cosmic-ray source material in proportion similar to their contributions to the solar system [22].

There is one well-established anomaly in the isotopic composition of galactic cosmic rays, the excess of ^{22}Ne. The ratio ^{22}Ne/^{20}Ne is enhanced by a factor of 4 compared with the solar reference value [23]. It can be explained only by the special conditions of nucleosynthesis. The enhancement of neutron rich isotopes would be expected in the highly evolved very massive stars in

their Wolf-Rayet stage when their surfaces contain large excesses of the products of core helium burning, including ^{22}Ne [24]. Increased cosmic ray ratio
C/O by a factor of 2 is also in favor of Wolf-Rayet stars.

The relatively high and close-to-the-solar value of the ratio ^{59}Co/^{56}Fe [25]
testifies that the major part of originally synthesized ^{59}Ni has decayed by the
K-capture of an orbital electron into ^{59}Co before the acceleration started. (This
test was suggested in [26].) Consequently, the delay between synthesis of this
material and acceleration is larger than 10^5 yr. This probably rules out the
models where a major portion of the observed mildly relativistic cosmic rays
is accelerated in young supernova remnants.

Acceleration by supernova blast waves

We come now to the discussion of acceleration mechanisms of galactic cosmic rays. The interpretation of data on the ultra-high energy particles suggests
that acceleration in galactic sources should go at least up to $10^{18} - 10^{19}$ eV. We
shall see that an appropriate mechanism in our Galaxy is not easy to identify.

The diffusive shock acceleration by supernova blast waves is the commonly
accepted process of cosmic ray acceleration in the Galaxy. This mechanism of
acceleration was suggested in [27], [28], [29], [30], see [31], [32] for review.
The acceleration of a fast particle diffusing near the shock front is a version of
Fermi type acceleration. Fast particles are scattered on the inhomogeneities of
magnetic field frozen into the background plasma and gain energy each time
they cross the shock where plasma is compressed.

A typical source of galactic cosmic rays is most probably associated with the
core collapse supernova, Type II supernovae, that is the final stage of evolution
for stars more massive than about 8 solar masses while on the main sequence.
Before the explosion, the massive star goes through the Main Sequence O-
star stage, the Red Supergiant stage, and, for the most massive progenitors
($> 20~M_\odot$), which give rise to the rare Type Ib/c supernovae, through the
Wolf-Rayet stage. The fast wind of a massive progenitor star on the main
sequence produces a big bubble of hot rarefied gas with a temperature of about
10^6 K in the surrounding interstellar medium, see [33], [8]. The typical Type
II supernova goes through the Red Super Giant phase before the explosion
and this process is accompanied by the flow of a low-velocity dense wind.
Thus, immediately after the supernova burst, the shock propagates through
the wind of a Red Super Giant star, then through the hot bubble, and finally
it enters the interstellar medium. A considerable fraction of cosmic rays is
probably accelerated in Type Ia supernovae (their explosion rate in the Galaxy
is about $1/4$ of that of type II supernovae). These supernovae are caused by the
thermonuclear explosions of compact white dwarfs following mass accretion.
The characteristic masses of the progenitor star and the ejecta are 1.4 solar

mass. The progenitor stars do not appear to have an observable amount of mass loss nor do they emit ionizing radiation that could modify the ambient medium.

The supernova explosion with a characteristic release of mechanical energy $W_{sn} = 10^{51}$ erg gives rise to the blast shock wave which propagates in the surrounding medium. At the ejecta-dominated stage during the first ~ 300 yr, the shock velocity is very high $u_{sh} = 5 \times 10^8...3 \times 10^9$ cm/s and the shock experiences only weak deceleration. At a later stage of adiabatic deceleration (the Sedov stage) when the mass of swept-up circumstellar gas significantly exceeds the mass of supernova ejecta, the shock velocity is falling with time as $u_{sh} \sim (W_{sn}/\rho)^{1/5}t^{-3/5}$ (here $\rho = $ const is the gas density) until about 3×10^4 yr when the shock enters into the stage of significant radiation cooling accompanied by the formation of dense slow envelope.

The characteristic time of particle acceleration at the shock can be estimated as D/u_{sh}^2. The distribution of accelerated particles on momentum has a power law form $f(p) \sim p^{-3r/(r-1)}$, where r is the gas compression ratio in the shock. The relation between the distribution function $f(p)$ and the cosmic ray intensity $I(E)$ which is a function of particle energy reads as $f(p)p^2 = I(E)$. The compression in strong shocks is $r = 4$ and hence the spectrum of accelerated test particles is $f(p) \sim p^{-4}$ that is close to the required source spectrum of galactic cosmic rays. This result is valid in the case of a step-like profile of the flow velocity at the shock or, more precisely, when the characteristic thickness of the shock L_{sh} is relatively small: $L_{sh} \ll D/u_{sh}$.

The large density of energetic particles in the vicinity of high Mach number shock where the diffusive shock acceleration occurs causes the modification of gas flow through the action of energetic particle pressure and initiates plasma instabilities produced by the current of energetic particles.

The modification of the flow by cosmic ray pressure leads to the violation of the thin shock approximation and results in a deviation of the spectrum of accelerated particles from the simple power-law form derived in a test particle approximation. Asymptotically at very high energies, the spectrum of accelerated particles flattens out to $p^{-3.5}$ instead of p^{-4}.

The plasma instabilities tend to increase the level of magnetohydrodynamic turbulence which scatters the cosmic-ray particles. The standard assumption is that cosmic ray streaming instability generate a strong turbulence with the amplitude of random magnetic field $\delta B \sim B_0$ where B_0 is the regular magnetic field upstream of the shock. The cosmic ray diffusion coefficient D decreases then to the Bohm value $D_{B0} = v r_g/3$, where v is the particle velocity, $r_g = pc/(ZeB_0)$ is the Larmor radius of a particle with momentum p and charge Ze.

The entire pattern of cosmic ray acceleration at supernova shocks critically depends on the assumption that the energetic particles themselves produce the turbulence needed to provide the anomalously slow diffusion at the site of acceleration. The dependence of diffusion on energy determines the maximum energy that particles can gain in the process of acceleration. The necessary condition of efficient acceleration at the shock is $D(E) \leq 0.1 u_{\rm sh} R_{\rm sh}$, where $R_{\rm sh}$ is the radius of spherical shock (the value of the numerical factor 0.1 is approximate here). Notice that a typical supernova burst with kinetic energy of ejecta $W_{\rm sn} = W_{51} 10^{51}$ erg in the interstellar gas with number density n_0 cm^{-3} gives the maximum value of the product $0.1 u_{\rm sh} R_{\rm sh} \sim 10^{27} (W_{51}/n_0)^{0.4}$ cm^2/s at the end of the free expansion stage of supernova remnant evolution when this product reaches its maximum value. At the same time, the typical value of cosmic ray diffusion coefficient in the Galaxy is close to 3×10^{28} cm^2/s and thus the necessary condition of acceleration cannot be fulfilled for relativistic particles unless their diffusion coefficient is anomalously small in the vicinity of the shock. The Bohm value, which is a lower bound of the diffusion coefficient along the magnetic field at $\delta B \leq B_0$, is equal to $D_B = 6 \times 10^{21} \beta R_{\rm m}$ cm^2/s at $B_0 = 5 \times 10^{-6}$ G. Here $\beta = v/c$ is the dimensionless particle velocity and $R_{\rm m} = pc/Z$ is the particle magnetic rigidity in units GV. This gives the maximum particle energy $E_{\rm max} \sim 2 \times 10^{14} Z (W_{51}/n_0)^{0.4}$ eV at the time of transition from the ejecta-dominated stage to the stage of adiabatic evolution of supernova remnants and with the scaling $E_{\rm max} \sim t^{-1/5}$ later on. The assumption about Bohm value of the diffusion coefficient in the vicinity of strong shocks is commonly used in the simulations of cosmic ray acceleration in supernova remnants, e.g. [34].

During the last few years, mainly in response to the new observations of high energy emission in supernova remnants, the standard scenario of cosmic ray acceleration in supernova remnants is undergoing a serious revision. As was said above, a number of young supernova remnants with ages less than about 3×10^3 yr were reported as cosmic ray accelerators of very high energy particles with energies up to 100 TeV based on the observations of nonthermal synchrotron X-rays or/and TeV gamma rays. Also, the presence of a strongly amplified magnetic field in young supernova remnants follows from the interpretation of data on their synchrotron X-ray emission, see e.g. [35] for review. On the other hand, the shell remnants at the age of more than a few thousand years are not prominent sources of TeV gamma rays produced by very high energy cosmic rays (see e.g. [36]), in spite of the previous optimistic theoretical estimates based on the assumption of Bohm diffusion [37], [38] and in spite of a few detections of gamma-rays from these sources at energies about 100 MeV [39].

The solution of this problem was suggested in [40], [41] where the strong nonlinear processes that accomany the cosmic-ray streaming instability in a precursor of a supernova shock was studied. The level of magnetohydrodynamic turbulence in this region determines the value of cosmic ray diffusion coefficient and the maximum energy of particles accelerated by the diffusive shock acceleration mechanism. The consideration was not limited by the case of weak turbulence and in this respect the investigation [40] extends the earlier work [42]. It was also assumed that the Kolmogorov-type nonlinear wave interactions together with the ion-neutral collisions restrict the amplitude of generated random magnetic field. It was found that the maximum energy of accelerated particles strongly depends on the age of the supernova remnant. The maximum energy can be as high as $10^{17} Z$ eV in young supernova remnants and falls to about $3 \times 10^{10} Z$ eV at the end of the Sedov stage. Thus the standard estimate of the maximum particle energy based on the Bohm limit calculated for the interstellar magnetic field strength turns out to be incorrect. The calculated diffusion coefficient is much less than D_{B0} in very young supernova remnants, and it considerably exceeds D_{B0} at the late Sedov stage. The accounting for strong non-linear effects in the instability that accompanies the acceleration of cosmic rays in supernova remnants simultaneously eliminates two difficulties of modern cosmic ray astrophysics. It raises the maximum energy of accelerated particles in young supernova remnants above the standard Bohm limit through the production of very strong random magnetic fields and thus helps to explain the origin of galactic cosmic rays with energies up to $\sim 10^{17} Z$ eV. It also decreases the maximum energy of particles in the late Sedov stage of supernova remnant evolution, which allows to explain why these objects are not bright in very high energy gamma rays.

The calculated average spectrum of cosmic rays injected in the interstellar medium in the course of adiabatic evolution of a supernova remnant (the Sedov stage) is approximately $I(E) \sim E^{-2}$ at energies larger than 30 Gev and with the maximum particle energy that is close to the position of the knee in the cosmic-ray spectrum observed at $\sim 4 \times 10^{15}$ eV. At an earlier stage of the supernova remnant evolution - the ejecta-dominated stage, the particles are accelerated to higher energies and have a rather steep power-law distribution. These results suggest that the knee may mark the transition from the ejecta-dominated to the adiabatic evolution of supernova shocks which accelerate cosmic rays. The results of calculations of the average spectra for Type II and Type Ia supernovae are close to each other [41]. The obtained source spectrum is consistent with the empirical model of cosmic ray propagation in the Galaxy. The breaks and cutoffs in the spectra of ions with different charges should occur at the same magnetic rigidity as for protons, i.e. at the same ratio p/Z (or E/Z for the ultrarelativistic nuclei). The data of the KASCADE experiment [Ulrich *et al.* 2004],[44] for the most abundant groups of nuclei (protons, helium, CNO group, and the iron group nuclei) are, in general, consistent with this concept.

At present, the main problem of the data interpretation centers around the second knee in the cosmic ray spectrum. The natural assumption that all individual ions have only one knee at $\sim 4 \times 10^{15} Z$ eV and that the knee in the spectrum of iron ($Z = 26$) expected at about 10^{17} eV explains the second knee in the all-particle spectrum does not agree with the observed position of the second knee at about 5×10^{17} eV. Of considerable promise is the approach [Sveshnikova 2003] where the dispersion of parameters of supernova explosions in the calculations of the knee position and the maximum particle energy was taken into account. This leads to a widening of the energy interval between the two knees in the overall all-particle spectrum. This analysis should be supplemented by the account of different chemical compositions of the progenitor star winds that determine the composition of accelerated cosmic rays [46].

The significant progress in our understanding of precesses with cosmic rays in supernova remnants is expected from the new generation of ground-based (CANGAROO-III, HESS, VERITAS, MAGIC) and space (GLAST) gamma-ray experiments in conjunction with new X-ray satellite observations.

Complementary processes of cosmic ray acceleration

The maximum energy which particles can reach in a course of diffusive shock acceleration in the supernova envelope is hardly sufficient to explain the extended cosmic ray spectrum up to the ultrahigh energies. The additional acceleration of very high energy particles by the regular electric field in pulsar driven supernova remnant was suggested in [47]. For a few msec pulsar with a proper (parallel) orientation of magnetic and rotation axes the energy gain can reach $10^{19}Z$ eV. The potential available for acceleration is the same whether the cosmic rays orbit the neutron star near the light cylinder or near the outer edge of the remnant. The scenario might include the diffusive shock acceleration for particles with insufficient energy to enter the nebula and the acceleration by the internal field for high-energy particles, which can drift into the nebula. It is not clear whether this scenario would provide the continuous energy spectrum and proper power law form of its high-energy part. Some further trajectory calculations are needed to determine the energy distribution of running out particles.

In principle, the relativistic winds of newly born magnetars (neutron stars with surface magnetic fields about 10^{15} G) can directly accelerate particles up to 10^{21} eV [48], [49] but they are very rare events in the Galaxy.

Also, the maximum particle energy gained at an individual shock wave may be increased after additional acceleration at a different shock. The acceleration can go in two stages [50]. An individual supernova envelope accelerates particle up to the knee (the first stage). The subsequent acceleration is furnished by the collective action of other supernova remnants (the second stage). This mechanism is probably the most efficient inside superbubbles. A superbubble may contain an ensemble of shocks generated by the winds of a few tens of massive bright OB stars and the supernova blast waves. The acceleration in the superbubble of the size 300 pc can gives the particle energy up to $10^{17}Z$ eV if the characteristic turbulent velocity is 3×10^8 cm/s and the magnetic field strength is as high as 3×10^{-5} G [51].

The largest shock, which one can suggest for cosmic ray acceleration in the Galaxy, is the (hypothetical) galactic wind termination shock [52]. The acceleration up to $3 \times 10^{18}Z$ eV was found for the shock radius 300 Kpc. It remains unclear how to provide a smooth conjugation between the part of the spectrum before the knee (generated in supernova remnants) and after the knee (generated at the termination shock) in this model. The particles can also gain energy in the interacting regions of the galactic wind flow [53].

It is worth noting that the knee observed in the cosmic ray spectrum at $\sim 4 \times 10^{15}$ eV may arise not in the sources but in the process of cosmic ray propagation in the Galaxy, e.g. as a result of interplay between ordinary and Hall diffusion in galactic magnetic fields [54], [Roulet 2004]. Of course, this explanation requires the existence of a power-law source spectrum which extends without essential breaks up to about 10^{18} eV or even further.

Acknowledgments The author is grateful to the personnel of the Ettore Majorana Centre in Erice for kind hospitality during the School.

References

[1] Baade, W., and F. Zwicky. (1934). "Remarks on super-novae and cosmic rays", Phys. Rev. 46, 76.

[2] Fermi E. (1949). "On the origin of the cosmic radiation", Phys. Rev. 75, 1169.

[3] Ginzburg, V.L., and S.I. Syrovatskii. (1964). *The Origin of Cosmic Rays*. Oxford: Pergamon Press.

[4] Berezinskii, V.S., S.V. Bulanov, V.A. Digiel, V.L. Ginzburg, and V.S. Ptuskin. (1990). *Astrophysics of Cosmic Rays*. North-Holland.

[5] Ptuskin, V.S. (2001). "Propagation, confinement models, and large-scale dynamical effects of galactic cosmic rays", Space. Sci. Rev. 99, 281

[6] Strong, A., and I.V. Moskalenko. (1998). "Propagation of cosmic- ray nuclei", ApJ 509, 212.

[7] Jones F.C. et al. (2001). "The modified weighted slab technique: models and results", ApJ 547, 264.

[8] Lozinskaya, T.A. (1992). *Supernovae and Stellar Wind: The Interaction with the Interstellar Medium*. AIP

[9] Jones, T. et al. (2003) "The identification of infrared synchrotron radiation from Cassiopeia A", ApJ 587, 227.

[10] Esposito, J.A. et al. (1996). "EGRET observations of gamma-ray emission from supernova remnants", ApJ 461, 820.

[11] Petre, R., U. Hwang, and G.E. Allen. (2001). "Evidence for cosmic-ray acceleration in supernova remnants from X-ray observatons". Adv. Space Res., 27, 647.

[12] Vink, J. (2003). "Shocks and particle acceleration in supernova remnants: observational features", Adv. Space Res. 35, 356.

[13] Muraishi, H. et al. (2000) "Evidence for TeV gamma-ray emission from shell type SNR RX J1713.7-3946", A&A 354, L57.

[14] Aharonian, F.A. et al. (2001). "Evidence for TeV gamma ray emission from Cassiopea A", A&A 370, 112.

[15] Tanimori, T. et al. (1998). "Discovery of TeV gamma rays from SN 1006: further evidence for the supernova remnant origin of cosmic rays", ApJ 497, L25.

[16] Hofmann, W. (2004). "Status of ground based gamma ray astronomy", *Intern. Symp. on High Energy Gamma Ray Astronomy*, Heidelberg (in press)

[17] Meyer, J.P. (1985). "Solar-stellar outer atmospheres and energetic particles, and galactic cosmic rays", ApJ Suppl., 57, 173.

[18] Epstein, R.I. (1980) "The acceleration of interstellar grains and the composition of the cosmic rays", MNRAS 193, 723.

[19] Meyer, J.P., L. O'C. Drury L., and D.C. Ellison. (1997). "Galactic cosmic rays from supernova remnants. I. A cosmic-ray composition controlled by volatility and mass-to-charge Ratio", ApJ 487, 182.

[20] Lingenfelter, R.E., R. Ramaty, and B. Kozlovsky. (1998). "Supernova grains: the source of cosmic-ray metals", ApJ 500, L153.

[21] Higdon, J.C., R.E. Lingenfelter, and R. Ramaty. (1998). "Cosmic-ray acceleration from supernova ejecta in superbubbles", ApJ 509, L33.

[22] Wiedenbeck, M.E. et al. (1999). "The isotopic composition of iron, cobalt, and nickel in cosmic ray source material", *26th ICRC*, Salt Lake City, 3, 1.

[23] DuVernois, M.A. et al. (1996). "The Isotopic Composition of Galactic Cosmic-Ray Elements from Carbon to Silicon: The Combined Release and Radiation Effects Satellite Investigation", ApJ 466, 457.

[24] Casse, M., and J.A. Paul. (1982). "On the stellar origin of the Ne-22 excess in cosmic rays", ApJ 258, 860.

[25] Mewaldt, R.A. et al. (1999). "The time delay between nucleosynthesis and acceleration based on ACE measurements of primary electron-capture nuclides", *26th Intern. Cosmic Ray Conf.*, Salt Lake City, 3, 45.

[26] Casse, M., and A. Soutoul. (1978). "Time delay between explosive nucleosynthesis and cosmic ray acceleration", ApJ 200, L75.

[27] Krymsky, G.F. (1977). "A regular mechanism for the acceleration of charged particles on the front of a shock wave", Soviet Physics-Doklady 22, 327.

[28] Axford, W.I., E. Leer, and G. Skadron. (1977). *15th Intern. Cosmic Ray Conf.*, Plovdiv, 11, 131.

[29] Bell, A.R. (1978). "The acceleration of cosmic rays in shock fronts", MNRAS 182, 147.

[30] Blandford, R.D., and J. Ostriker. (1978). "Particle acceleration by astrophysical shocks", ApJ 221, L29.

[31] Drury, L.O'C. et al. (2001). "Test of galactic cosmic-ray source model", Space Sci. Rev. 99, 329.

[32] Malkov, M.A., and L.O'C. Drury. (2001). "Nonlinier theory of diffusive acceleration of particles by shock waves", Rep. Progress in Physics 64, 429.

[33] Weaver, R. et al. (1977). "Interstellar bubbles. II- Structure and evolution", ApJ 218, 377.

[34] Berezhko, E.G., V.K. Elshin, and L.T. Ksenofontov. (1996). "Cosmic-ray acceleration in supernova remnants", JETP 82, 1.

[35] Vink, J. (2003). "Shocks and particle acceleration in supernova remnants: observational features", astro-ph/0304176.

[36] Buckley, J.H. et al. (1998). "Constraints on cosmic-ray origin from TeV gamma-ray observations of supernova remnants", A&A 329, 639.

[37] Drury, L. O'C., F.A. Aharonian, and H.J. Völk. (1994). "The gamma-ray visibility of supernova remnants. A test of cosmic-ray origin", A&A 287, 959.

[38] Naito, T., and F. Takahara. (1994). "High energy gamma-ray emission from supernova remnants", JPhG 20, 477.

[39] Esposito, J.A. et al. (1996). "EGRET observations of gamma-ray emission from supernova remnants", ApJ 461, 820.

[40] Ptuskin, V.S., and V.N. Zirakashvili. (2003). "Limits on diffusive shock acceleration in supernova remnants in the presence of cosmic-ray streaming instability and wave dissipation", A&A 403, 1.

[41] Ptuskin, V.S., and V.N. Zirakashvili. (2004). "On the spectrum of high-energy cosmic rays produced by supernova remnants in the presence of strong cosmic-ray streaming instability and wave dissipation", A&A, in press; astro-ph/0408025.

[42] Bell, A.R., and S.G. Lucek. (2001). "Cosmic-ray acceleration to very high energie through the non-linear amplification by cosmic rays of the seed magnetic field", MN-RAS 321, 433.

[43] Ulrich, H. et al. (2003). 'Spectra of cosmic rays in the knee region", Nucl. Phys. B (Proc. Suppl.) 122, 218.

[44] Hörandel, J.R. (2003). "On the knee in the energy spectrum of cosmic rays", Astroparticle Physics 19, 193.

[45] Sveshnikova, L.G. (2003). "The knee in galactic cosmic ray spectrum and variety in supernovae", A&A 409, 799.

[46] Silberberg, R., C.H. Tsao, M.M. Shapiro, and P.L. Biermann. (1991). *Cosmic Rays, Supernovae, Interstellar Medium*, ed. M.M.Shapiro, R. Silberberg & J.P. Wefel, (Kluwer Ac. Publ.), p. 97.

[47] Bell, A.R., and S.G. Lucek. (1996) "Cosmic-ray acceleration in pulsar-driven supernova remnants: the effect of scattering", MNRAS 283, 1083.

[48] Arons, J. (2003) "Magnetars in the Metagalaxy: an origin for ultra-high energy cosmic rays in the nearby Universe", ApJ 589, 871.

[49] Blasi, P., R.L. Epstein, and A.V. Olinto. (2000). "Ultra-high-energy cosmic rays from young neutron star winds", ApJ 533, L123.

[50] Axford, W.I. (1994). "The origin of high-energy cosmic rays", ApJS 90, 937.

[51] Bykov, A.M., and I.N. Toptygin. (2001). "A model of particle acceleration to high energies by multiple supernova explosions in OB associations", Astron. Let. 27, 625.

[52] Jokipii, J.R., and G.E. Morfill. (1985). "On the origin of high-energy cosmic rays", ApJ 290, L1.

[53] Völk, H.J., and V.N. Zirakashvili. (2004). "Cosmic ray acceleration by spiral shocks in the galactic wind", A&A 417, 807.

[54] Ptuskin, V.S. et al. (1993). "Diffusion and drift of very high energy cosmic rays in galactic magnetic fields", A&A 268, 726.

[55] Roulet, E. (2003). "Astroparticle theory, some new insights into high energy cosmic rays", astro-ph/0310367.

RADIAL DISTRIBUTION OF GRBS IN HOST GALAXIES

D.I. Kosenko

Sternberg Astronomical Institute, Universitetskij pr. 13, 119992 Moscow, Russia
Institute for Theoretical and Experimental Physics, Bolshaya Cheremushkinskaya, 25, 117218 Moscow, Russia

lisett@xray.sai.msu.ru

S.I. Blinnikov

Institute for Theoretical and Experimental Physics, Bolshaya Cheremushkinskaya, 25, 117218 Moscow, Russia
Sternberg Astronomical Institute, Universitetskij pr. 13, 119992 Moscow, Russia

blinn@sai.msu.su

K.A.Postnov

Sternberg Astronomical Institute, Universitetskij pr. 13, 119992 Moscow, Russia

pk@sai.msu.ru

O.S. Bartunov

Sternberg Astronomical Institute, Universitetskij pr. 13, 119992 Moscow, Russia

oleg@sai.msu.su

Abstract   We investigate correlation of gamma-ray burst radial distribution in host galaxies with distributions of other phenomena in galaxies (such as supernovae, dark matter, X-ray binary systems). Various statistical methods are used to test the hypothesis that gamma-ray bursts and some of other objects are distributed similarly. The analysis shows that among all considered objects the distribution of type Ib/c supernovae in galaxies has features which are close to properties of gamma-ray burst distribution.

Keywords: gamma-ray bursts, distribution in host galaxies, statistical analysis

M. M. Shapiro et al. (eds.), Neutrinos and Explosive Events in the Universe, 143–147.

Introduction

Since INTEGRAL era an opportunity for precise measuring of gamma-ray bursts (GRB) localization in their host galaxies as well as of the hosts radii has emerged. So it becomes possible to investigate GRB's radial distribution in their hosts and to compare it with distribuitons of other objects in galaxies. If one can find this class of objects, it could be a candidate for GRB relatives.

Tsvetkov et al. (2001) have shown that surface density distribution of GRBs in their hosts is close to surface brightness distribution in spirals (Kolmogorov-Smirnov test gives the probability $P_{KS} = 68\%$) and to surface brightness of ellipticals ($P_{KS} = 40\%$). The authors did not find significant correlation of GRB distribution and distribution of OB associations ($P_{KS} = 4\%$), also there is no correlation between GRB and supernovae (SN) Ib/c distributions ($P_{KS} = 9\%$).

Bloom et al. (2002) have collected a good GRB galactocentric offests data, took into account errors of GRBs localizations, and estimates of host galaxies radii. Comparison of GRBs offsets distribution with merging NS-NS and NS-BH binaries ($P_{KS} < 2 \times 10^{-3}\%$) and with distribution of massive star forming regions in galaxies ($P_{KS} = 45\%$) has been performed in the paper.

In this study we compare distribution of GRB galactocentric offsets with radial distributions of supernovae of types Ia and Ib/c, low mass and high mass X-ray binaries, and with models of dark matter (DM) halo density profile. We compare first and second moments of distributions and also their median values, moreover we apply visual technique for comparing a pair of empirical distributions — quantile-quantile plot.

Data

We consider the following data sets: GRB, SN Ia, SN Ib/c, X-ray binaries, models of DM distribution in galaxies. Each data set of galactocentric offsets is normalized by typical galaxy radius ($r = R/R_{galaxy}$), which is usually equal to optical radius ($R_{galaxy} \simeq R_{opt}$). Data on GRB offset errors also avaible.

Most of the GRB data was taken from Bloom et al. (2002). Others are presented in table 1. Supernova data sets was taken from D. Tsvetkov and N. Pavluk SN online catalogue (http://virtual.sai.msu.ru/~pavlyuk/distrib/radial.htm). Radial distribution of X-ray binaries was taken from Grimm et al. (2001).

We consider two models of DM density distribution. These are the model by Burkert (1955) and NFW model by Navarro et al. (1997). The first one is described by formula $\rho_B(r) \propto 1/((r/r_{core} + 1)((r/r_{core})^2 + 1))$ and has one scale paramter $r_{core} \simeq r_{opt}$, optical radius $r_{opt} \simeq 15$ kpc for nearby galaxies. The second one is $\rho_{NFW}(r) \propto 1/((r/r_s)(1 + r/r_s)^2)$ and has a scale parameter $r_s \simeq 30$ kpc for local galaxies.

Table 1.   GRB galactocentric offset data, z – the redshift of the burst, r – estimate of galacto-centric offset (in host galaxy radius units), σ_r – error of an offset estimate.

GRB	z	$r. \pm \sigma_{r_0}$	Source
000131	4.5	1.09 ± 0.30	J.S. Bloom et al. (GCN notice#1133)
000210	0.846	1.50 ± 0.99	L. Piro et al. (2002)
000911	1.058	0.23 ± 0.95	P.A. Price et al. (2002)
000926	2.038	0.13 ± 0.01	S. Castro et al. (2001)
010222	1.477	0.77 ± 0.92	D. A. Frail et al. (2001)
010921	0.45	1.28 ± 0.44	J. S. Bloom et al. (GCN notice #1135)
011121	0.36	0.86 ± 0.12	J. S. Bloom (GCN notice #1260)
011211	2.14	1.15 ± 0.38	D.W. Fox et al. (GCN notice #1311)
020405	0.69	2.21 ± 0.62	N. Masetti et al. (GCN Circular #1375)
021004		0.00 ± 0.94	http://www-int.stsci.edu/~fruchter/GRB/021004/
021211	1.01	2.32 ± 1.07	(GCN notice #1758)
031203	0.1055	0.03 ± 0.015	A. Gal-Yam et al. (astro-ph/0403608)

Cumulative surface density profiles are presented in figure 1. All distances are normalized by host galaxies radii, where $R_{galaxy} \simeq 2.3$ kpc for distant galaxies in which gamma-ray bursts are observed, $R_{galaxy} \simeq 15$ kpc for nearby galaxies, these are hosts for supernovae and X-ray binaries. $F(r) = P(x < r) = N(x < r)/N_{\text{total}}$ — p.d.f. of sources in galaxies. Smoothed GRB offset distribution (black solid curve on figure 1) is a sum of normalized p.d.f of each observation with account for offset error (see Bloom et al. (2002)).

Comparison of the distributions

Estimators of two first moments of the avaible empirical distributions. are presented in table 2 (median values of galactocentric offsets also shown). All distances are normalized by hosts optical ridii. Estimators of a first (mean) an second (variation) moments in the table were calculated by

$$\langle r \rangle = \mu^1(r) = \frac{1}{N}\sum_{i=1}^{N} r_i, \qquad D(r) = \mu^2(r) = \frac{1}{N-1}\sum_{i=1}^{N}(r_i - \langle r \rangle)^2. \quad (1)$$

Median values estimated so that $F(r^{med}) \simeq 1/2$, where F — empirical cumulative probability distribution function of r. Table 2 shows that mean and median values of galactocentric offsets of SNIb/c are very close to ones of GRBs.

Figure 2 shows quantile-quantile plots (QQ-plots) of GRB distribution compared with distributions of supernovae. QQ-plot shows quantiles of one data set against quantiles of another data set. If two populations come from one dis-

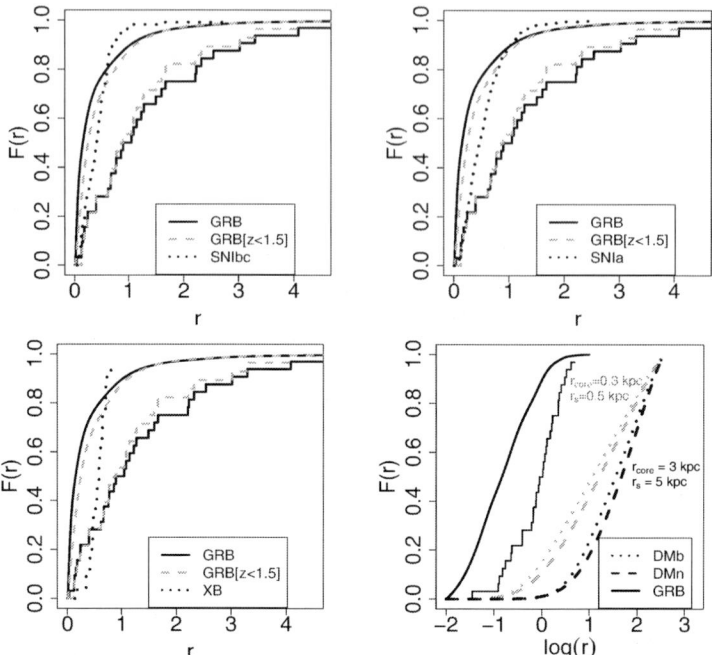

Figure 1. Cumulative p.d.f. (see insets): GRB – SNIb/c, GRB – SNIa, GRB – XB, GRB – two DM models. Smooth solid black curve is a GRB surface density profile with localization errors taken into account.

Table 2. Estimators of mean, variance and median offset values of surface density distributions with standart deviations GRB, SNe, XB: $\mu^{\cdot}(r)$ – mean distance, $\mu^{\cdot}(r)$ – variations, r^{med} – median value.

$\mu^i(r),\ i=1,2$	$\mu^{\cdot}(r) \pm \sigma_{\mu^1 \cdot r \cdot}$	$\mu^{\cdot}(r)$	r^{med}
GRB	0.260 ± 0.045	0.068	0.164
Nearby GRB, $z < 1.5$	0.320 ± 0.040	0.046	0.392
SNIbc	0.238 ± 0.017	0.031	0.175
SNIa	0.324 ± 0.012	0.053	0.245
LMXB and HMXB	0.543 ± 0.026	0.019	0.545

tribution function, then all data on the plot will be on the diagonal (dashed on the figures) line. Deviations from this line reveal differences in distributions.

Discussion and Results

In the study statisticlal analysis of GRB surface density distribution in host galaxies has been performed. The distribution was compared to those of SN Ib/c

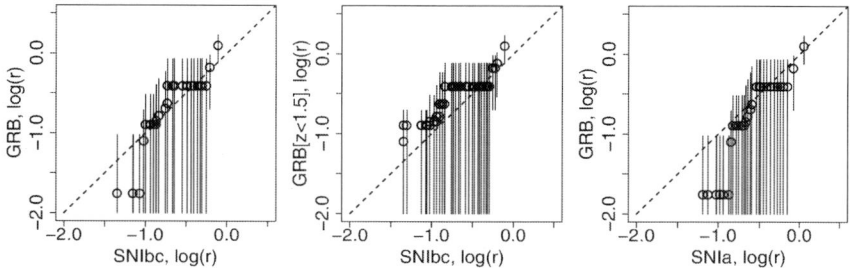

Figure 2. Quantile-Quantile plots. From left to right: GRB — SNIbc, GRB at $z < 1.5$ — SNIbc, GRB — SNIa. Vertical lines — error bars of GRB localization.

and SN Ia, X-ray binaries and DM theoretical profiles. Estimators of mean, variance and median values of galactocentric offsets were found. All considered distribution profiles have rather different shapes. Only the surface density profile of supernovae Ib/c shows some similarity to the profile of gamma-ray burst surface density distribution. This can be seen from quantile-quantile plots (figure .2) and from the comparison of means and medians: they are $\langle r \rangle_{GRB} = 0.260$, $\langle r \rangle_{SNIbc} = 0.238$ and $r_{GRB}^{med} = 0.164$, $r_{SNIbc}^{med} = 0.175$.

Almost all gamma-ray bursts are cosmological, thus their host galaxies could have structure and morphology rather different from those of local galaxies. The typical size of a GRB host galaxy is about 2-3 kpc, and nearby galaxies (hosts of SNIb/c) have size about 15 kpc. If we neglect evolution of galaxies from redshift $z \simeq 1.5$ and consider only GRB which have a redshift $z < 1.5$, then the surface density distribution behavior becomes closer to the SNIb/c profile (see figure 2).

References

[1] J. S. Bloom, S. R. Kulkarni, S. G. Djorgovski (2002). "The Observed Offset Distribution of Gamma-Ray Bursts from Their Host Galaxies: A Robust Clue to the Nature of the Progenitors", The Astronomical Journal, 123, p. 1111-11148 and references therein.

[2] A. Burkert (1995). "The structure of dark matter halos in dwarf galaxies", Astrophysical Journal, v.447, pp.25-28.

[3] H.-J. Grimm, M. Gilfanov, and R. Sunyaev (2001). "The Milky Way in X-rays for an outside observer Log(N)-Log(S) and Luminosity Function of X-ray binaries from RXTE/ASM data Astronomy and Astrophysics", astro-ph/0109239.

[4] Navarro, Julio F., Frenk, Carlos S., White, Simon D. M. (1997). "A universal density profile from hierarchical clustering", Astrophysical Journal, v.490, pp.493-508.

[5] D.Yu. Tsvetkov, S.I. Blinnikov, N.N. Pavlyuk (2001). "Gamma-ray Bursts, Type Ib/c Supernovae and Star-forming Sites in Host Galaxies", Astronomy Letters, v.27, issue 7, pp.411-415.

BLACK HOLES: PHYSICS AND ASTROPHYSICS

Stellar-mass, supermassive and primordial black holes

Jacob D. Bekenstein

Racah Institute of Physics, Hebrew University of Jerusalem, Givat Ram, Jerusalem 91904, Israel

Jefferson Physical Laboratory, Harvard University, Cambridge, MA 02138

bekenste@vms.huji.ac.il

Abstract   I present an elementary primer of black hole physics, including its general relativity basis, all peppered with astrophysical illustrations. Following a brief review of the process stellar collapse to a black hole, I discuss the gravitational redshift, particle trajectories in gravitational fields, the Schwarzschild and Kerr solutions to Einstein's equations, orbits in Schwarzschild and in Kerr geometry, and the dragging of inertial frames. I follow with a brief review of galactic X-ray binary systems with known black holes, stressing the QPO phenomenon in particular.

I then discuss the evidence from AGN for the existence of supermassive black holes in galaxy nuclei, as well as evidence for such black holes in ordinary galaxy nuclei. I use the free motion of gas parcels to illustrate aspects of accretion disks around black holes, showing how to calculate energy efficiency and surface emissivity of disks, and the rate of black hole spin-up. I recall the primary methods for black hole mass determination, the correlation of black hole mass with the stellar velocity dispersion of its neighborhood, and implications for the origin of supermassive black holes. Finally, I consider the formation of primordial black holes, and calculation of their mass spectrum at present in the case of scale invariant primordial inhomogeneities.

Keywords: black holes, Schwarzschild, Kerr, supermassive, primordial

Introduction

From the Newtonian standpoint a black hole (BH) is a compact object whose surface gravitational potential approaches the square of the speed of light. Thus in contrast with other astrophysical objects, the radius R of a BH scales linearly

M. M. Shapiro et al. (eds.), Neutrinos and Explosive Events in the Universe, 149–173.
© 2005 *Springer. Printed in the Netherlands.*

with its mass M:

$$R = GM/c^2 \approx 1.47(M/M_\odot) \text{ km}. \tag{1}$$

General Relativity (GR) *requires* the existence of BHs. Not only radius but several other BH properties scale simply with mass. For example the average mass density $\bar{\rho} \equiv 3M/4\pi R^3$ scales as M^{-2},

$$\bar{\rho} \approx 1.5 \times 10^{20}(M_\odot/M)^2 \text{ Kg m}^{-3}, \tag{2}$$

so that massive BHs are tenuous and light BHs are dense. Classical BH physics lacks a special scale; thus only in specific astrophysical surroundings are the physics of small and large BHs qualitatively different.

I have chosen to discuss three separate categories of black holes: those that arise from stellar collapse, those that may come about from merging and accretion processes and are termed supermassive black holes (SMBH), and those which may be relics of the dense highly inhomogeneous medium in the early universe (primordial black holes—PBH).

No longer can it be claimed that "for astrophysical purposes a BH is a Newtonian point mass"; GR has become indispensable to understand fine points of the observations relating BHs. Yet GR is considered by many opaque and hard to wield. Thus many model builders use GR formulae and results without really understanding where they come from. I try to alleviate the situation by providing in these lectures a little primer to GR together with some example calculations which are both easy to do and have important consequences. I also discuss some concrete issues in BH astrophysics, using whenever feasible GR results here obtained.

How do black holes form from stars ?

A good review is that of Brown, et al. [1]. A normal star is stable so long as nuclear burning in it provides thermal pressure to support it against gravity. But nuclear burning gradually converts the star core's hydrogen to helium, and for massive cores ($M > 5M_\odot$) the helium is subsequently burned to carbon, and carbon to heavier elements until we reach the iron group. The core contracts as each type of fuel runs out because it temporarily loses pressure until the heating resulting from the contraction ignites the next type of fuel. Since nucleosynthesis of heavy nuclei from lighter ones raises the mean particle mass, it contributes to pressure loss.

In the end the endothermic disintegration of iron-group nuclei, the tightest bound of nuclei, precipitates core collapse. Evolutionary calculations [1] show that single stars with initial masses $20 - 30M_\odot$ leave cores of more than $1.8M_\odot$. These promptly collapse to "massive BHs" ($M > 1.8M_\odot$) with no optical display, but intense neutrino emission from neutronization. For initial mass $18 - 20M_\odot$ the core implosion (also accompanied by neutronization) engenders a shock wave which can blow part of the star out (type II supernova). A core of mass $1.5M_\odot < M < 1.8M_\odot$, momentarily neutron rich, is left; after deneutronization it collapses into a "low mass BH". Stars with initial masses $10 - 18M_\odot$ would seem to make supernovae which leave neutron star remnants. The fate of single stars in the initial mass range $\sim 35 - 80M_\odot$ is uncertain.

Binaries develop differently. According to various authors, a star with initial mass $20 - 35M_\odot$ in a binary can loose its H envelope by overflowing its Roche lobe to become a "naked" helium star, which upon going supernova leaves a low mass BH or a neutron star. Since a goodly fraction of stars occur in binaries, this is a way to make normal star-BH binaries, provided the BH formation does not disrupt the binary, e.g. by copious gravitational waves emission [2]. The neutron stars in binaries provide a two-step path for black hole binary (BHB) formation following Roche lobe overflow by the the companion star as it evolves to a giant. Accretion may take the neutron star's mass over the critical mass, whereupon it will implode to a BH. Of course, the residual normal star can itself go supernova forming a two-compact object binary. BH-neutron star binaries will be more common than neutron star-neutron star ones, but harder to see. Stars with $M > 80M_\odot$ in binaries will make high mass BHs (as defined above). This is the origin of BHBs like Cyg-X1 whose ultimate fate is supernova explosion of the massive normal star and merger of the compact binary.

Evidently late stellar evolution yields BHs. It is estimated that the galaxy contains some 3×10^8 stellar-mass BHs [3]. To discuss BHs we need full GR.

The gist of General Relativity

GR conceives gravity not as a force between masses but as a change in the geometry of *spacetime* due to neighboring energy. Consider Euclidean space which may be written in the familiar forms

$$ds^2 = dx^2 + dy^2 + dz^2 \tag{3}$$
$$ds^2 = dr^2 + r^2(d\vartheta^2 + \sin^2\vartheta\,d\varphi^2), \tag{4}$$

or generically as

$$ds^2 = \sum_{i,j=1}^{3} g_{ij}\,dx^i\,dx^j, \tag{5}$$

where the g_{ij} coefficients are called the metric coefficients. In principle there are six of them (g_{ij} counts the same as g_{ji}), but in the popular versions of the Euclidean line element, three vanish. If the six g_{ij} are arbitrary functions, then (5) does not necessarily describe Euclidean space, but some curved space in three dimensions, a Riemannian space. A clear example is the line element $ds^2 = d\vartheta^2 + \sin^2 \vartheta \, d\varphi^2$ for two-dimensional space. It describes the geometry of the surface of an ordinary ball of unit radius. This is a curved space (try flattening a world globe without distorting it) and it is so in two-dimensions.

In special relativity we deal not just with space, but with flat four-dimensional spacetime, Minkowski spacetime, with metric

$$ds^2 = -c^2 dt^2 + dx^2 + dy^2 + dz^2. \tag{6}$$

If we replace the space part here by expression (5), we still deal the Minkowski spacetime. In fact

$$ds^2 = \sum_{\alpha,\beta=0}^{3} g_{\alpha\beta} \, dx^\alpha \, dx^\beta, \tag{7}$$

now with Greek indices like α and β running over $0, 1, 2, 3$ (x^0 represents the time coordinate, x^i a space one), is still the spacetime of special relativity, provided we came by the $g_{\alpha\beta}$ coefficients by transforming the coordinates from t, x, y, z in the line element (6). (Henceforth we use the Einstein convention: any product or single factor where a pair of up and down indices are equal means a sum over this index. Thus the last formula will be written $ds^2 = g_{\alpha\beta} \, dx^\alpha \, dx^\beta$). However, for a collection of ten arbitrary functions $g_{\alpha\beta}$ ($g_{\alpha\beta}$ counts the same as $g_{\beta\alpha}$), the spacetime will be curved or (pseudo-)Riemannian spacetime.

Einstein proposed that each curved spacetime represents a particular gravitational field. Thus writing the equations of nongravitational physics, e.g. Maxwell's, in such a spacetime rather than in Minkowski spacetime (7) automatically takes care of the effects of gravity on the physics. This is now called the Einstein equivalence principle. Einstein got this revolutionary idea from the universality of free fall: all objects that fall freely in a given gravitational field from identical initial position and velocity move on identical worldlines regardless of differences in structure and composition. This fact is now tested experimentally to precision of one part in 10^{13}) (upcoming satellite experiments should improve the number by 3-4 orders).

Contrast this with motion in an electromagnetic field: a positive particle moves one way, a negative one another way, and a neutral particle with a dipole in yet another. What happens in gravity seems a big coincidence from this perspective. Einstein elevated the universality of free fall in gravity to the status of a principle: the *weak* equivalence principle. He realized that the irrelevancy of structure and composition can be understood if particle worldlines are some universal trajectories in a curved geometry, so that gravity is no longer a force but a reflection of the curvature of spacetime. Then he generalized the notion to all nongravitational physics—this is the Einstein equivalence principle. Obviously the two principles are consistent with each other.

What are the universal trajectories ? In special relativity a particle not subject to forces moves inertially: if $x^\alpha(\tau)$ is the trajectory $\{x, y, z, t\}$ as functions of the proper time τ, then

$$\frac{d^2 x^\alpha}{d\tau^2} = 0. \tag{8}$$

Now, a theorem in Riemannian geometry tells us that *locally* any metric (7) with the correct signature can be rewritten as (6) by an appropriate change of coordinates. At different points we use different transformations of coordinates, but always end up with the Lorentz metric in the new coordinates. So the equation (8), when written in terms of the coordinates for which the metric looks like (7), must describe the trajectory in the gravitational field. This is the geodesic equation (sum over β, γ)

$$\frac{d^2 x^\alpha}{d\tau^2} = -\Gamma^\alpha_{\beta\gamma} \frac{dx^\beta}{d\tau} \frac{dx^\gamma}{d\tau} \tag{9}$$

with the functions $\Gamma^\alpha_{\beta\gamma}$ determined uniquely by ($\partial_\alpha \equiv \partial/\partial x^\alpha$)

$$g_{\mu\alpha} \Gamma^\alpha_{\beta\gamma} = \frac{1}{2}(\partial_\gamma g_{\mu\beta} + \partial_\beta g_{\gamma\mu} - \partial_\mu g_{\beta\gamma}). \tag{10}$$

Note that the "gravitational force" [r.h.s. of (9)] is velocity dependent.

But is Newtonian gravitation grossly wrong ? Not at all. It is relevant for nonrelativistic motion, namely one with $|dx^i/d\tau| \ll c$. This means the terms on the r.h.s. of (9) with a factor $dx^i/d\tau$ or two must be negligible. We are left with (no sum; nonrelativistically $\tau \to t$)

$$\frac{d^2 x^i}{d\tau^2} \approx -\Gamma^i_{tt} \frac{dt}{d\tau} \frac{dt}{d\tau} \approx -\Gamma^i_{tt}, \tag{11}$$

where the residual force is velocity independent. Now whenever $g_{\alpha\beta}$ is time independent, e.g. field of the sun or a galaxy, and close to the Minkowski metric (6), we have from (10) $\Gamma^i_{tt} = -\frac{1}{2}\partial_i g_{tt}$. Thus (11) looks like the Newtonian equation for motion in a potential Φ_N provided $g_{tt} = -2\Phi_N +$ const. Comparing with the special relativistic metric (6) we see that nonrelativistic motion is Newtonian if we can write

$$ds^2 = -(c^2 + 2\Phi_N)dt^2 + 2\sum_i g_{ti}dt\, dx^i + \sum_{ij} g_{ij}dx^i\, dx^j. \tag{12}$$

Thus Newtonian dynamics fixes the form of g_{tt}, at least to first order in Φ_N which is determined by Poisson's equation. We may rely on the above so long as $|\Phi_N| \ll c^2$; otherwise the condition $|dx^i/d\tau| \ll c$ would break down. Newtonian dynamics gives no information about g_{ti} and g_{ij} save that they cannot be too large.

Let us use (12) to compute the gravitational redshift, the reduction in the frequency of waves as they climb out of a gravitational potential well. Recall that we obtained (12) by assuming $g_{\alpha\beta}$ is time independent. Imagine an oscillator (decaying atom, radar device) produces a wave train of sharp frequency ν_1 at a point x_1. This means that $N \equiv \nu_1 \Delta\tau_1$ is the number of cycles of the wave in an interval $\Delta\tau_1$ of the (proper) time ticked by a clock at rest at x_1. But by (12) we have the relation $\Delta\tau_1 = (c^2 + 2\Phi_N(x_1))^{1/2}\Delta t$ with the interval of t time spanned by the train. Thus the number of cycles can be written $N = \nu_1(c^2 + 2\Phi_N(x_1))^{1/2}\Delta t$. Now the metric is not changing, so Δt is also the interval of t time spanned by the wave train anywhere, in particular at the destination x_2. But wave cycles cannot get lost, so using *his* clock, the observer at x_2 must assign the wave a frequency ν_2 such that $\nu_2(c^2 + 2\Phi_N(x_2))^{1/2}\Delta t = N$. We thus have

$$\frac{\nu_1}{\nu_2} = \frac{(c^2 + 2\Phi_N(x_2))^{1/2}}{(c^2 + 2\Phi_N(x_1))^{1/2}} \approx 1 + \Delta\Phi_N/c^2 \qquad (13)$$

with $\Delta\Phi_N \equiv \Phi_N(x_2) - \Phi_N(x_1)$. Now the redshift z suffered by the wave is defined by $1 + z = \lambda_2/\lambda_1 = \nu_1/\nu_2$; hence Einstein's celebrated result,

$$z \approx \Delta\Phi_N/c^2. \qquad (14)$$

The redshift is positive (negative) if the waves propagated uphill (downhill) in the potential, and its magnitude is just the change in Newtonian potential measured in units of c^2. Formula (14) applies to the frequency of any repetitive phenomenon, e.g. the rate of arrival of pulses from a pulsar. Applications are many. The ratio of mass to radius of white dwarf stars can be deduced from the redshift of the spectral lines using (3). Of course one has to first correct for Doppler shift coming from motion along the line of sight. In the approximation which we work in, Doppler shift and gravitational redshift are additive. Another example more germane to this lecture: positronium annihilation gives rise to a gamma ray line at 511 KeV. But if the annihilation occurs in the near environment of a gravitating body, neutron star or black hole, those photons will be received on Earth with a lower energy, the defect measuring the depth of the potential well.

How big can the gravitational redshift get ? A neutron star of $1.5M_\odot$ has a radius of ≈ 10 km giving a formal surface Newtonian potential $\approx 0.22c^2$. This already calls for nonlinear corrections. A BH, being totally collapsed, is more extreme. We observe from (14) that $\nu_2 \to 0$ for any ν_1 when $\Phi_N(x_1) \to -c^2/2$. We are obviously pushing our formulae too far because they were obtained from nonrelativistic arguments, and when Φ_N is of order unity, motion is relativistic. Nevertheless the prediction that the formal Newtonian potential can only reach down to $-c^2/2$ turns out to be correct, and so is the conclusion that the redshift is then infinite (see below).

The deflection of light by gravitating bodies is another famous phenomenon. Unlike the gravitational redshift, it does depend on g_{ti} and g_{ij}. The reason is that photons move at speed c, so the terms in Eq. (9) involving Γ^i_{tj} and Γ^i_{jk} are no longer small compared to that with Γ^i_{tt}, which was the only one implicated in nonrelativistic motion. Γ^i_{tj} and Γ^i_{jk} both involve spatial derivatives of g_{ij}, so g_{ij} must be known to calculate light deflection. Newtonian arguments cannot help us here.

We now turn to Einstein's full gravitational equation. There being ten metric components, there are ten partial differential equations to determine them. One is a fanciful elaboration of Poisson's equations with the relativistic energy density—as opposed to rest mass density—as source. Pressure and energy fluxes become the sources of the others. If we are mostly interested in the *external* gravitational field of a spherically symmetric body, then the sources can be dropped and the unique exact solution is Schwarzschild's metric (not Martin Schwarzschild but his dad Karl Schwarzschild, also the father of photographic photometry):

$$ds^2 = -(c^2 + 2\Phi_N)dt^2 + (1 + 2c^{-2}\Phi_N)^{-1}dr^2 + r^2(d\vartheta^2 + \sin^2\vartheta\, d\varphi^2) \quad (15)$$

As usual, $\Phi_N = -GM/r$ where M is the mass of the object and r that radial coordinate for which $4\pi r^2$ is the area of a $r = $ const. surface.

The black hole concept in astrophysics

Schwarzschild's metric describes the exterior of any spherical mass-energy distribution as well as the simplest BH. Obviously the g_{tt} of this metric is the same as for the approximate metric (12). Hence the discussion about gravitational redshift of a source at rest goes as before, and we again have

$$z = \frac{\nu_1}{\nu_2} - 1 = \frac{(c^2 + 2\Phi_N(\mathbf{x}_2))^{1/2}}{(c^2 + 2\Phi_N(\mathbf{x}_1))^{1/2}} - 1 \quad (16)$$

But now we can trust the result even when Φ_N is not small on scale c^2. We confirm that if a source lies at $r = r_h$, where $\Phi_N = -c^2/2$, then the redshift of radiation it emits is infinite. Thus the surface at radius

$$r_h \equiv 2GM/c^2 = 2.94\, M/M_\odot \text{ km}, \quad (17)$$

and more generally any surface where $g_{tt} = 0$, is called the surface of infinite redshift. A fixed light source there is invisible from a distance because its radiation gets shifted to infinite wavelengths.

BHs and other compact objects are frequently surrounded by accretion disks in which gas confined close to a plane moves about the object in nearly circular orbits. With the Schwarzschild BH in mind we can now understand the phenomenon of asymmetric iron line profiles in the X-ray spectra of BHBs. Some of these like GRS 1915+105 and V4641 Sgr observed with *BeppoSaX* [4] show the unmistakable Fe Kα emission line (from the 2P—-1S transition in Fe and its ions) with a broad red tail. Now, if the disk is not face on, the Doppler shift and relativistic beaming will convert the narrow line into two, the blueshifted component being the stronger. The second order (relativistic) Doppler effect in conjunction with the gravitational redshift will shift all frequencies down by a fraction $\mathcal{O}(v^2/c^2)$ (because $v^2/c^2 \sim \Phi_N/c^2$). But light from the inner parts of the disk is shifted more strongly than that from further out, thus also broadening the line. The three effects together thus broaden asymmetrically about the line's center. As we shall mention, the Schwarzschild accretion disk can extend down to $r = 6GMc^{-2}$ so that formula (16) attests to a respectable gravitational redshift range quite sufficient to explain some of the mentioned observations. In other cases, like XTE J1650-500 observed with *XMM Newton* [5], a rotating BH (see next section), whose accretion disk can reach further in, seems to be called for.

We notice that the Schwarzschild g_{rr} metric component blows up at $r = r_h$. This does not signal physically harsh conditions there, but only that the coordinates t and r do not retain their intuitive meaning at points inside $r = r_h$. For instance, in that region the path $\vartheta = $ const., $\varphi = $ const. and $t = $ const. with r varying has $ds^2 = -c^2 d\tau^2 < 0$, so that it represents a physically possible motion. But who ever saw motion with time at a standstill ? Thus in the interior region t is *not* time, that role being usurped by r. In these lectures we do not care about the interior, so we can use metric (15) with the usual meanings provided we are careful at $r = r_h$. The surface $r = r_h$ where g_{rr} blows up is the boundary of the BH, and is generally called the *horizon*. In the Schwarzschild case the roles of infinite redshift surface and horizon are played by the same surface, but this is not a fast rule.

The horizon can be crossed by physical entities only inward. To see this look at the curve $\vartheta = $ const., $\varphi = $ const. and $r = $ const. It has $ds^2 = -c^2 d\tau^2 = -(c^2 + 2\Phi_N)dt^2$. With $r = r_h$, $d\tau = 0$, so this is the track of a "hovering" photon, and a photon is forever on the lightcone. But the lightcone can only be crossed inward. We conclude that in Schwarzschild spacetime gravity bends and blends the local light cones at $r = r_h$ into a spherical surface—the horizon. Anything inside it is in limbo for it cannot send signals out. In fact, no characteristic trace of it remains. A large number of calculations have made it clear that when a nonrotating electrically neutral object totally collapses, the BH, when it settles down, is a Schwarzschild BH with M as its only parameter. All other information about the object, its chemical composition, its thermal state, etc. is veiled from the exterior. A BH is the prefect shredder.

Now we apply all this to elucidate the structure of accretion disks near Schwarzschild BHs. Ignoring fluid and magnetohydrodynamic effects, we consider free particles in circular orbits on the equatorial plane of the Schwarzschild metric. A key question is what is the relation between the azimuthal frequency ν_a and the radius r of the orbit. For example, the frequencies of the quasiperiodic oscillations (QPOs) seen in some galactic X-ray sources are thought to reflect the ν_a of close-in orbits.

An easy procedure to $\nu_a(r)$ is as follows. We set r a constant and $\vartheta = \pi/2$ (circular orbit in the plane). The r component of Eq. (9) gives

$$\Gamma^r_{tt}\left(\frac{dt}{d\tau}\right)^2 + \Gamma^r_{\varphi\varphi}\left(\frac{d\varphi}{d\tau}\right)^2 + 2\Gamma^r_{t\varphi}\left(\frac{dt}{d\tau}\right)\left(\frac{d\varphi}{d\tau}\right) = 0. \tag{18}$$

The factor 2 comes about because $\Gamma^r_{t\varphi} = \Gamma^r_{\varphi t}$. Now since the metric (15) is φ and t independent, and $g_{\alpha r} \propto \delta^r_\alpha$, it is evident from Eq. (10) that $\Gamma^r_{t\varphi} = 0$, while $\Gamma^r_{tt}/\Gamma^r_{\varphi\varphi} = \partial_r g_{tt}/\partial_r g_{\varphi\varphi}$. Thus $\nu_a = (2\pi)^{-1}\Omega$ where

$$\Omega = d\varphi/dt = (-\Gamma^r_{tt}/\Gamma^r_{\varphi\varphi})^{1/2} = (-\partial_r g_{tt}/\partial_r g_{\varphi\varphi})^{1/2}. \tag{19}$$

Since $\sin\vartheta = 1$, a simple calculation using Eq. (15) gives

$$\nu_a = (2\pi)^{-1}\left(\frac{GM}{r^3}\right)^{1/2} = 5.79 \times 10^4 \left(\frac{M}{M_\odot}\right)^{\frac{1}{2}} \left(\frac{1\ \text{km}}{r}\right)^{\frac{3}{2}} \text{Hz}. \tag{20}$$

Coincidentally, this has the the same form as the Newtonian azimuthal frequency. Other (usually smaller) frequencies associated with the orbit are the radial epicyclic frequency κ (the frequency of a radial perturbation of Eq. (9) about the circular orbit just discussed), and the vertical frequency $\nu_\perp$ (of perturbations off the exactly circular planar orbit). The orbit must lie at $r > r_h$; in addition all circular orbits with $r_h < r < 3r_h$ are known to be unstable: $\kappa \to 0$ at $r = 3r_h$ which is thus known as the innermost stable circular orbit (ISCO). Periodic phenomena associated with circular motion must take place outside the ISCO. Thus Eq. (20) implies for the frequency of a periodic phenomena,

$$\nu < 2.2(M_\odot/M)\,\text{kHz}. \tag{21}$$

Interestingly enough, QPOs in galactic X-ray sources have frequencies from a few Hz reaching up to half a kHz, corresponding to the largest frequency allowed by (21) for a BH with 1.5 up to a few solar masses, or to a solar mass neutron star but with r well outside ISCO (neutron star radii are a few times r_h). These high frequencies are one more proof of the existence of compact objects in nature: even for a white dwarf ν_a for r outside the star is way too low to fit many of the observed QPOs.

Rotating black holes

The Schwarzschild BH does not rotate. BH physics tells us that a collapsed neutral rotating star gives a Kerr BH. Its line element [6] is parametrized by mass M and angular momentum *per unit mass* a:

$$ds^2 = g_{tt}dt^2 + 2g_{t\varphi}dtd\varphi + g_{\varphi\varphi}d\varphi^2 + \Sigma\Delta^{-1}dr^2 + \Sigma d\vartheta^2 \tag{22}$$

$$g_{tt} = -(c^2 - 2GMr\Sigma^{-1}) \tag{23}$$

$$g_{t\varphi} = -2GMac^{-2}\Sigma^{-1}r\sin^2\vartheta \tag{24}$$

$$g_{\varphi\varphi} = [(r^2 + a^2c^{-2})^2 - a^2c^{-2}\Delta\sin^2\vartheta]\Sigma^{-1}\sin^2\vartheta \tag{25}$$

$$\Sigma \equiv r^2 + a^2c^{-2}\cos^2\vartheta \tag{26}$$

$$\Delta \equiv r^2 - 2GMc^{-2}r + a^2c^{-2}. \tag{27}$$

Note that the element $g_{t\varphi}$ no longer vanishes. The Kerr parameter ac^{-1} has dimensions of length. The larger the ratio of this scale to GMc^{-2} (the *spin parameter* $a_* \equiv ac/GM$), the more aspherical the metric. Schwarzschild's BH is the special case of Kerr's for $a = 0$.

The infinite redshift surface, this time oblate, is still where $g_{tt} = 0$:

$$r = r_\infty(\vartheta) \equiv GMc^{-2} + [(GMc^{-2})^2 - a^2c^{-2}\cos^2\vartheta]^{1/2}. \tag{28}$$

Again, radiation reaching a distant observer from an emitter at rest there has its frequency gravitationally redshifted to zero.

The horizon, that surface which cannot be crossed outward, is delineated by the condition $g_{rr} \to \infty$. It lies at $r = r_h$ where

$$r_h \equiv GMc^{-2} + [(GMc^{-2})^2 - a^2c^{-2}]^{1/2}. \tag{29}$$

Indeed, the track $r = r_h$, $\vartheta = \text{const.}$ with $d\varphi/d\tau = a(r_h{}^2 + a^2)^{-1} dt/d\tau$ has $d\tau = 0$ (it represents a photon circling azimuthaly *on* the horizon, as opposed to hovering at it). Hence the surface $r = r_h$ is tangent to the local lightcone. The horizon meets the infinite redshift surface at the poles, but is otherwise inside it. Eq. (29) should not fool us; the (Boyer-Lindquist) coordinates used in Eq. (22) make the horizon *look* spherical, but it is not because the angular part of the Kerr metric cannot be put in the form $f(r)(d\vartheta^2 + \sin^2\vartheta\, d\varphi^2)$ reflecting the full symmetry of a sphere. However, all properties of the metric, including the horizon's shape, are *axially* symmetric, and of course, stationary. The horizon radius r_h is well defined only for $a_* \leq 1$; a BH's angular momentum has a maximum value that rises as M^2.

Inside the so called *ergosphere*, $r_h < r < r_\infty(\vartheta)$, no particle, whether free-falling or propelled, can stay at fixed r, φ and ϑ: such track would have $ds^2 > 0$. At the very least the particle must circulate constantly in the angular directions. This reflects the phenomenon of "frame dragging", common to all axially symmetric metrics with $g_{t\varphi} \neq 0$. Contrary to expectations from Mach's principle, the local Lorentz inertial frames are rotating with respect to the inertial frame at infinity (the frame of the stars), and when this phenomenon gets intense enough, it does not let the particles stay in one place. For large r, $g_{t\varphi} \sim r^{-1}$ while both $g_{\vartheta\vartheta}$ and $g_{\varphi\varphi}$ grow as r^2, so that the dragging becomes rapidly imperceptible with growing r. It is strong only in or near the ergosphere.

To get a better feeling for the nature of the frame dragging, let us find the angular velocity $\Omega \equiv d\varphi/dt$ of a free particle in a circular orbit $r = $ const. and $\vartheta = \pi/2$ about the BH. Symmetry tells us such an orbit will not get out of the equatorial plane. The geodesic equation (9) again gives Eq. (18). In view of definition (10) we now have

$$(\partial_r g_{tt} + 2\Omega\partial_r g_{t\varphi} + \Omega^2\partial_r g_{\varphi\varphi})(dt/d\tau)^2 = 0, \tag{30}$$

which quadratic condition for Ω has the solutions

$$\Omega = \frac{(GM)^{1/2}}{(GM)^{1/2}ac^{-2} \pm r^{3/2}} \cdot \tag{31}$$

As $a \to 0$, the $+$ solution here asymptotes to Schwarzschild's (19)-(20), as expected. The $-$ solution obviously corresponds to a retrograde circular orbit since it formally gives $\Omega < 0$. For $a = 0$ both prograde and retrograde orbits for the same r have the same $\nu_a = (2\pi)^{-1}|\Omega|$. But for $a \neq 0$ the retrograde orbit has a larger ν_a than the prograde one with like r ! This effect, totally foreign to Newtonian theory, is a consequence of the frame dragging phenomenon. With (31) goes the formula

$$\nu_{a\pm} = \frac{3.24 \times 10^4(M_\odot/M)}{(c^2r/GM)^{3/2} \pm a_*} \text{ Hz.} \tag{32}$$

Just as in the Schwarzschild case, a Kerr BH has an ISCO corresponding to the vanishing of the epicyclic frequency κ. (Both κ and $\nu_\perp$ are plotted against r in Ref. [7]; they are smaller than ν_a.) The formulae for the innermost stable orbit radius r_{ISCO} for both prograde and retrograde orbits are given by Bardeen et al. [8], and are here plotted in Fig. 1 together with the horizon radius r_h, all as a function of a_*. The large value of r_{ISCO} for retrograde orbits is what prevents the formal pole in Eq. (32) from showing up as $a_* \to 1$ when $r_h \to GMc^{-2}$.

Case $+$ of formula (32) is plotted in Fig. 2 for the stable circular orbits around a Schwarzschild and extreme Kerr ($a_* = 1$) BH. It may be seen that in range of r where they overlap, the curves are quite similar. However, for a rapidly spinning BH the curve extends much further in.

r_{ISCO} or r_h

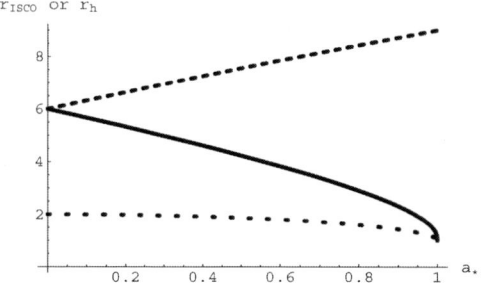

Figure 1.   **The ISCO.** The solid (broken) curve gives the radius of the ISCO of a Kerr BH as function of $a_* \equiv ca/GM$ for prograde (retrograde) orbits, while the dotted line is the radius of the horizon. All radii are expressed in units of $GMc^{-\cdot}$.

QPOs in X-ray binary black holes

In a detailed review of BHB, McClintock and Remillard [3] list some 40 well observed X-ray binary sources in the galaxy with candidate BHs, about half of the number with high confidence. These BHBs, often called micro-quasars, exhibit five distinct spectral/temporal states, each source flipping among several of these: (1) the thermal–dominant (TD) or high/soft (HS) state, a high-intensity state dominated by thermal emission from the accretion disk; (2) the low/hard (LH) state, a low-intensity state dominated by power-law emission spectrum with a photon index ≈ 1.7, rapid variability and frequently accompanied by a radio jet; (3) the quiescent state, of very faint luminosity $3 \times 10^{30} - 3 \times 10^{33}$ erg s^{-1}, and also dominated by power–law emission (index 1.5–2.1) whose very faintness is regarded as good evidence that the accreting surface is a horizon and no solid boundary; (4) the very high (VH) or steep power law (SPL) state with index 2.5 or so, and (5) the intermediate state. The best candidates for BHB have been seen in all of the HS, LH and VH states. The transition between two different states is thought to be related to changes in the rate of accretion to the disk.

A respectable minority of BHB candidates exhibit bumps in the power spec-trum of the X-ray count rate; these are called quasiperiodic oscillations (QPOs; as opposed to periodic which would imply a sharp frequency). Here we are not interested in the low frequency (0.1–few Hz) QPOs seen for days or weeks during the VH state with amplitudes of order 10% of the total X-ray counts, and whose median frequencies often change with the X-ray flux. Rather we shall dwell on the high frequency (40–450 Hz) HFQPOs seen in seven of the BHB candidates during the VH state. These QPOs usurp 1–3% of the X-ray flux. Three of them exhibit single frequency QPOs: 4U163047 (184 Hz), XTE J1859+226 (190 Hz), XTE J1650-500 (250 Hz). Three others have each a

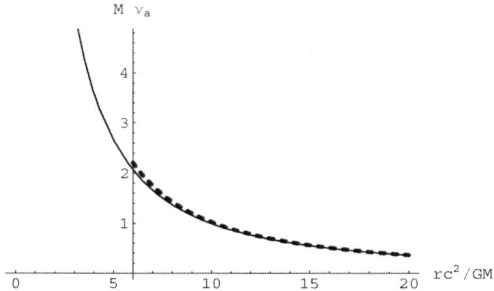

Figure 2. **The azimuthal frequency.** The azimuthal frequency ν_a in kHz multiplied by $M/M_\odot$ for prograde circular orbits outside ISCO in the equatorial plane of a Schwarzschild BH (dashed) and extreme ($a_* = 1$) Kerr BH (solid curve) as function of $rc^{-\cdot}/GM$.

pair of HFQPOs: H 1743-322 (240 Hz + 160 Hz), GRO J1655-40 (450 Hz + 300 Hz) and XTE J1550-564 (276 Hz + 174 Hz). In addition the source GRS 1915+105 possesses *two* pairs of HFQPOs (168 Hz + 113 Hz and 67 Hz + 41 Hz) which actually show up during the HS state. These QPO frequencies do not vary with the X-ray flux; they are a fingerprint of the system.

The first four pairs of frequencies stand in the ratio of 3:2; in H 1743-322 and GRO J1655-40 the accuracy of this ratio is exquisite, in GRS 1915+105 it is better than 1%, and in XTE J1550-564 about 5%. Even more striking, the lower QPO-pair frequency has been found to follow the empirical law $M\nu_0 \approx 931 M_\odot$ Hz (ν_0 is half that frequency and is identified as the "fundamental" frequency) with respect to reliable determinations of the BH masses [3]. What is this telling us ?

A suggestive analog is resonances in our planetary system, e.g. the fact that the orbital period of a Saturnian satellite and that of a subset of the particles in the planet's rings stand in the ratio of small integers makes gaps in the ring. Abramowicz and colleagues [9] suggest that in the twin QPOs one is seeing resonances of the frequencies κ and $\nu_\perp$ for orbits at a fixed value of rc^2/GM for various systems. Recall that the Kerr metric has only one scale, GM/c^2, and one dimensionless parameter, a_*, and that ν_a's dependence on a_* is weak (Fig. 2). Orbits at fixed rc^2/GM around various BHs are thus similar in a geometric sense, and phenomena involving just them could be expected to be alike despite the range of M involved.

Now whereas in Newtonian gravity ν_a, κ and $\nu_\perp$ are identical, they differ for orbits around (Kerr) BHs. In fact almost by definition κ becomes relatively small as the orbit approaches the ISCO. So, for example, there are stable orbits at a fixed rc^2/GM around Kerr BHs for which $\nu_\perp/\kappa$ is exactly $3/2$; the res-onance could amplify radial and off-plane perturbations, and the consequent "rattling" might well get imprinted on the X-ray power spectrum. Although it is not clear what the physics of the modulation is, the repetition of the 3/2 strongly suggests that the same resonance shows up in all four BHBs. Now since by Fig. 2 $M\nu_a$ is fixed for given rc^2/GM, and κ and $\nu_\perp$ are then definite fractions of ν_a, we see that the McClintock-Remillard $\nu_0 \propto M^{-1}$ law is clean evidence that we are dealing with QPOs of BHs.

If these ideas are correct, then the azimuthal frequency associated with each QPO pair must obey $M\nu_a > 3 \times 0.931 M_\odot$ kHz (recall that κ and $\nu_\perp$ lie below the corresponding ν_a). Then Fig. 2 discloses that the QPO source must lie at $r < 5GMc^{-2}$: QPO pairs are associated with compact objects, and they are inner disk phenomena. In fact, $5GMc^{-2}$ falls below r_{ISCO} for Schwarzschild, so that some QPO pairs are Kerr BH phenomenon. Abramowicz et al. [9] have estimated $a_* \approx 0.9$ for GRO J1655-40: rapidly rotating BHs exist in nature (more in Sec. 4).

Supermassive black holes in galactic nuclei

Soon after the discovery of QSOs, Salpeter [10] and Zel'dovich [11] suggested accretion onto a supermassive black hole (SMBH) as the QSO energy source. Lynden-Bell [12] solidified this understanding by stressing that the large energy in a QSO radio halo, $\sim 10^{54}$ J, were it produced by nuclear reactions whose maximal efficiency is .007, would require the "burning" of $10^9 M_\odot$. This massive a refuse, concentrated within 10^{13} m (a dimension required by the $\sim$day timescale of QSO X-ray flux variability) speaks for 10^{55} J of gravitational binding energy. So whether nuclear power has any role in fueling a QSO or not, gravitational accretion power *must* be involved.

The involvement of deep gravitational wells in QSO's was further exposed by the observations of seemingly "superluminal" jets emerging from a number of QSOs, jets which are straight for up to a Mpc. The "superluminal'expansion" implies speeds that approach c, so the relativistic parameter GM/Rc^2 in the accelerating region must approach unity [13]. Rotation of a massive "gyroscope" is the logical way to stabilize the emission direction so that the jets are straight, and, of course, BHs can be massive rotators [14]. A good review of SMBHs in AGNs is given by Laor [15]. By the 1980's it was accepted that QSO's are ephemeral displays in nuclei of galaxies, so that if they implicate SMBHs, then SMBHs may be left in all galaxies which once harbored QSOs.

More recently direct evidence for SMBHs in not obviously active galaxies has emerged. The evidence for some 40 of them is discussed in an extensive review by Kormendy and Gebhardt [16]. The strongest case is linked with the radio source Sgr A* at the center of the Galaxy. A cluster of stars is observed within 0.02 pc of it, and motions of some of them with velocities up to 1350 km s^{-1} are seen to change in time; the stars orbit the radio source in tens of years ! Well determined acceleration vectors for several of the stars point at

Sgr A* and the data allow a determination of the central mass at $(2.6 \pm 0.2) \times 10^6 \, M_\odot$ [17]. This massive object cannot itself be a cluster of smaller objects: $10^6 \, M_\odot$ of ordinary stars would be visible; were it brown dwarfs, they would be so densely packed that they would collide, merge and become luminous beyond observable limits. Finally a $10^6 \, M_\odot$ cluster of collapsed stars would be rapidly whittled away by evaporation [18].

Another strong case is in the nucleus of the spiral NGC 4258 where a number of H_2O masers are observed. The radial (Doppler) velocities can be fit by rotation of a circumnuclear disk of pc scale with a Keplerian rotation velocity profile reaching up to 1080 km s^{-1}. This means there is a "point" mass at the center to the tune of $4 \times 10^7 \, M_\odot$. Once again we would be hard put to squeeze this massive a cluster of brown dwarfs or collapsed objects inside the disk's inner radius at 0.13 pc [18]. It must thus be a BH. The Seyfert nuclei of NGC 1068 and the Scd NGC 4945 also have (more modest) SMBHs discovered in them by the maser method.

Supermassive black holes as energy sources

Salpeter, Zel'dovich and Lynden-Bell all stressed the great efficiency with which an accreting Schwarzschild SMBH produces radiation. Later Bardeen [19] emphasized that BHs in nature are likely to be rotating, so that one should consider the energy efficiency of an accreting Kerr BH. Let us use the BH physics we have learned to compute the maximum such efficiency. The physical picture is of an accretion disk in which matter slowly spirals inward. As we saw, the spiraling stops at r_{ISCO} where an instability develops, circular motion is no longer possible, so that the disk ends there, if not sooner.

First we establish that provided gas dynamics and MHD effects are negligible, the correct expression for relativistic total energy per unit mass of a gas parcel is $E = -g_{tt} dt/d\tau - g_{t\varphi} d\varphi/d\tau$. In flat spacetime $g_{tt} = -c^2$ while $g_{t\varphi} = 0$, so $E = c^2 dt/d\tau$. But from special relativity $dt/d\tau$ is the Lorentz γ, so $E = \gamma c^2$, indeed the relativistic energy of a unit mass free particle. Returning to curved spacetime we now show that E is conserved in free motion, so that it correctly includes the gravitational energy.

In fact, whenever a metric is independent of some coordinate x^μ (time or a spatial coordinate), the quantity $K = g_{\mu\alpha} dx^\alpha/d\tau$ is a constant of the motion for a free particle (no pressure, dissipation, etc.). For

$$\frac{dK}{d\tau} = \partial_\beta \, g_{\mu\alpha} \frac{dx^\beta}{d\tau} \frac{dx^\alpha}{d\tau} - g_{\mu\alpha} \Gamma^\alpha_{\beta\gamma} \frac{dx^\beta}{d\tau} \frac{dx^\gamma}{d\tau}, \tag{33}$$

where we have used the geodesic equation (9) to get the second term. But Eq. (10) and the assumed symmetry tell us that

$$g_{\mu\alpha} \Gamma^\alpha_{\beta\gamma} = \frac{1}{2}(\partial_\beta \, g_{\mu\gamma} + \partial_\gamma \, g_{\beta\delta}). \tag{34}$$

And because of the symmetry between dummy indices β and γ, the Γ term in Eq. (33) exactly cancels the first term. We did not need to know $g_{\alpha\beta}$ explicitly to show this, only that it does not depend on x^μ. Since Kerr's metric is t independent, and $g_{tr} = g_{t\vartheta} = 0$, $K = -g_{tt}dt/d\tau - g_{t\varphi}d\varphi/d\tau$ is conserved in free motion. By analytical mechanics it must be proportional to the energy, and indeed it is just E, the proposed expression for energy.

In reality as a gas parcel orbits, dissipation and radiative losses cause its E to decrease, so that it spirals inward. Thus the $E(r)$ for circular motion in Kerr as given above means the specific energy *remaining* by the time the parcel reaches r. To reduce $E(r)$ to a practical form recall that $ds^2 = -c^2 d\tau^2$, so that for circular motion

$$c^2 d\tau^2 = -g_{tt}dt^2 - 2g_{t\varphi}dtd\varphi - g_{\varphi\varphi}d\varphi^2 = -g_{tt}dt^2 \left(1 + \frac{2g_{t\varphi}\Omega}{g_{tt}} + \frac{g_{\varphi\varphi}\Omega^2}{g_{tt}}\right).$$
(35)

Substituting $dt/d\tau$ from here, $d\phi/dt = \Omega$ as given for prograde orbits by Eq. (31), and the metric elements (23-27), we get after some labor

$$E = c^2 \frac{1 - 2GM/c^2 r + a_*(GM/c^2 r)^{3/2}}{[1 - 3GM/c^2 r + 2a_*(GM/c^2 r)^{3/2}]^{1/2}}.$$
(36)

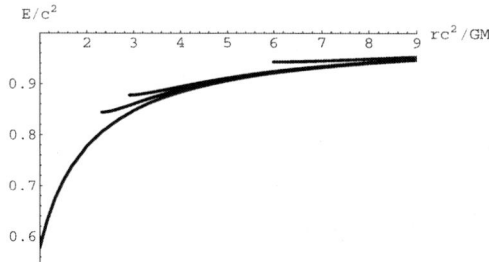

Figure 3.   **Specific energy.** $Ec^{-\bullet}$ per unit mass of a parcel in circular orbit of radius r around a Kerr BH with (in ascending order) $a_* = 1, 0.9, 0.8, 0$. Each curve is shown down to the corresponding ISCO. Radii are expressed in units of $GMc^{-\bullet}$.

Fig. 3 plots Ec^{-2} for a range of radii down to r_{ISCO} for BHs with several spin parameters. E starts very near c^2 in the outer disk edge because $|\Phi_N| = GM/r \ll c^2$ there: the parcel's energy is then its full rest energy. In line with our previous remark, the distribution of E with radius is a good predictor of how much energy is deposited at any particular ring of the disk; this in turn sets a bound on the emissivity. The steep slope of the curve for a_* shows that much radiation will come from near the ISCO (requiring corrections for losses into the hole).

At ISCO E reaches its lowest value, for thereafter it cannot spiral in, but must take a plunge into the hole so rapidly that gas dynamical processes make little difference, and E is very nearly conserved thereafter. Thus $c^{-2}E_{ISCO}$ signifies the fraction of the original parcel rest mass finally accreted by the hole. Then $1 - c^{-2}E_{ISCO}$ is the peak efficiency ("peak" since some of the disk's radiation may get emitted into the BH). This is plotted in Fig. 4 by taking Fig. 1 into account in Eq. 36. For a Schwarzschild BH ($a_* = 0$) the

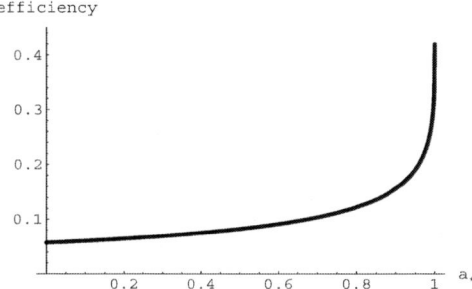

Figure 4. **Efficiency.** The peak efficiency for conversion of mass to radiation by a disk around a Kerr BH with spin parameter a_*.

efficiency is 0.0572, while for an extreme Kerr BH ($a_* = 1$) it reaches 0.42. Thus a BH accretion disk can convert mass into energy up to two orders of magnitude more efficiently than nuclear burning: Einstein's $E = mc^2$ at its best !

The origin of black hole spin

But why to expect a central SMBH to be surrounded by an accretion disk ? Gas collecting in the hole's deep gravitational well will come from some distance out. Whatever angular momentum it has will prevent it from immediately plummeting into the BH, and it will rather swirl around. Any rotation of the galaxy (even ellipticals and bulges rotate somewhat) will impose coherent swirling of different gas streams, and collisions among them will align the divers angular momentum vectors leading to formation of a thin disk.

There is direct evidence for accretion disks in AGNs from observations by the X-ray satellite ASCA of the Fe Kα line coming from many such sources (and lately also from *Chandra* and XMM *Newton* [20]). The line profile shows the asymmetry discussed in Sec. 2. Models which assume a profile for the disk emissivity distribution consistent with the specific energy profile of Fig. 3 are successful in reproducing the observed shape of the line from the Seyfert nuclei of NGC 3516 and NGC 3227 [21]. In the latter case there is some indication that a rapidly rotating BH is involved because the strong asymmetry seen requires very relativistic conditions found only if the disk reaches close to the horizon.

On this note it is easy to extend the calculations of the preceding section to show why, as Bardeen expected, a BH in an AGN spins rapidly. The physical reason is that the disk feeds the hole with a high angular momentum to mass ratio. To make matters simple I assume the BH is initially Schwarzschild with mass M_0. We need an expression for specific angular momentum about the axis of the disk. For nonrelativistic motion in Euclidean space this would be $(r \sin \vartheta)^2 \, d\varphi/dt = g_{\varphi\varphi}\Omega$ (see Eq. (3)). For relativistic free motion in Kerr spacetime we should replace this by something similar which is conserved due to the rotational symmetry about the axis (as should angular momentum). The rule stated just ahead of Eq. (33) suggests the conserved quantity

$$\ell = g_{\varphi t} dt/d\tau + g_{\varphi\varphi} d\varphi/d\tau. \tag{37}$$

In the limit of flat spacetime ($g_{\varphi t} \to 0$) this indeed reduces to the nonrelativistic expression when $\tau \to t$, so ℓ is the specific angular momentum.

When a unit mass in the disk reaches ISCO, it plummets into the BH without further changes to its energy E and angular momentum ℓ. Thus its accretion causes changes $\delta M = E$ and $\delta(aM) = \ell$. Meanwhile, a simple calculation gives for an increment of $a_* = ca/GM$,

$$\delta a_* = [c(GM)^{-1} d(aM)/dM - 2a_*]M^{-1}\delta M. \tag{38}$$

Putting $d(aM)/dM = \ell/E$ and substituting the explicit formulae for E (from Sec. 2), ℓ from Eq. (37) and Ω from Eq. (31), and evaluating at $r = r_{ISCO}$, we obtain the differential equation for the growth of a_* with respect to M:

$$da_* = \left[\frac{\varrho(a_*)^2 - 2a_*\varrho(a_*)^{1/2} + a_*^2}{\varrho(a_*)^{1/2}[\varrho(a_*) - 2] + a_*} - 2a_* \right] d\ln M. \tag{39}$$

Here $\varrho(a_*) \equiv c^2(GM)^{-1} r_{ISCO}$ is that complicated function of a_* used to make Fig. 1. Integrating this equation numerically from $a_* = 0$ at $M = M_0$ shows that the spin parameter reaches $a_* = 0.3058$ ($a_* = 0.7689$) when M has grown by 10% (40%). By the time the BH has doubled its initial mass, $a_* = 0.9875$. Since $a_* \leq 1$, we see that if a SMBHs has changed its mass substantially by accretion from its disk, it must be rotating rapidly. Clearly the idealization of an initially Schwarzschild BH is not crucial. An originally mildly rotating BH would be transformed into an extreme Kerr BH by the time it gained significant mass.

How are SMBH masses measured ?

Most SMBHs have been identified in early type galaxies [16] (exceptions include the mentioned maser galaxies, the Galaxy and the two nearby Sbs M31 and M81). No SMBHs have been found in pure disk pr Ir galaxies. Detecting a BH means finding kinematic evidence (velocities of emission lines from a circumnuclear ionized gas disk, or velocities of absorption lines from bulge stars) for a large mass which is unresolved (more on this below). Only 10-15% of early type galaxies have well formed central gas disks, but more than ten have yielded evidence for SMBHs, including the large E0 galaxy M87 which sports a $3 \times 10^9 \, M_\odot$ hole as well as the famous jet which is perpendicular to the gas disk. Gas disk derived masses are regarded as accurate to $\sim 30\%$. Mass determination via stellar velocity dispersion is hampered by the long integration times required (ellipticals and bulges have low surface brightness), as well as by ambiguities of interpretation, e.g. is the velocity distribution anisotropic ? Yet more than 25 SMBH masses have been measured by this method. The range of SMBH masses is $10^6 - 10^9 \, M_\odot$.

The detailed tables [16] are less important than two correlations that emerge from them. First, the BH mass M is almost proportional to the bulge blue luminosity L_b, or more precisely [22, 16]

$$M \approx 7.8 \times 10^7 (L_b/10^{10} \, L_\odot)^{1.08}. \tag{40}$$

And the BH mass is nearly proportional to the fourth power of the bulge velocity dispersion σ_e within de Vaucoleurs' effective radius [23, 16]:

$$M \approx 1.3 \times 10^8 (\sigma_e/200 \, \mathrm{km \, s^{-1}})^{3.65}. \tag{41}$$

This last has a scatter not much bigger than the mass measurement errors can account for (see graph in ref. [16]), so it seems to be the more fundamental. The SMBH accounts for about 0.0013 of the galaxy's or bulge's mass, but deviations of an order of magnitude are known.

As a rule SMBHs show up in every galaxy (or bulge) observed with enough resolution to find a BH. In what sense ? Suppose that from the observed L_b and σ_e an estimate is made of a central SMBH's M. Then the radius of its sphere of influence (region where its mass disturbs stellar orbits drastically) $r_* = GM/\sigma_e^2$ is determined. Wherever this sphere can be resolved, evidence for a SMBH has always turned up. Hence the widely accepted belief that most early type galaxies have a central SMBH. SMBHs in disk galaxies with small bulges—if they exist—fail to comply with the $M - \sigma_e$ relation. For example, for the Scd M33 a bound of $10^3 \, M_\odot$ has been put on any central BH [16], yet the $M - \sigma_e$ relation would predict one four orders heavier. SMBHs go naturally with early type galaxies and big bulges, not with disks.

AGNs are generally too far for the BH sphere of influence to be resolvable. For them the method of reverberation mapping has been useful. In a AGN there are rapid optical and UV fluctuations. Reverberation mapping interprets the time delays between variations in the AGN continuum (thought to come from the accretion disk) and its broad emission lines (coming from enveloping gas clouds where the continuum is reprocessed into lines) as the light travel times between BH and the clouds, and hence equal to the linear size of the region R divided by c. The width of the emission lines gives a velocity dispersion σ. Then, assuming that the BH is the main mass in the region one has the virial estimate $M \approx \sigma^2 R/G$. This method provides no direct evidence for a point mass. And it is somewhat model dependent. Yet, after some initial confusion, it has been agreed that central SMBH reverberation masses lie on the same $M - \sigma_e$ relation as SMBH masses in normal galaxies as determined by the gas disk and stellar dynamical methods.

Where did the SMBHs come from ?

Like questions of heritage in everyday life, the origin of the SMBHs abounds in controversy. Did SMBHs precede their galaxies and act as nuclei for their formation [24, 14] ? Or did they grow from humble seeds to their present proportions by accretion from an already developed bulge or elliptical ? Or did each develop together with its host galaxy ? The dearth of SMBHs in pure disks suggests that a BH is not required for galaxy condensation, thus militating against the first hypothesis. On the other hand, their presence in late-type systems is reasonable if SMBHs form by accretion over time; in disks rotational support would impede their growth, and so their absence is also reasonable.

Early type disk galaxy *bulges*, like ellipticals, obey the Faber-Jackson relation [25] $M \sim \sigma^4$ between mass and velocity dispersion. Relation (41) has very much this form; hence as hinted at earlier, a SMBHs tends to contain a fixed mass fraction of its elliptical or bulge host. This again seems to support the joint growth hypothesis; for how could the full blown SMBH manage to control its host's final mass despite the small dimension of the former's sphere of influence. SMBHs have been found in AGNs out to $z = 6$, some with masses $\sim 10^9 \, M_\odot$ [26]. Evidently these too militate against the idea of late SMBH formation in fully blown galaxies. As mentioned, these early SMBHs obey the same rule (41) as "normal" SMBHs at the present epoch, suggesting that they too developed together with their hosts.

Accordingly, Kormendy and Gebhardt [16] reach the verdict that AGN activity, bulge formation and SMBH gestation are facets of one and the same primeval process. This granted, there are remain unsolved problems regarding the path by which BHs grow to SMBH size, as well as regarding the nature of the seed BHs at the outset of the evolution. Two categories of growth mechanism stand out in the literature: accretion by BHs [27] and mergers of BHs [28]. Seed candidates include stellar-mass BHs, intermediate-mass $(10^2 - 10^5 M_\odot)$ black holes (IMBHs) and PBHs. Any seed-growth method combination has to be judged by its success in producing the $10^9 \, M_\odot$ SMBHs in AGNs at $z > 6$.

A stellar-mass BH accretes slowly. Early type galaxies are gas poor, but have stars. To accrete a star the relatively tiny BH must first capture it, and then tidally disrupt it in order to ingest it, all the time respecting the Eddington limit. It is known that this process acting in early galaxies would have trouble producing $10^9 \, M_\odot$ SMBH by $z = 6$ [29]. But stellar-mass black holes could play out their role in a massive star cluster. Such assembly would tend to drop to the bottom of the galactic potential well by dynamical friction on the background, lighter, stars. Simultaneously distant encounters among its constituent stars would cause it to develop a tightly bound core and a loosely bound envelope; the last would slowly be lost to evaporation.

In the dense core stellar collisions would be common and would lead to stellar amalgamation, and the consequent formation of massive short-lived stars which would evolve rapidly, some leaving BHs [30, 31]. Of course collisions between these last and the remaining stars should make the holes larger. In the meantime some of the lighter objects of either kind would be expelled from the cluster by Newtonian slingshot. Consolidation of the cluster into an IMBH by runaway evolution would follow [31]. Other studies [32] are more pessimistic regarding the possibility of producing IMBHs by cluster evolution in one Hubble time.

Elliptical galaxies (and bulges) are thought to have grown by mergers of smaller building blocks at $z \sim 3$. Being very massive on stellar scale, the clusters (or their IMBH remnants) would be caused to congregate at the merged galaxy's center by dynamical friction. Their mergers would thus be sped up ultimately forming baby SMBH [30]. However, this mechanism faces a challenge. Violent formation of a black hole can lead to its strong recoil by the emission of gravitational waves; this process can expel fresh stellar-mass BH's from the Galaxy [2]. It was recently realized that it can also expel newly merged SMBHs from their host bulges or galaxy center's [33], thus impeding their growth.

Primordial black holes

These problems have riveted attention on the notion that BHs originating in the early universe are the seeds from which SMBHs grow [34]. PBHs, thus far conjectural, offer so much that is novel that they deserve a look in their own right. Because there is as yet no observational confirmation of PBHs, I treat them in brief. Carr [35] and Kiefer [36] are recent reviews on PBH astrophysics (the second on BH physics as well).

According to Eq. (2) the lighter a BH, the denser: a $10^9 \, M_\odot$ SMBH has a density like snow's; a stellar mass one is above nuclear density. And a BH the mass of a mountain (5×10^{12} Kg) has a whopping density 2.5×10^{55} Kg m^{-3}. Intuitively, to form a BH of such high density should require matter or radiation about as dense. Only in the early universe do we meet densities orders above nuclear density. Hence when tiny BHs are the issue, attention turns to cosmology, and they are termed PBHs.

High density is not enough; a highly dense homogeneous distribution of matter/radiation is perfectly consistent with cosmology. To make BHs one needs inhomogeneities. And obviously collapse cannot take place until the inhomogeneity's radius R has fully entered the (particle) horizon (radius ct with t the cosmological time). Now during inflation the exponential expansion stretches every R outside the horizon (which grows more placidly); hence PBHs cannot form then. In addition, any PBHs from former eras are rapidly diluted. The post-inflation radiation era is the first relevant one in connection with PBHs formation.

As an example take flat space ($k = 0$) radiation dominated cosmology; we know that the expansion factor $a \propto \sqrt{t}$. Thus the Hubble parameter is $H = \frac{1}{2}t^{-1}$, and the condition is $R < \frac{1}{2}cH^{-1}$. But for gravitation to overpower pressure, R must exceed Jeans' length $c_s(4\pi G\rho)^{-1/2}$, where c_s is the sound speed and ρ the mass density. Due to radiation dominance $c_s = c/\sqrt{3}$, and since Friedmann's equation gives $H^2 = (8\pi/3)G\rho$, we have $R > \frac{1}{3}\sqrt{2}\,c\,H^{-1} = 0.47c\,H^{-1}$. Comparing we see that the collapse is possible only briefly after horizon crossing; thereafter R becomes very small compared to cH^{-1} and the second condition fails.

From Eq. (17) for R and $H \approx \frac{1}{2}t^{-1}$ we have for the time of formation of a PBH of mass M ($M_* \equiv 10^{12}$ Kg):

$$t \approx 4.9 \times 10^{-24}(M/M_*)\,\text{s} \qquad (42)$$

As mentioned this is relevant for post-inflation, that is $t > 10^{-35}$ s or $M > 1$ Kg. Since the smaller the PBH, the earlier it forms, the smallest PBHs detectable would probe the earliest cosmological times. Once the PBH is formed, M is not increased sizably by accretion [35]. On the other hand, M can decrease by Hawking radiation.

According to BH thermodynamics [37, 38], a Schwarzschild BH is endowed with a quantum temperature

$$T_{BH} = \hbar c^3(8\pi GkM)^{-1} = 1.23 \times 10^{11}(M_*/M)\,^0\text{K} \qquad (43)$$

where k stands for Boltzmann's constant. This is manifested by the BH emitting into space any kind of particles in nature, each with thermal spectrum and statistics corresponding to the Hawking temperature T_{BH}. We see that for $M \gg M_*$, only photons and neutrinos will be emitted. Anyway, this radiation is paid for by a decreasing Mc^2. Estimating the energy loss with the

Stefan-Boltzmann law $P = (4\pi r_h{}^2)\sigma T^4$ using T from Eq. (43), we get for each massless (or very light) species

$$dM/dt \approx -4.0 \times 10^{-9}(M_*/M)^2 \,\mathrm{Kg\,s}^{-1}. \qquad (44)$$

In reality, the BH radiation crossection is larger than $4\pi r_h^2$ by a factor of ~ 5. And since there exist very light neutrinos and antineutrinos, the total dM/dt is an order of magnitude above Eq. (44). Integrating the fixed-up equation from $M = M_0$ to $M \approx 0$ gives the lifetime $\tau < 27 \times 10^{10}(M/M_*)^3$ y, the inequality appearing because as M approaches M_*, the PBH begins emitting massive particles and M falls more quickly than just estimated. Evidently PBHs with $M \ll M_*$ have evaporated by now. Nevertheless, as we shall see, PBH searches say a lot about the inflationary era (at whose end PBHs of only $M \sim 1\,\mathrm{Kg}$ would form).

PBH formation is bound up with the primordial density fluctuations spectrum. Inflation produces a scale invariant spectrum: the root mean square fluctuation ϵ for an inhomogeneity just inside the horizon is M independent. When the propitious time (42) comes, a PBH can form only if the density contrast $\delta \equiv \delta\rho/\rho$ at horizon scale is large enough; the rule is that for equation of state $p = w\rho$, δ must exceed w. Evidently, the fraction β of horizon-size volumes which actually collapse (the probability that the inhomogeneity makes a PBH), is M independent.

To calculate the present space density of PBH formed during the radiation, we look at a fixed comoving volume of space; it expands as $a^3 \propto t^{3/2}$. By contrast the volume contained in a horizon is $\propto t^3$: the large volume thus contains a number of horizon-size volumes that varies as $t^{-3/2}$. Then Eq. (42) shows that the number N of PBHs of mass M formed out of the big volume scales like $\beta M^{-3/2}$. Now the collapse takes about a dynamical time, GMc^{-3}, and only thereafter can the process repeat. Thus N refers to a range of masses $\sim M$ so that the differential distribution is $dN/dM \propto \beta M^{-5/2}$. The power law reflects the lack of a special scale in the inflation spectrum. The distribution is preserved over time except for dilution by expansion and the dying out due to Hawking evaporation which only leaves PBHs with $M > M_*$. Accordingly at the present epoch the differential space density is [35]

$$dn/dM = \frac{1}{2}M_*{}^{1/2}M^{-5/2}\,\Omega_{PBH}\,\rho_c\,\Theta(M - M_*) \qquad (45)$$

where ρ_c is the present critical density and Ω_{PBH} is the fraction of it in PBHs. We adjusted the numerical factor so that the integral of $M(dn/dM)$ over M—the mass density—is precisely $\Omega_{PBH}\,\rho_c$.

To calculate Ω_{PBH} focus, for example, on the end of radiation dominance, $t = t_r$, $z = 10^4$, when the horizon volume was $\approx 4(ct_r)^2$. By today's time t_0 that volume has expanded by a factor $(10^4)^3$. The present space density of

PBHs formed then is obviously $n = \beta\, 10^{-12}/(4c^3 t_r{}^3)$. Eq. (45) gives the alternative estimate $n \approx \frac{1}{2} M_*{}^{1/2} M_r{}^{-3/2} \Omega_{PBH}\, \rho_c$. Equating the two, replacing M_r by $c^3 t_r/2G$, and remembering the relation $8G\rho_c \approx H_0{}^2$ between ρ_c and Hubble constant, we get

$$\Omega_{PBH} \approx 2.5 \times 10^{-12} \beta (M_r/M_*)^{1/2} (H_0 t_r)^{-2} = 1.3 \times 10^{18}\, \beta. \qquad (46)$$

The second equality comes from two observations: for $t > t_r$, $a \propto t^{2/3}$ so that $H_0 = \frac{2}{3} t_0^{-1}$; and $t_0/t_r = (10^4)^{3/2}$ (one computes directly $M_r/M_* = 6 \times 10^{34}$). Within the scale invariant paradigm (46) is the total PBH density parameter, and Eq. (45) has predictive power.

Because we must have $\Omega_{PBH} < 1$ we obtain a stiff inequality on β. Using Carr's result (based on a Gaussian fluctuation spectrum)

$$\beta \sim \epsilon \exp(-1/18\epsilon^2), \qquad (47)$$

we get $\epsilon < 0.038$. This is the accepted bound on primordial fluctuations from inflation [35] in the range $M < 10^{27}$ Kg. It should be stressed that the scale invariance assumption is a strong one, and even a small M dependence of ϵ would make β strongly mass dependent. Thus, for example, Carr derives a M dependent version of Eq. 46 [Eq. (7) of Ref. [35]], from which he sets constraints on $\epsilon(M)$ separately for each mass range from a variety of astronomical data. Among these are the known γ ray background (which would be affected by just now dying black holes), and the entropy in the CMB which receives a contribution from Hawking radiation. All in all the search for PBHs makes us much wiser about conditions in the very early universe.

References

[1] Brown, G.E., Lee, C.-H., Wijers, R.A.M.J. & Bethe, H.A. 2000, Physics Reports, 333-334, 471 (astro-ph/9910088).

[2] Bekenstein, J.D. 1973, Ap. J., 183, 657.

[3] McClintock, J.E. & Remillard, R.A. 2004, in *Compact Stellar X-ray Sources*, eds. W.H.G. Lewin and M. van der Klis, Cambridge: Cambridge University Press (astro-ph/0306213).

[4] int Zand, J.J.M., Markwardt, C.B. et al. 2002, A&A 390, 597; int Zand, J.J.M., Miller, J.M., Oosterbroeck, T. & Parmar, A.N. 2002, A&A, 394, 553.

[5] Miller, J.M., Fabian, A.C., Wijnands, R. et al. 2002, Ap. J., 570, L69.

[6] Misner, C.W., Thorne, K.S. and Wheeler, J.A. 1973, *Gravitation*, San Francisco: Freeman.

[7] Psaltis, D. 2003, in *X-Ray Timing: Rossi and Beyond*, ed. P. Kaaret, F.K. Lamb, & J. H. Swank (astro-ph/0402213).

[8] Bardeen, J.M., Press, W.H., & Teukolsky, S.A. 1972, Ap. J., 178, 347.

[9] Abramowicz, M.A., Kluzniak, W. et al. 2001, A&A, 374, L19.

[10] Salpeter, E.E. 1964, Ap. J., 140, 796.

[11] Zeldovich, Ya. B. 1964, Sov. Phys. Dokl., 9, 195.

[12] Lynden-Bell, D. 1969, Nature, 223, 690 and 1978, Physica Scripta, 17, 185.

[13] Blandford, R.D., McKee, C.F., & Rees, M.J. 1977, Nature, 267, 211.

[14] Rees, M.J., 1984. Ann. Rev. Astron. Astrophys. 22, 471.

[15] Laor, A., 1999, Phys. Reports, 311, 451.

[16] Kormendy, J., & Gebhardt, K., 2001, in *The 20th Texas Symposium on Relativistic Astrophysics*, ed. H. Martel & J.C. Wheeler, New York: AIP (astro-ph/0105230).

[17] Genzel, R., Eckart, A., Ott, T., Eisenhauer, F. 1997. Mon. Not. R. Astron. Soc. 291, 219.

[18] Maoz, E. 1995, Ap. J., 447, L91 and 1998, Ap. J., 494, L18.

[19] Bardeen, J. M. 1970, Nature 226, 64.

[20] T.J. Turner, R.F. Mushotzky, T. Yaqoob et al. 2002, Ap. J. 574, L123.

[21] Nandra, K., George, I.M., Mushotzky, R.F., Turner, T.J., Yaqoob, T. 1997, Ap. J. 477, 602.

[22] Kormendy, J. 1993, in *The Nearest Active Galaxies*, ed. J. Beckman, L. Colina, & H. Netzer, Madrid: CSIC, 197; Kormendy J. & Richstone, D.O. 1995. Ann. Rev. Astron. Astrophys, 33, 581.

[23] Ferrarese, L., & Merritt, D. 2000, Ap. J., 539, L9.

[24] Ryan, M., 1972, Ap. J. 177, L79.

[25] Faber, S. M. & Jackson, R. E. 1976, Ap. J., 204, 668.

[26] Fan, X. et al. 2001, Astron. J, 122, 2833.

[27] Rees, M. J. 1984, Ann. Rev. Astron. Astrophys., 22, 471.

[28] Barkana, R., Haiman, Z. & Ostriker, J.P. 2001, Ap. J., 558, 482.

[29] Yoo, J. & Miralda-Escudé, J. 2004 (astro-ph/0406217).

[30] Rees, M. J. 2003, in *Future of Theoretical Physics and Cosmology* ed. G.W. Gibbons, et al., Cambridge: Cambridge University Press, 217 (astro-ph/0401365).

[31] Quinlan, G.D. & Shapiro, S.L., 1990, Ap. J., 356, 483.

[32] Baumgardt, H., Makino, J. & Ebisuzaki, T. (astro-ph/0406227).

[33] Madau, P., Rees, M.J., Volonteri et al. 2004, Ap. J. 604, 484-494; Favata, M., Hughes, S.A., & Holz, D.E., 2004, Ap. J. 607 (2004) L5-L8; Merritt, D., Milosavljevic, M., Favata, M. et al., 2004, (astro-ph/0402057).

[34] Bean, R. and Magueijo, J. 2002, Phys. Rev., D66, 063505; M.Yu. Khlopov, S.G. Rubin, A.S. Sakharov 2004, (astro-ph/0401532).

[35] Carr, B. J. 2003, in *Quantum Gravity: From Theory to Experimental Search*, ed. D. Giulini, C. Kiefer & C. Lammerzahl, Dordrecht: Kluwer (astro-ph/0310838).

[36] C. Kiefer 2002, in *The Galactic Black Hole*, eds. H. Falcke and F. W. Hehl, Bristol: IOP Publishing (astro-ph/0202032).

[37] Bekenstein J.D. 1973, Phys. Rev. D, 7, 2333.

[38] Hawking, S.W. 1975, Commun. Math. Phys., 43, 199

THE EFFICIENCY OF USING ACCRETION POWER OF KERR BLACK HOLES

Ioana Dutan[1,3] and Peter L. Biermann[2,3]

[1] *University of Bucharest, Faculty of Physics*
Str. Atomistilor, 405, RO-76900, Bucuresti-Magurele, Romania

[2] *University of Bonn*
Regina-Pacis-Weg, 3, D-53113, Bonn, Germany

[3] *Max-Planck-Institute for Radioastronomy*
Auf dem Hügel, 69, D-53121, Bonn, Germany

idutan@mpifr-bonn.mpg.de; plbiermann@mpifr-bonn.mpg.de

Abstract   The efficiency of a rapidly spinning Kerr black hole to turn accretion power into observable power can attain 32 percent for the photon emission from the disk, as is well known, following the work of Novikov-Page-Thorne. But many accretion disks are now understood to be underluminous ($L < L_{Edd}$), while still putting large amounts of energy into the jet. In this case, the apparent efficiency of jets driven by the innermost accretion disk of a highly rotating Kerr black hole ($a_* = 0.999999$) can reach 96 percent.

Keywords: black holes, accretion, energy extraction, efficiency

Introduction

The idea of extracting energy from black holes has been proposed by Penrose thirtyfive years ago and followed by a cascade of energy extraction models. Nowadays it is known that from almost all Active Galactic Nuclei (AGNs) which harbour a supermassive ($M > 10^5 \times M_\odot$) Kerr black hole, focused jets of hot gas shoot into space at relativistic speed. "How much energy can these jets get from the black hole or its accretion disk?" is a question which arises by itself.

Energy extraction from Kerr black holes

The Kerr solution (1963) of the Einstein's vacuum field equations describes the gravitational field of a material source at rest having mass and angular momentum. The angular momentum determines a physically significant di-

M. M. Shapiro et al. (eds.), Neutrinos and Explosive Events in the Universe, 175–180.

rection, the axis of symmetry; consequently the field cannot be spherical as in the Schwarzschild solution but axial symmetric. The geometrical units are $G = c = 1$, the metric has the signature (-+++), the Greek indices range from 0 to 3, and the asterisk "*" refers to the dimensionless parameters (divided by black hole's mass) throughout the paper.

The metric equation $ds^2 = g_{\mu\nu}dx^\mu dx^\nu$ in the case of a stationary, axial symmetric field takes the form of

$$ds^2 = g_{tt}dt^2 + 2g_{t\phi}dtd\phi + g_{rr}dr^2 + g_{\theta\theta}d\theta^2 + g_{\phi\phi}d\phi^2 \qquad (1)$$

The *Boyer-Lindquist coordinates* (t, r, θ, ϕ) form a basis of the Kerr spacetime, such that the metric becomes

$$ds^2 = -\left(1 - \frac{2Mr}{\Sigma}\right)dt^2 - \frac{4Mar\sin^2\theta}{\Sigma}dtd\phi + \frac{\Sigma}{\Delta}dr^2$$
$$+\Sigma d\theta^2 + \left(r^2 + a^2 + \frac{2Ma^2r\sin^2\theta}{\Sigma}\right)\sin^2\theta d\phi^2 \qquad (2)$$

where M is the mass of the black hole, $a = J/M$ is its angular momentum per unit mass ($0 \leq a \leq M$) and the functions Δ and Σ are defined by $\Delta = r^2 - 2Mr + a^2$ and $\Sigma = r^2 + a^2\cos^2\theta$.

Since the metric coefficients in Boyer-Lindquist coordinates are independent of t and ϕ, both $\zeta^\mu = \partial_t$ and $\eta^\mu = \partial_\phi$ are the Killing vectors of the metric, the timelike and axial, respectively, with Boyer-Lindquist components (1,0,0,0) and (0,0,0,1), respectively.

The Kerr solution becomes a singular one, when both $\Delta = 0$ ($g_{rr} \to \infty$) and $\Sigma = 0$. There are three possibilities: $M^2 < a^2$, $M^2 = a^2$, and $M^2 > a^2$. The first two cases are not of physical interest (Carroll 2004) and for the third case the equation $\Delta = 0$ has two solutions $r_\pm = M \pm \sqrt{M^2 - a^2}$, where both radii are null surfaces which will turn out to be *horizons*, the outer r_+ and the inner r_- ones. In the case of the Kerr metric the Killing vector is spacelike at the outer horizon, except at the north ($\theta = 0$) and south ($\theta = \pi$) poles, where it is null.

The locus of points where $\zeta^\mu\zeta_\mu = 0$ is known as the *Killing horizon* (also called the *static limit*), $g_{tt} = 0$.

The region between the Killing horizon and outer horizon is called the ergoregion. Inside the ergoregion no particle can stay at r, θ, ϕ fixed. They are dragged (the Lense-Thirring effect) in the same direction as the black hole rotates, and follow a timelike world line $ds^2 > 0$. These particles have access to negative energy trajectories (Penrose process) which extract energy from the black hole.

Let us consider the geodesic motion of a particle in the Kerr spacetime. Since the Kerr metric is independent of t and ϕ coordinates, there are some conserved quantities for the free motion of a particle, which are associated with ζ^μ and η^μ; they are the specific energy ($E^\dagger = E/m$) and specific angular momentum ($L^\dagger = L/m$) of a particle with mass m orbits on the equatorial plane of a Kerr black hole. But, the circular orbits do not exist for all values of r and also the bound circular orbits are not all stable. There is a peculiar radius, called the radius of the innermost stable orbit r_{ms} from which the orbits of the particles are no longer stable. The value of $r_{ms*} = r_{ms}/M$ depends only on the spin parameter ($a_* = a/M$, $-1 \leq a_* \leq +1$) of the black hole. For more details see (Bekenstein 2004).

Accretion onto black holes. The accretion onto Kerr black holes differs from the accretion onto other objects; the absence of the stable circular orbits between the innermost stable orbit and outer horizon implies a rapid flow of matter in black hole, and the presence of the frame-dragging in the ergoregion such that the spin of black hole becomes a source for energy extraction.

The *Novikov-Page-Thorne model* describes the conversion of accreting mass into outgoing photon energy (Novikov and Thorne 1973, Page and Thorne 1974, Thorne 1974) in the case of a black hole with a classical thin accretion disk lies in the equatorial plane of the hole, such that as the material of the disk spirals inward, releasing gravitational energy, a negligible amount of this energy is stored internally. Almost all energy is radiated away.

In this model a canonical black hole is the one of spin parameter limited at the value 0.998, by the effect of the photons, which buffer the spin away from the extreme $a_* = 1$ Kerr value.

The efficiency of converting the accreting mass into radiation measured at infinity, and ignoring small corrections due to capture of photons by the hole, is

$$\eta = 1 - E^\dagger_{ms} = \left(\begin{array}{l} \text{the specific binding energy of} \\ \text{the last stable circular orbit} \end{array} \right) \qquad (3)$$

where $E^\dagger_{ms}$ is the specific energy of a particle at the innermost stable orbit.

The black hole rotates faster, the higher efficiency of black hole, such that for the canonical black hole the value of the efficiency parameter is around 0.32, lower than 0.42, the maximum value of the efficiency parameter obtained for an extreme Kerr black hole ($a_* = 1$), by this model.

Jets driven by accretion onto Kerr black hole. Quasars are one of the most energetic astrophysical objects. They live at the centers of galaxies and are manifestation of more general phenomenon of AGNs, which also include Seyfert galaxies, Blazars, Radio galaxies. As is well-known the engines of AGNs are the supermassive black holes and their jets (loud Radio galaxies, Blazars) are associated to the high spin black holes ($a_* \sim 1$). Even though there is no opportunity to watch what is happening when the jets are forming close to the black hole, there are two possible energy sources, the "pure" rotational energy of black hole (Semenov et al. 2004) and the accretion by releasing the gravitational energy of the disk very close to the black hole. Let us concentrate on the last possibility.

The low luminosity of some AGNs (Yuan et al. 2002) could be explained by considering the innermost part of the accretion disk to be non-radiant (Donea and Biermann 1996), such that the only energy-flow out of the innermost disk is along the jets. The magnetic fields of the hole produce a torque on the disk and angular momentum is transported from the accretion flow to the jets through the magnetic stresses. The mass flow rate into the jets is $\dot{M}_{jet} = q_m \dot{M}_D$, where $\dot{M}_D$ is the accretion mass onto the hole and $q_m \simeq 0.05$. In the following calculations the disk is considered to be nearly Keplerian.

By these considerations, the conservation of the total energy-momentum tensor (Page and Thorne 1974), in the covariant sense $\nabla_\mu T^{\mu\nu} = 0$, associated with the Killing vectors for the Kerr metric in Boyer-Lindquist coordinates provide the following conservation laws:

The angular-momentum conservation law

$$\frac{d}{dr}[(1 - q_m)\dot{M}_D L^\dagger] = 4\pi r(JL^\dagger - H) \qquad (4)$$

where, $L^\dagger$ is the specific angular-momentum of a particle in the disk, J is the flux of energy-flow along the jet, H is the flux of angular momentum (Li 2001) transferred from the black hole to the disk by the magnetic fields, which is given by the magnetic torque produced by the black hole on the disk $T_{HD} = 2 \int_{r_*}^{r_*} 2\pi r H dr$, where the factor 2 accounts for the fact that the disk has two surfaces.

The energy conservation law

$$\frac{d}{dr}[(1 - q_m)\dot{M}_D E^\dagger] = 4\pi r(JE^\dagger - H\Omega_D) \qquad (5)$$

where, $E^\dagger$ is the specific energy of a particle in the disk and $\Omega_D = 1/M(r_*^{3/2} + a_*)$ is the angular velocity of the accretion disk .

The total energy flow along the jets to infinity is given by

$$E_{jet} = 2 \int_{r_*}^{r_*} 2\pi J E^\dagger r dr = \int_{r_*}^{r_*} [\frac{d}{dr}((1 - q_m)\dot{M}_D E^\dagger) + 4\pi r H\Omega_D]dr \quad (6)$$

here, the factor 2 from the first right-hand side of the equation accounts for the fact that they are two jets. Considering the jets are formed close to the black hole where the frame dragging process takes place, the inner and outer radii of the above integral could be taken as innermost stable orbit and $r_2 = 2r_g$, the radius of the ergosphere, respectively, where $r_g = GM/c^2$ is the gravitational radius.

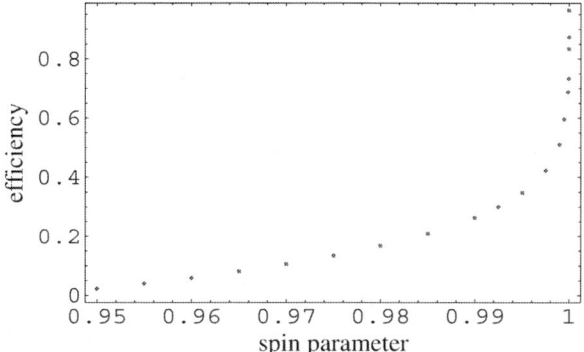

Figure 1. The apparent efficiency η of jets driven by accretion disk as function of the spin parameter a_* of a Kerr black hole.

Assuming the disk is perfectly conducting ($Z_D \rightarrow 0$), the flux of angular momentum transferred from the black hole to the disk by the magnetic fields is given by $H = \frac{1}{8\pi \cdot r} \left(\frac{d\Psi_{HD}}{dr} \right)^2 \frac{\Omega_H - \Omega_D}{(dZ_H/dr)}$ (Li 2002).

The *efficiency of the jets driven* by accretion disk η is defined as the ratio of the energy flow along the jets to the rest energy of the accreting mass

$$\eta = \frac{E_{jet}}{\dot{M}_D} \tag{7}$$

Therefore,

$$\eta = (1 - q_m)[E^\dagger(r_2) - E^\dagger(r_{ms})] + \frac{4\pi}{\dot{M}_D} \int_{r_{ms}}^{r_\bullet} H\Omega_D r \, dr \tag{8}$$

Here, the second term describes the transfer of energy from the black hole to the disk, and then we assume it can go into jets; and so tapping the energy of the black hole's rotation implies additional energy for the jets, so that the apparent efficiency can be high.

For an accretion rate $\dot{M}_D$ below the Eddington ($\dot{M}_{Edd} = 1.39 \times 10^{15} (M/M_\odot)$ kg s^{-1}) rate, $\dot{m} = \dot{M}_D/\dot{M}_{Edd} \simeq 0.13$, as an example, the value of efficiency of jets driven by the disk is maximum $\eta_{max} = 0.9638$ in the case of a highly rotating Kerr black hole of a spin parameter $a_* = 0.999999$ (see fig. 1).

Conclusions

In this short paper we have tried to describe the energy, mass and angular momentum of a black hole - accretion disk - jet system. We find that the efficiency of such a system putting energy from the accretion into the jet is limited by Thorne's (Thorne 1974) bound of 42 percent, but including the possible transfer of angular momentum and energy from the black hole to the inner disk and then to the jets can temporarily increase the apparent efficiency to near 100 percent. Thus, observed jets might represent such stages of apparent high efficiency.

Acknowledgments I.D. would especially like to thank the Organizers for the financial support. I.D. was supported by the Erasmus/Socrates programme of the EU, the AUGER-project and now by VIHKOS, as well as the MPIfR.

References

[1] Bekenstein, J.D. (2004), "Black Holes: Physics and astrophysics" in Proc. of the ISCRA 14th Course, Eds. Shapiro, M., Stanev, T. and Wefel, J.

[2] Carroll, S.M. (2004), "Spacetime and Geometry. An Introduction to General Relativity", San Francisco, CA, USA: Addison Wesley, and gr-qc/9712091.

[3] Donea, A.-C. and Biermann, P.L. (1996), "The symbiotic system in quasars: black hole, accretion disk and jet.", A&A 316, 43.

[4] Li, L.-X. (2002), "Toy model for the magnetic connection between a black hole and a disk", Phys. Rev. D65, 084047.

[5] Novikov, I.D. and Thorne, K.S. (1973), "Astrophysics of Black Holes" in Black Holes. Les Astres Occlus, Eds. DeWitt, C. and DeWitt, B.S., Gordon & Breach, New York, p. 343.

[6] Page, D.N. and Thorne, K.S. (1974), "Disk-accretion onto a black hole. I. Time-averaged structure of accretion disk", Ap. J. 191, 499.

[7] Semenov V., Dyadechkin S., Punsly, B. (2004), "Simulation of Jets Driven by Black Holes Rotation", Science 305, 978, and astro-ph/0408371.

[8] Thorne, K.S. (1974), "Disk-accretion onto a black hole. II. Evolution of the hole", Ap. J. 191, 507.

[9] Yuan, F., Markoff, S., Falcke, H. and Biermann, P.L. (2002), "NGC 4258: A jet-dominated low-luminosity AGN?", A&A 391, 139.

DARK MATTER: PAST, PRESENT, AND FUTURE

Virginia Trimble

Department of Physics and Astronomy, University of California, Irvine, CA 92697-4575, USA
vtrimble@uci.edu

Abstract   The words "dark matter" are a shorthand for an enormous range of evidence indicating (a) that various astronomical mass-to-light (M/L) ratios are larger than can be accounted for by visible stars and gas at any temperature and (b) that M/L systematically increases as one measures it on larger and larger distance scales. The evidence is reviewed historically and attention given to the range of possible gravitating substances (collectively "dark matter candidates") that might make up all or part of the stuff. A handful are currently taken seriously, but the total inventory is at least several dozen. Dark energy comes at the end, but is also to be taken seriously.

1.   INTRODUCTION

Both the concept of dark matter ("unenlightened stars" according to Edward Pigott (1) in 1805) and evidence for some forms of it (companions of Sirius and Procyon and outlying gas giants in the solar system, associated with the names of Bessell, Adams, and LeVerrier in the 1840s) are older than any reader of this paper. Jeans wrote of dark stars (outnumbering light stars 3:1) in 1922, and Kapteyn wrote of dark matter ("quantity not excessive") in the same year. The first observation generally mentioned in dark matter reviews is Fritz Zwicky's 1933 analysis of his own measurements of velocities of galaxies in the Coma cluster (2), which lead (via a virial theorem) to a ratio of mass to light near 100 in solar units, using a modern distance scale. He wrote of "dunkel materie" and supposed it to consist of gas and faint stars. The present author does not quite remember the era of Kapteyn and Jeans but knew Zwicky near the end of his career and the beginning of hers (1964-74).

Forty years later, measurements of masses of galaxies from rotation curves, binary pairs, cluster velocity dispersions, and other indicators had accumulated to the point where two brief 1974 reviews by an Estonian trio (3) and an American trio (4) tipped the consensus of the community in favor of

M. M. Shapiro et al. (eds.), Neutrinos and Explosive Events in the Universe, 181–199.

large quantities of matter that neither emitted nor absorbed its fair share of light and that was much less concentrated toward the centers of galaxies than the luminous stars and gas. An extrapolation of their M/L vs. scale relations reached the density needed to close the universe at somewhere around the Hubble radius.

An expectation that the total density would be the closure one, but with only 5-10% of the matter luminous, also took hold, particularly in light of the predictions of inflation theory early in the 1980s that space should be flat. James Gunn, Richard Gott, David Schramm, and Beatrice Tinsley (5) in the same year made a strong case for the total density being only 20-30% of the critical value, setting up a sort of observers vs. theorists (yes, they counted as observers in this context!) confrontation which held for a couple of decades. Their arguments pertained partly to baryons and partly to any sort of matter, and, with a current value of the Hubble constant, their considerations of big bang nucleosynthesis, the age of the universe, and so forth still apply. Reconciliation of the two views has come through the (gradual) acceptance of a non-zero cosmological constant, or its variants, quintessence, dark energy, etc., which contribute to flattening space but not to velocity dispersions or to nuclear reactions. Not reaching the critical matter density can be associated with (but not attributed to) the lack of structures on distance scales large than 150-200 Mpc. The universe is not fractal beyond superclusters, filaments, and voids (6).

An early theoretical argument for dark matter was the need for massive spherical halos to suppress bar instabilities in disks (7). This has been re-evaluated from time to time, but in any case, modern opinion is far more swayed by the need for dark matter (and indeed dark energy) in order to arrive at a satisfactory scenario for the formation of galaxies and clusters from the very small density fluctuations present at the time of recombination. Crudely, the idea is that linear perturbations can grow only linearly with $(1+z)^{-1}$, so we would need something like 10^{-3} fluctuations in density at $z = 1000$ to grow to non-linear ones now, but the near-isotropy of the CMB (Cosmic Microwave Background) requires that the actual fluctuations be only parts in 10^5 not parts in 10^3. The literature of galaxy formation is simply enormous, and (8) is just a random recent paper to get you into the system.

The total topic of dark matter transcends enormity in its accumulated (and exponentially increasing) literature. I have provided earlier snapshots in 1987 (9), 1988 (10), 1993 (11), and 2002 (12) and more extensive historical material in 1990 (13) and 1995 (14). In addition, from 1997 onward, each of the reviews "Astrophysics in 1997" to "Astrophysics in 2003" (15) has had a section of dark matter candidates and cosmological models, including the momentary favorites (which have changed with time), new promising candidates (like self-interacting dark matter, which came and went very quickly), and some suggestions about which one can only say, "remarkable!". These secondary sources will be relied upon a good deal, lest the reference section of this discussion overflow the total page limits. They are abbreviated as Ap97 etc.

2. DATA THEN AND NOW

How early could someone have compiled the sort of M/L trends reported in (3) and (4) and shown in Table 1? In principle, shortly before the outbreak of World War II, using the work of Jeans and Kapteyn, or Oort a decade later for the solar neighborhood, Hubble's 1934 discussion of the inner parts of galaxies (16), a rotation curve for the outer parts of M31 that formed part of the PhD dissertation of the late Horace W. Babcock (17), the binary galaxies presented by Holmberg in 1937 (18), also part of his PhD dissertation, and a 1936 study of the Virgo cluster by Sinclair Smith (19), in case you happened not to like Fritz Zwicky, as indeed some of his contemporaries did not.

TABLE 1. MASS-TO-LIGHT RATIOS (SOLAR UNITS) AS A FUNCTION OF SIZE SCALE ON WHICH THEY ARE MEASURED AND CONTRIBUTIONS TO THE COSMIC DENSITY

LENGTH SCALE	TYPICAL, OBJECTS, TECHNIQUES, REFERENCES	M/L	CONTRIB. TO Ω_M
few - 100 pc	star clusters, solar neighborhood: star velocity dispersions (23, 50)	0.2-10	0.0002 - 0.01
1-10 kpc	optically bright parts of galaxies: stellar velocity dispersions, inner rotation curves (M/L systematically larger for Es than Ss) (24, 25)	few-20	0.003 - 0.02
10-100 kpc	whole galaxies: rotation curves, velocities of globular clusters & satellite galaxies, X-rays, strong lensing (24, 26)	10-50	0.01 - 0.05
0.3 - 3 Mpc	binary galaxies, Local Group, other small groups: Virial theorem, X-rays, velocity dispersions of outer satellites, pair-wise velocity differences, approach of M31 (27, 28, 29)	20 - 100	0.02 - 0.1
10 - 30 Mpc	groups, clusters: Virial theorem, X-rays, gravitational lensing (30)	100	0.1
100-300 Mpc	large clusters and superclusters of galaxies: weak gravitational lensing, X-ray temperatures, Virial theorem (31, 32, 33, 34)	100 - 300	0.1 - 0.3
Global	universe out to the Hubble radius and beyond: Type Ia supernovae, cosmic shear, largest structures, fluctuations in 3K background (35, 36, 37, 38)	270	0.27

Note that the conversion between the last two columns assumed a cosmic luminosity density of 1000 solar luminosities per cubic mega-parsec. Only for the last line (where the 0.27 is known to better than 10%) does the uncertainty on this matter to any of the numbers indicated.

How early did someone actually provide such a table? The first I've found came from Martin Schwarzschild in 1954 (20) and had the key numbers for inner galaxies, outer parts, and whole clusters.   He also noted that elliptical galaxies nearly always have larger M/L's than spirals. We would now say that this is mostly because their bright young stars are gone, so that L is smaller, rather than M being bigger. Indeed ellipticals of various sorts include both the very smallest dwarf spheroidal galaxies of 10^6 solar masses and giants of up to 10^{13} solar masses. Schwarzschild suggested that old white dwarfs were an important part of dark matter, which would account for the differences between ellipticals and spirals.

In order to compare numbers across the decades, it is essential to allow for changes in the best estimate of the cosmic distance scale or the Hubble constant (H). In most cases, the L that you deduce for a galaxy (etc.) from its brightness will be proportional to the square of its distance, d, or to H^{-2}, and the mass you calculate, from some form of $M = V^2R/G$, will be proportional to d or H^{-1}. Thus M/L scales as d^{-1} or H, and a velocity dispersion for a cluster of galaxies plus its angular size on the sky that led to M/L = 1000 for H = 500 km/sec/Mpc now, with H = 70 km/sec/Mpc, corresponds to M/L = 140.   When Schwarzschild produced his table, the community was just incorporating the first of the large drops in H, from about 500 to 250 km/sec/Mpc.

What did the numbers look like when I first reviewed the subject in 1987 (9)? Pretty much like Schwarzschild's, after allowance for the continued decline in H, and with a few additional data points for the Local Group, for galaxies and clusters with X-ray sizes and temperatures (hence independent Virial-type masses), and for large galaxies (including our own) from the dispersion of velocities of globular clusters and satellite galaxies.

Those numbers, in turn, apart from being given for H = 100 km/sec/Mpc, were very much like those of Table 1. The references indicated in the table are a very small subset of recent ones, not always the most comprehensive, but they provide a representative view of details of the determinations and the major potential sources of errors. Divergent views exist, and recent papers in the ApXX series (15) each include a few data sets for which a matter density close to the closure value (9.5×10^{-30} g-cm^{-3}) is a good a fit as, or better than, the consensus value of 27% of this. Some special mention must be made of Jan H. Oort, whose early papers (21, 22) drew attention to dark matter in our own and other galaxies, which he took to be very faint stars and/or gas at temperatures not then observable, and who continued to consider related issues, including black holes at galactic centers, for another half century.

3. THE CANDIDATES YOU COULD TAKE HOME TO MOTHER

You are going to meet a very large number of these, beginning with a summary of categories ordered chronologically. Then will come some discussion of those that have been taken seriously over a reasonable length of time. The next section will include a take-out menu of more recent suggestions, some to be taken seriously, others perhaps not.

Baryons were first, from the dark, eclipsing stars or planets of Edmund Pigott and John Goodricke (1) in the 1780s via the faint stars, gas, and faint galaxies of Oort (21) and Zwicky (2). As far as I can tell, we then skip 30 years to the possibility of deviations from Newtonian or Einsteinian gravity that will allow the luminous matter to bind galaxies and clusters, as put forward by Arigo Finzi (39). Candidates belonging to the realm of modern theoretical physics started slowly with neutrinos of non-zero rest mass from Gershtein and Zeldovich in 1966 (40) and primordial black holes the same year, also a Russian invention (41). And then the flood gates opened up on both sides of the iron curtain. I would not swear to absolute priority for these papers, but at any rate the indicated candidates were not lost again afterwards.

- Topological singularities from spontaneous symmetry breaking at phase transitions, according to Kirzhnits and Linde in 1972 (42).
- Primoridal gravitational radiation, Grischuk 1974 (43), associated with scenarios we would now call inflation, Starobinsky 1979 (44). And if you declare that inflation had not yet been invented, you will have the givers of several major prizes on your side.
- Non-topological solitons, Friedberg et al. 1976 (45), of which Q-balls, quark nuggets, and soliton stars are later variants.
- Axions arising from Peccei-Quinn symmetry (1977) as a cure for the strong CP violation problem (46).
- And, the name that eventually led all the rest, supersymmetric (and other symmetries) partners of the particles you know and love, put in initial order by Lee and Weinberg (47). The particles have been called inos (as in gravitino), spartners (as in sneutrinos), and WIMPs (weakly interacting massive particles).

Within the year, all the best people (48, 49) were including cold dark matter (CDM) in their models for galaxy formation. From then to now, speakers and writers of the CDM words generally mean WIMPs and their ilk, though axions and a few other candidates behave in much the same way during structure formation. "Bias," the idea that luminous baryons will be more tightly clustered than the CDM (presumably because they are dissipative) dates from the same period.

Now, what has become of all these?

Baryons still exist (for which we are grateful) and will continue to do so for at least the 10^{32} year (lower limit) half life of the proton. That they do not make up most of the dark matter is a joint and concordant conclusion from (a) data collected by the WMAP satellite (Wilkinson Microwave Anistropy Probe) and (b) comparisons of calculations with data for big bang nucleosynthesis (38, 51, 52). The official number is $\Omega_b = 0.044 \pm 0.004$. At large redshift, most of the baryons were in moderate-contrast structures responsible for producing assorted kinds of absorption features in QSO spectra. At present, about half are in visible stars and X-ray-emitting gas. The locus of the rest is generally thought on both observational and theoretical grounds to be a filamentary WHIM (warm-hot intergalactic medium, Ap02, Sect. 12.6.3). This has not kept colleagues from putting forward additional baryon collections (next section).

Electromagnetic radiation, closely allied with baryons, now contributes only 0.1% of the closure density, nearly all of it from the CMB (53), though of course radiation dominated all forms of matter at redshifts larger than about 1200.

Non-standard descriptions of gravity designed to avoid the need for dark matter now number half a dozen or so. None has had the full range of its consequences worked out for comparison with data (meaning the solar system tests, evolution of binary pulsars, time scales of QPOs in X-ray binaries, etc., as well as large scale gravitational lensing, structure formation, and so forth). MOND, (MOdified Newtonian Dynamics [54] and many earlier papers) comes closest. It continues to have supporters (who are younger than we and so should outlive us), but the tide is against it (55, 56). The general idea is a minimum allowed acceleration, below which gravity turns over to a 1/R force. The difficulties arise in making structure formation, reionization, and gravitational lensing all come out right with the same value of that minimum.

Neutrinos no longer count as "dark matter candidates" since they are known experimentally to oscillate among flavors and so to have non-zero rest masses. The implied masses are such that the ones we know and love contribute less than $\Omega_v = 0.01$ (Ap00, Sect. 12.4.2). Observed large scale structure and models for its formation are happiest with $\Omega_v = 0.006$ or thereabouts. (57)

Primordial black holes (PBH) cannot be entirely ruled out at present, provided that they formed early enough not to have been baryonic during nucleosynthesis. One can say that either the universe is not closed by black holes near 10^{15}g, or Hawking radiation doesn't happen, or both (Ap02, Sect. 12.7). Planet-to-star masses that would gravitationally lens stars behind them (called MACHOs for MAssive Compact Halo Objects) are excluded, and so are (a) blacks holes of 10^{5-6} solar masses, which would mess up galactic dynamics and (b) ones of 10^{6-11} solar masses, which would act as gravitational lenses of background QSOs and GRBs (Ap01, Sect. 12.5). A bit of phase space still remains (Ap00, 12.4.2, Ap03, 3.6, Ap99 Sect. 12.6).

Topological singularities or defects have had their good years and bad years. They come in all possible numbers of dimensions: zero (monopoles, and that these turned out to be the same as the Dirac magnetic monopole was a surprise) with mass near the decoupling scale, one (strings), two (domain walls), three (textures), and convoluted (vortons, hedge-hogs, etc). The finite-dimension ones can have masses at least as large as a small galaxy. That the universe was not entirely overrun with monopoles (limits on which can be set from the persistence of large scale magnetic fields) was one of the original motivations for inflation. Too many strings or domain walls would make recognizable patterns in the CMB. All act more or less like cold dark matter and can perhaps also arise without phase transitions (Ap00, 12.4.3). It does not seem possible to close the universe with such singularities, but they may have a role in seeding galaxy formation (Ap00) or in generating primordial magnetic fields (Ap01, Sect. 12.5).

Primordial gravitational radiation means the sort that comes from stuff sloshing around in the early universe. As with the case of PBHs, there are limits well below $\Omega_{gr} = 1$ in many regimes (wavelength rather than mass in this case, [58]), but also still some for which limits are not very tight, except in the generic sense that the universe acts like most of its positive pressure stuff is matter, with density proportional to $(1+z)^3$ rather than radiation with density proportional to $(1+z)^4$.

Non-topological solitons bring us to the first of the candidates about which one is left a little uncertain about which things belong to which class, given that, for instance, Q-balls turned up in another list of "non-topological extended objects that may or may not be supersymmetric and may or may not be stable," while quark nuggets also appeared as "super-heavy entities, produced by non-thermal processes in the early universe." They are, therefore, all relegated to the next section.

Axions come out of a sort of Bose-Einstein condensation in the early universe rather than from a state of thermal equilibrium, and so will be cold no matter how small their masses. Kolb and Turner in their 1990 duograph (59) gave them a whole chapter (chapter 10), and nothing seems to have happened since to make them a less serious candidate, though the masses must now be in the smaller of the two ranges then possible, near 10^{-5} eV, to keep decay products below detectability (Ap01, Sect. 12.5). Laboratory searches are in progress, and if they find something persuasive, you won't need me to tell you about it.

WIMPs are in the same condition, with searches in progress. Indeed there has been one positive report, but also contradictory evidence from other experiments (60), suggesting that the annual variation in flux reported may have arisen from something other than our annual change in velocity of motion relative to the galactic halo rest frame. Only one conventional WIMP can be truly stable, the lowest-mass sypersymmetric particle (LSP for short) which is quite likely to be a linear combination of Higgsino, photino, and Zino (for instance). Back in 1987, the LSP could have had any mass from 10-100 eV up to a TeV. The low end is now ruled out by the requirement that the stuff be cold during structure formation, and the laboratory limits have gradually squeezed in as well. Some very strange particle physics also begins to happen for dominant stuff above about half a TeV (60). Thus, current laboratory searches focus on masses of 100-400 GeV and cross sections smaller than 10^{-5} picobarns. Recall that "pico" = 10^{-12}, and a barn was what traditional nuclear physicists couldn't hit the broad side of (10^{-24} cm^2). A shed was smaller.

Some of the more popular variations on classical WIMPs do not require inventing any new words. These include (a) decaying CDM (61), admittedly the authors start with charged dark matter and end with something like a superWIMP or massive Kaluza-Klein graviton, but you get the idea, (b) annihilating dark matter (62), and (c) self-interacting dark matter. This last has some significant cross section for scattering its own particles, e.g. 4×10^{-25} cm^2/GeV, but much smaller, or zero, cross sections for annihilation or dissipation or other interactions with baryons. It was a surprise "princess" candidate (Ap00, 12.4.2) to which more and more objections were found over the next few years (Ap00-Ap03), the most serious perhaps that it didn't actually solve the problem for which it had been invented, that of turning a sharp density cusp of a simulated galaxy halo into the flat core of a typical observed galaxy halo.

4.   CANDIDATES TO LEAVE BEHIND AT THE SINGLES BAR

Some of these are simply rather new (but promising), and there is no use distressing the family prematurely. Others you should perhaps not take home no matter how long you have known them. Will I always tell you which is which? No; why offend colleagues unnecessarily (but you could chase down the journals in which each originally appeared as an indicator). Alternative titles for this section might have been NIBY ("not in my back yard", like nuclear reactors) or NIH ("not invented here").

I begin with a handful of names that appeared in early lists (9, 10) but seem to have dropped from the recent literature. Some have probably changed names (in the way that shadow matter is now often called mirror matter); others may be indistinguishable for astrophysical purposes from better motivated candidates. And others may have been ruled out by non-astrophysical considerations.

- majoron and goldstone boson: 10^{-5} eV, like the axion, arising out of QCD
- paraphoton and right-handed neutrino, keV particles from modified QED and super-weak interaction theory
- cosmion, flatino, and magnino: MeV to GeV particles from SUSY and supergravity
- preons: multi-TeV particles from composite models
- pyrgons, maximons, perry poles, newtorites, and Schwarzchilds: Planck-mass particles in higher-dimensional theories
  Baryonic candidates were traditionally brown dwarfs, old white dwarfs, neutron stars, and stellar mass black holes. Both they and the new ones are subject to the big bang nucleosynthesis constraint above; others might accrete matter, thus radiating, absorb at some wavelengths, or otherwise reveal themselves.
- solid hydrogen (Ap97, Sect. 12.2)
- dense, cold molecular gas clouds in galaxies (Ap97, 00, 01, 03), fairly firmly ruled out by absence of absorption (63)
- high velocity clouds of neutral hydrogen (Ap00 Sect. 12.4.1)
- white dwarfs supposedly detected in the MW halo (Ap00, 01)
- stellar mass black holes (or quark stars or boson stars) because early accretion on them would produce reionization too soon (Ap01 - this should perhaps be re-thought given the WMAP evidence for reionization beginning near $z = 20$)
- white dwarfs that were never luminous and never produced carbon (etc.) that was blown out (overcoming two of the objections to the usual ones, Ap99, 01)
- large numbers of halo BDs, WDs, or anything else (stellar mass quark nuggets?) that would contribute more MACHO events than seen (Ap03, Sect. 3.6)

Some non-standard theories of gravity have seen less play in the astrophysical literature than has MOND, perhaps only because fewer people have felt passionately that they want to disprove these theories.

- time dependent G (Ap97)
- conformal gravity (Ap97, 02)
- a vector-based theory of gravity (Ap97)
- quantized rotation curves for galaxies (Ap99)
- a scalar-tensor theory in which masses of PBHs grow with time (Ap00, Sect. 12.5.2)
- a second gravitational potential of the Yukawa form (64), i.e. a finite rest mass for the carrier. Secretly we have always rather liked this, as a modern-sounding version of Finzi's (39) idea, but have also always

suspected that, if you invoke one non-zero-rest-mass graviton or gravitino to strengthen gravity as you go from local to galactic scales, you will need additional ones to move outward to larger scales, ending with 10^{-63}g or less at 10 Mpc or more, a number we first heard from Fritz Zwicky in about 1965. He was, of course, in favor of dark matter, but against superclustering.

Additional neutrinos beyond the three you know and love would have to be sterile (that is, not get involved in big bang nucleosynthesis or in neutrino oscillations as currently observed). They could be massive enough to be warm, rather than hot, DM (Ap01 Sect. 12.5). One sort would be made cold from the regular neutrino flavors by MSW oscillation (Ap99, Sect. 12.6). Decaying neutrinos had a certain wild beauty when it seemed their masses might be just above 2 x 13.6 eV, so that the product photons could have contributed to (re)ionization in various contexts (65). Most combinations of properties that would make them interesting can now be ruled out (Ap97, 99, 01), because of their propensity to mess up the CMB early on and to yield detectable UV photons later.

A couple more "warm" candidates probably go about here, including mirror or shadow neutrinos or majorons (Ap01, Sect. 12.5). These must have masses in excess of 0.25 - 0.4 keV or something bad will come down the chimney (probably reionization that smears out the Lyman alpha forest clouds). A 40 eV neutrino popped up in Ap97 (Sect. 12.2), and it was not clear how to avoid having so many of these that the universe would fold up into Pauli's pocket. (See Ap02 for generic objections to warm DM of any sort.)

We haven't found any new sorts of PBHs or sources of gravitational radiation that seem less likely than those in Sect. 3, so here is the lightest gluino-containing baryon (73) to keep the paragraph from feeling unwanted. A couple of more items were published in mainstream journals by people at mainstream institutions, or you might otherwise have supposed that they should go at the screwy end of the paper (whichever that is). Three examples: Rydberg matter (Ap03), which means atoms in highly excited orbits like n = 109, not something at 13.6 eV; dust with charge = mass in the c = G = 1 units of general relativity (Ap97); and clusters of MACHOs with cluster mass around 4×10^4 solar masses, so as to evade the limits from both gravitational microlensing and from stellar dynamics (Ap97).

The topological and non-topological boxes both seem to have a few left-overs. Indeed they have left-overs in common, which would probably bother us less if we knew precisely what topological meant in this context. Imprecisely, those are the ones that also get described as singularities, while the non-topological also get described as solitons (standing waves, we think, in something, but not water or galactic gas). The T-box still holds decaying domain walls (Ap98 Sect. 11.4), which sweep the monopoles out of the way first, and several sorts of topological solitons, including instantons, Skyrmions, and 't Hooft-Polyakov monopoles (Ap02, Sect. 12.7).

The non-T box is perhaps the most crowded. Early ones live in chapter 7 of (59) and include soliton stars, Z-balls, non-topological cosmic strings, neutrino balls, and quark nuggets. The soliton stars and quark nuggets probably have masses in the MACHO range, and so are not the answer for the halo of the Milky Way and, by extension, other galaxies. "Moving right along" as speakers often say when they have already exceeded their allotted time by 15 minutes, we find Q balls still alive and well in Physical Review Letters (Ap02), while the B balls are decaying in the same journal (Ap99).

Axions need a big box by particle standards. Have you ever worried about a Heisenberg uncertainty size for something with a mass of 10^{-5} eV at T = 3K? We get 2.2 mm. And they have such a box, a big chunk of a Reviews of Modern Physics (66).

Most of the recent variants of WIMP-ish CDM (decaying, annihilating, self-interacting) live in the previous section, but here are LIMPS (which do their weak interacting only with leptons, (67), and have masses of 1 - 10 TeV); hot, low-mass WIMPs from the decay of scalar particles (Ap99); the CDM that comes from B-ball decay (Ap99); and axinos as the product of LSP decay (Ap99). Since no decaying or annihilating WIMPs have been seen in the lab, limits on their masses, cross-sections, and half-lives sometimes come from astrophysical constraints (Ap00, Ap03). Repulsive dark matter (meaning the shape of its potential, not any personal characteristics) popped up a couple of times (Ap00, Ap01) and may or may not belong in this section.

At this point, we could branch off in several directions and must (unlike Lord Ronald in Stephen Leacock's "Gertrude the Governess") resist the urge to jump on our horse and ride madly off in all of them. The territory galloped over can still be described as "respectable but new or less familiar," and we encounter:

- WIMPzillas (Ap03) of more than 10^{15} GeV, somewhere around the energy scale where the color force breaks loose from the electroweak force. These could (Ap99, Ap00) even be capable of nuclear and electromagnetic interactions, if the masses are large enough and the charges small enough. Somewhere nearby live the cryptons of D. V. Nanopoulos and the strange quark nuggets of E. Witten (Ap03).
- A generic scalar field will have an associated generic scalar boson, and with it you can do most of the things that dark matter is supposed to do (75), and some (like making 10^{10} solar mass stars) that it is not (Ap00). The fields and particles can be traced back to 1968 (68), which perhaps entitles them to Section 3 status.

Additional spatial dimensions beyond the three we move in (anywhere from one extra, as in traditional Kaluza-Klein theory, up to 11 or thereabouts) imply the possibility of extra symmetries, extra conserved quantities, and a lowest-mass particle that cannot decay without violating that conservation law and which is, therefore, a possible DM candidate. Some names we caught were:

- branons (Ap03), with masses of 10^2 - 10^4 GeV, the usual WIMP range and with cross sections less than 10^{-43} cm^2 at the low-mass end
- the lightest Kaluza-Klein gauge boson (Ap03), with potentially observable decay products (69, 70)
- the Kaluza-Klein gravitino (76) which might get trapped inside neutron stars

Mirror or shadow matter also belongs to the extra-dimension territory. The particles are just like ours (hadrons, leptons, and all), but can interact with us only gravitationally across a slightly extended fifth dimension of a particular sort of superstring theory. The standard early reference is (71), and a recent one, whose authors would like to use the stuff for supernovae is (72).

"Other" you might suppose would be a superfluous category after all the items above, but the following have also been swept up in the yearly overviews. They are ordered chronologically.

Ap97: Short-lived particles, created continuously from vacuum quantum fluctuations in the gravitational fields of galaxies; 10^{-34} eV mass, spin zero, neutral bosons with a time-dependent scalar field.

Ap99: Ether.

Ap00: PBHs that are also responsible for ball lightning, because the presence of the Earth catalyzes radiation emitted by gravitational tunneling. DAEMONS (DArk Electric Matter Objects) are particles for which thermodynamic time runs backwards, preventing electromagnetic interactions.

Ap01: A fluid with negative energy density (which would permit an oscillating universe without a singularity). Vacuum energy of a simple, quantized free scalar field of low mass.

Ap02: Stars made of WIMPs. Point-like particles of stellar mass with gas halos around them, so that the gravitational lensing events they produce are non-grey and so not recognized. Planck mass relics of the evaporation of PBHs.

At the transition to the next section, we find (1) naturally soft bosons that could also be responsible for cosmic acceleration (Ap00), and (2) a unified "dark sector" in which the energy density of a scalar field acts like a cosmological constant and the value of the field sets the DM particle masses via Yukawa coupling (77). Current data can be fit, but some of the predictions are different from those of standard CDM, including the long range future of the universe (though this is not a testable prediction for people my age). Two dark matter families are possible.

5.   QUINTESSENCE, DARK ENERGY, AND ALL THAT?

For something that began life as an integration constant, "in order to obtain the most general expression with vanishing covariant derivative," (78), Λ has received some remarkable bad press over the years. "Einstein's infamous cosmological constant," "Einstein's greatest blunder," "whose desirability is otherwise itself controversial" (This last from Dennis W. Sciama, who was no stranger to controversy), and so forth.

Let us skip at once to 2004 and observe that the "consensus cosmology," in place at least since 1997 (the Kyoto General Assembly of the International Astronomical Union, for instance) has 70% or a bit more of the gravitating stuff in the universe (whose total adds up to the critical, flat density) present in the form of something with negative pressure. The simplest possible incarnation (consistent with all available data at present) is indeed constant in both time and space, and, like cold dark matter, has no new interactions or forces (79).

If you write the Friedman equation in one of several standard forms as

$$H^2 = \frac{8\pi G}{3}\rho - \frac{k}{a^2} + \frac{\Lambda}{3} \tag{1}$$

then the consensus parameter set has k = 0 (flat space-time), and the other terms of roughly comparable size, with units of t^{-2} implied. Others tuck in a c^2 so that Λ has units of (length)$^{-2}$. A universe in which it is the dominant term will expand exponentially with time, as in (a) the de Sitter solution, (b) steady state cosmology, and (c) an inflationary epoch.

It is generally said that Einstein kept the Λ term in his early papers to permit a static universe. Alexander Friedman showed in 1922-24 that it could also be part of an expanding (or contracting) universe. Many relativists through the 1930s held on to the constant, often because it permitted an age larger than H^{-1}, and that motivation has recurred more or less continuously (80, 81). It eased off a bit when estimates of the Hubble constant dropped from 1952 into the 1970s, and crept back when the oldest stars and radioactive elements began to push the age on past 10 Gyr. The worry that the Earth and stars might be older than the universe was one (of several) motivations also for the 1948 steady state universe.

There was a separate burst of enthusiasm for non-zero Λ in the late 1960s, arising from a seeming excess of QSOs with redshifts very close to 1.95, which could be interpreted as the coasting epoch of a universe dominated early on by matter of some sort, from which Λ had begun to take over. Notice that flat space, where $\Omega_M + \Omega_\Lambda = 1$ has only an inflection point in R(t), indeed somewhere around z = 2, not a coasting period. The excess of z = 1.95 redshifts has, in any case, been gradually erased by additional samples with different selection effects.

With a little effort, one could probably find at least one paper every five years, from 1917 to the present, in which a cosmological constant was taken reasonably seriously, so the idea has never really been lost. Until recently however, most textbooks set it equal to zero, at least from chapter 2 onward. Authors could claim various, Occamal, philosophical reasons for this. But, perhaps more important, if you keep it, the equations relating angular diameters, observed brightnesses, look back time, etc. to redshift become very much more complex and not always amenable to analytic solutions.

A separate objection came from the particle physicists when they began to take an interest in cosmology in the 1970s and to think of Λ as the zero-point energy of one or more fields. The natural value in the universe of Eq. (1) would be $(\tau_p)^{-2}$, where τ_p is the Planck time, about 10^{-43}s. Thus it would be bigger than the other terms in Eq. (1) by about 10^{120} (see 82), and theorists tended to feel that it would be easier to achieve exact cancellation and $\Lambda = 0$ than cancellation so nearly, but not quite, complete. Many astronomers, on the other hand, have remained content to say that zero, versus small but non-zero, was an observational issue. Gerard de Vaucouleurs was among them, and the number grew when it began to be clear that there were observations that could actually separate Λ from the other cosmological parameters (83). Some of these are still producing only upper limits, some of which hit hard against the consensus number (15, sect. 3.5), but there is now a positive result from the integrated Sachs-Wolfe effect (84), which is the expected correlation between relatively small angular scale fluctuations in the CMB and the distribution of gas in protoclusters that are still expanding, so that photons passing through come out blueshifted. It is there in a comparison of the Sloan Digital Sky Survey of galaxies and clusters with WMAP.

The observational camp being on top, at least for the time being, theoretical physicists have turned their attention to new candidates, and new names, for the negative pressure stuff. The inflaton was first. If inflation results from the behavior of a scalar field, someone was bound to call it an inflaton field and the associated scalar boson the inflaton. This was already common by 1990 (59), with one version regarding the present exponential expansion as just a leftover bit of inflation.

Ap97 was the first annual report to record a majority of papers favoring non-zero Λ. Ap98 registered both the word "quintessence" and the idea of a more general equation of state, with P = -wρ, and w potentially a function of time (or something else). The stuff is also allowed to be clumpy (85), and to participate in structure formation. The case w = 1 is equivalent to the invariant Λ. [The name "quintessence" derives from the classical elements: earth, air, fire, and water for terrestrial objects/phenomena, and quintessence for celestial ones.] X-matter appeared the next year (Ap99) with roughly the same meaning as quintessence and fairly short half-life. An upper limit to -w near -0.6 came from the Type Ia supernova data as they then stood. That point has proven fairly robust. None of the assorted relevant data prefer w ≠ 1, and the WMAP limit is 0.78 (38).

The term dark energy acquired majority status in Ap00, and theorists were working out the consequences of w being a function of time. For instance, if the dark energy melted away, our accelerating universe might yet contract back to high enough density that any surviving observers would cease to take an interest in whether a singularity would result. Some versions of the scalar field required for current dark energy were motivated by string theory, though this did not guarantee agreement with observations (86 and many slightly later papers).

The 0.7/0.3 ratio of dark energy to dark matter has fairly small error bars if you take all the observations seriously, so theorists began, in the new century (Ap01), to try to identify mechanisms that could lock in such a ratio (vs. zero or infinity) long enough for life to evolve and all. Some of the associated concepts are a Brans-Dicke (scalar-tensor) field, a false vacuum, a scalar field that locks onto negative pressure, and, of course, ideas connected with strings and branes. Up to this point, at least, most of the ideas first surfaced in Physical Review Letters, Physics Letters, and other reasonably respectable venues, in contrast to some of the more imaginative candidates for dark matter. This happy state of affairs probably won't last.

Observable consequences of the various candidates typically appear in the fine structure and polarization of the CMB, probably undetectable until the Planck Satellite Mission or beyond, though a more interesting one would be a bubble of negative potential energy expanding at the speed of light, which you would presumably notice (very briefly) as it overtook you. Some of the ideas are likely to send the reader scurrying back to textbooks. Let a holographic quantum contribution that should always keep Λ smaller than the dominant matter component (87) stand for many appearing in Ap02 and Ap03.

It would be rash indeed to be on the "dark flavor of the month" (except perhaps bitter chocolate). A unified dark sector proposal (77) was noted at the end of the previous section. Raman Sundrum (88) has given the graviton a finite size and urged that searches for deviations of the gravitational force from $1/R^2$ be continued down to the 20 μm scale. The difficulty with such experiments (89) is in getting the centers of two very large masses very close together. Samples of neutron star material would not help, since the gravitational potential confines the stuff only for masses larger than about 0.1 solar masses.

New darkness leaks continuously into the astronomical literature from the physics literature and astro-ph. The past year has seen a positive (or rather negative) spigot of Chaplygin gas (90, 91, 92). In light (or darkness) of my fruitless search for Prof. Cardass of Cardassian expansion (93), I was pleased to discover that Chaplygin (Sergei Alekseevich, 1869-1942), whose village of birth was renamed for him, was a real person, to be found in the Dictionary of Scientific Biography and other (mostly Russian) sources. Most of his work was in what would now be called fluid dynamics, and the crucial 1904 paper invoked a compressible fluid with an equation of state given by $P = -A/\rho$ for use in considering the lift of a rigid wing. His work in general has not ever been forgotten, with regular citations in SCI, but that particular equation of state had to be independently rediscovered several times, for instance by Theodore von Karman in 1941, also in an aeodynamical context. I mention von Karman of several rediscoverers because he is the anti-hero of a story of which the punchline is, "Mr. Bird, where is Maryland?"

Kamenshchik et al. (94) seem to have been the first to fill a universe with a Chaplygin gas. They also mention its connections with brane theory and supersymmetry, discovered by others. A generalized Chaplygin gas has $P = -A/\rho^{\alpha}$ where α can be a function of time, to carry the universe across from a pressure-free state to a negative pressure one.

The long range future of a universe with a dark energy sector (whatever you call it) is a lonely one (Ap02, Ap03). It is today already too late to send a message to a galaxy we see with redshift greater than 1.7-1.8 and have it ever arrive (unless the dark energy decays away), and we will never see a galaxy with $z = 5$ at any age large than 6 Gyr. Come back in a few Hubble times, and nothing will remain within our horizon (stationary at about 5 Gpc away) but the merged product of our own Local Group and its nearest neighbors.

The phrase "phantom energy" snuck into glossaries a couple of years ago to describe dark energy with $w > 1$. Observationally (Ap03, 9.5) it cannot be much larger, or $-w$ much smaller than -1, but even a few % will take Ω_{Λ} to infinity in finite time, tearing up clusters and galaxies, then the solar system and stars, the Earth (at half an hour before the end), and by then it hardly matters that atoms are going to vanish 10^{-19} seconds before the instant of doom!

REFERENCES

1. E. Pigott 1805. Phil. Trans. Royal Society 95, 131
2. F. Zwicky 1933. Helv. Phys. Acta 6, 110
3. J. Einasto, A. Kaasik, & E. Saar, 1974. Nature 250, 309
4. J.P. Ostriker, P.J.E. Peebles, & A. Yahil. 1974. ApJ Lett. 193, L1
5. J.R. Gott, J. E. Gunn, D. N. Schramm, & B.M. Tinsley 1974. ApJ 194, 341
6. B.J.E. Jones, V. Martinez, E. Saar, & V. Trimble. 2004. Rev. Mod. Phys. (in press).
7. J.P. Ostriker & P.J.E. Peebles 1073. ApJ 186, 467
8. D.H. Weinberg et al. 2004. ApJ 601, 1
9. V. Trimble 1987. Ann. Rev. A&A 25, 415
10. V. Trimble 1988. Contemp. Phys. 29, 373
11. V. Trimble 1993. in N.S. Hetherington, (ed.), *Encyclopedia of Cosmology,* Garland Publishing, p. 148
12. V. Trimble 2004. in W. Freedman et al. (eds.), *Carnegie Symposium, Measuring and Modeling the Universe,* Cambridge Univ. Press (in press)
13. V. Trimble 1990. in B. Bertotti (ed.), *Modern Cosmology in Retrospect,* Cambridge Univ. Press. p. 355
14. V. Trimble 1995. in S.S. Holt & C.L. Bennett, (eds.), *Dark Matter,* AIP Conf. Ser. 336, 57
15. V. Trimble & M.A. Aschwanden 2004. Publ. Astron. Soc. Pacific 116, 187
16. E.P. Hubble 1934. ApJ 76, 44
17. H.W. Babcock 1939. Lick Obs. Bull. 19, 41
18. E. Holmberg 1937. Lund Obs. Annals 6, 1
19. S. Smith 1936. ApJ 83, 23
20. M. Schwarzschild 1954. AJ 59, 273
21. J.H. Oort 1932. BAN 6, 249
22. J.H. Oort 1940. ApJ 91, 273.
23. J. Kovalevsky 1998. ARA&A 36, 99
24. Y. Sofue & V. C. Rubin 2001. ARA&A 39, 137
25. V. de Bruyne et al. 2004. MNRAS 349, 440
26. F. Prada et al. 2003. ApJ 598, 260
27. J. Junquiere & R. Chan 2002. Astrophys. Sp. Sci. 279, 271
28. I.D. Karachentsev et al. 2001. A&A 389, 812 and 383, 812
29. C. Gallarco et al. 2001. AJ 121, 2572
30. N. A. Bahcall & P. Bode 2003. ApJ 588, L1
31. L. Ciotti & S. Pellegrini 2004. MNRAS 350, 609
32. J.G. Cohen & J.-P. Kneib 2002. ApJ 573, 524
33. K.-Y. Chae 2003. MNRAS 346, 746
34. F.C. van den Bosch et al. 2003. MNRAS 345, 889
35. R.A. Knop et al. 2003. ApJ 598, 102
36. B.J. Barris et al. 2004. ApJ 602, 571
37. D.J. Bacon et al. 2003. MNRAS 344, 673
38. C.L. Bennett et al. 2003. ApJS 148, 1
39. A. Finzi 1963. MNRAS 127, 21
40. S.S. Gerstein & Ya. B. Zeldovich 1966. JETP Lett. 4, 175
41. Ya. B. Zeldovich & I.D. Novikov 1966. Astron. Zh. 43, 758 (Sov. Astron. AJ 10, 602)
42. D.A. Kirzhnits & A. Linde 1972. Phys. Lett. 42B, 471
43. L.P. Grischuk 1974. JETP 40, 409
44. A.A. Starobinsky 1979. JETP Lett. 30, 682
45. R. Friedberg et al. 1976. PR D15, 1964
46. R.D. Peccei & H.R. Quinn 1977. PRL 38, 1440

47.   B.W. Lee & S. Weinberg 1977. PRL 39, 165
48.   J.E. Gunn 1977. ApJ 218, 592
49.   S.D.M. White & M.J. Rees 1978. MNRAS 183, 341
50.   V.I. Korchagin et al. 2003. AJ 126, 2896
51.   A. Coc et al. 2004. ApJ 600, 545
52.   D. Kirkman et al. 2003. ApJS 149, 1
53.   J.A. Zurita Heras et al. 2003. A&A 411, 71
54.   M. Milgrom 2002. ApJ 577, L75
55.   H. Hoekstra et al. 2004. ApJ 606, 67
56.   A. Lue & G.D. Starkman 2003. PRL 92, 131102
57.   S.W. Allen et al. 2003. MNRAS 346, 593
58.   J.W. Armstrong et al. 2003. ApJ 599, 806
59.   E.W. Kolb & M.S. Turner 1990. *The Early Universe.* Addison Wesley
60.   G. Chardin 2003. in R. Bandiera et al. (eds.) *Texas in Tuscany.* World Scientific,
        p. 167
61.   K. Sigurdsson & M. Kamionkowski 2003. PRL 92, 171302
62.   C. Boehn et al. 2004. PRL 92, 101301
63.   T.E. Clarke et al. 2003. ApJ 601, 798
64.   R.H. Sanders 1986. A&A 154, 135
65.   D.W. Sciama 1993. *Modern Cosmology and the Dark Matter Problem,* Cambridge
        University Press
66.   R. Bradley et al. 2003. Rev. Mod. Phys. 75, 777
67.   E.A. Baltz & L. Bergstrom 2003. PR D67, 043516
68.   D.J. Kaup 1968. Phys. Rev. 172, 1331
69.   H.-C. Cheng et al. 2003. PRL 89, 211301
70.   E. Hooper & G.D. Kribs 2003. PRD67, 055003
71.   E.W. Kolb et al. 1985. Nature
72.   R. Foot & Z. Silagadze. 2004. Astroph 040415
73.   S. Balberg et al. 2001. ApJ 548, L179
74.   J. Feng et al. 2003. PRL 91, 011302
75.   S. Fay 2003. A&A 413, 799
76.   M. Casse et al. 2004. PRL 92, 111102
77.   G.R. Farrar & P.J.E. Peebles 2003. ApJ 604, 1
78.   R.C. Tolman, 1934, *Relativity, Thermodynamics, and Cosmology,* Oxford Univ.
        Press
79.   S. Carroll 2004. Nature 429, 27
80.   K. Krisciunas 1993 in N.S. Hetherington, (ed.) *Encyclopedia of Cosmology,* Garland
        Publishing, p. 218
81.   P. Kerszberg 1993, in N.S. Hetherinton, (ed.) *Encyclopedia of Cosmology,* Garland
        Publishing, p. 566
82.   S.W. Weinberg 1989. Rev. Mod. Phys. 61, 1
83.   C.A. Alcock & B. Paczynski 1978. Nature 281, 356
84.   P. Michal et al. 2003. Nature 421, 608
85.   R.R. Caldwell et al. 1998. PRL 80, 1582
86.   A.A. Starobinsky, 1998. JETP 68, 757
87.   S. Thomas 2002. PRL 89, 081301
88.   R. Sundrum 2004. PR D69, 044014
89.   J. Chiaverini et al. 2003, PRL 90, 151101
90.   U. Alam et al. 2003. MNRAS 344, 1057
91.   P.T. Silva & O. Bertolami 2003. ApJ 599, 829
92.   A. Dev et al. 2004. A&A 417, 847

93.   K. Freese & M. Lewis 2002. Phys. Lett. B540, 1
94.   A. Kamenshchik et al. 2000. Phys. Lett. B511, 265
95.   Z.-H. Zhu et al. 2003. A&A 417, 835

ON DYNAMICS OF RELATIVISTIC SHOCK WAVES WITH LOSSES IN GAMMA-RAY BURST SOURCES

E.V. Derishev, Vl.V. Kocharovsky, K.A. Martiyanov
Institute of Applied Physics
46 Ulyanov st., 603950 Nizhny Novgorod, Russia
mca1@appl.sci-nnov.ru

Abstract   Generalization of the self-similar solution for ultrarelativistic shock waves (Bland-ford & McKee, 1976) is obtained in presence of losses localized on the shock front or distributed in the downstream medium. It is shown that there are two qualitatively different regimes of shock deceleration, corresponding to small and large losses. We present the temperature, pressure and density distributions in the downstream fluid as well as Lorentz factor as a function of distance from the shock front.

Keywords: relativistic shock waves, gamma-ray bursts

Introduction

The progenitors of gamma-ray bursts (GRBs) are believed to produce highly relativistic shocks at the interface between the ejected material and ambient medium (see, e.g., Meszaros, 2002; Piran, 2004 for review). Non-thermal spectra and short duration of GRBs place a firm lower limit to the bulk Lorentz factor of radiating plasma, which must exceed a few hundred to avoid the compactness problem (e.g., Baring & Harding, 1995).

Consider a relativistic spherical blast wave expanding into a uniform ambient medium with the Lorentz factor $\Gamma \sim 300$. The average energy per baryon in the fluid comoving frame behind the shock front is of the order of $\Gamma m_p c^2$ (Taub, 1948), where m_p is proton mass, and the plasma in the downstream presumably forms a non-thermal particle distribution extending up to very high energies. Under these conditions, medium downstream is subject to various loss processes. The non-thermal electrons produce synchrotron radiation, which accounts for GRB afterglow emission, and (at least partially) for the prompt emission. Apart from the synchrotron radiation of charged particles there is another mechanism of energy and momentum losses connected with

M. M. Shapiro et al. (eds.), Neutrinos and Explosive Events in the Universe, 201–205.

inelastic interactions of energetic protons with photons. These reactions cause proton-neutron conversion as a result of charged pion creation. It should be noticed that for typical interstellar density the Coulomb collisions are inefficient and the charged particles instead interact collectively through the magnetic field. This allows to describe plasma motion using hydrodynamical approach, though it can break for a small fraction of the most energetic particles.

When a proton turns into a neutron or another neutral particle is born, it does not interact with the magnetic field and hence the energy spent for its creation is lost from the hydrodynamical point of view. The synchrotron and inverse Compton emission, as well as energetic photons, neutrinos and neutrons produced via photopionic reactions, escape from downstream giving rise to non-zero divergence of the energy-momentum tensor. The creation of energetic neutrons is also a first step in the production of highest-energy cosmic rays through the converter mechanism (Derishev et al., 2003).

Ejection from the GRB progenitor of a mass M_0 with initial Lorentz factor Γ_0 results in two shocks propagating asunder from the contact discontinuity. The forward shock moves into the external gas and has a much greater compression ratio at its front than the other, reverse shock, which passes through the ejected matter. As the shocked external gas has a temperature much higher than that in the vicinity of the reverse shock, we neglect the losses in the ejecta.

We discuss two models. In the first one we assume the energy losses to be localized close to the shock front, whereas the matter downstream the shock is considered lossless. In another model the shock front is treated as non-dissipative and the losses are distributed all over the shocked gas.

Following the recipe of Blandford and McKee (1976) we generalize their well-known self-similar solutions for relativistic blast waves for the case, where the energy and momentum of the relativistic fluid is carried away by various species of neutral particles.

Self-similar solutions

We start from the energy-momentum continuity equations, where in the case of distributed losses a non-zero r.h.s. is included:

$$\frac{\partial T^{00}}{\partial t} + \frac{1}{r^2}\frac{\partial(r^2 T^{0r})}{\partial r} = -\varphi_0\, T^{00}, \quad \frac{\partial T^{0r}}{\partial t} + \frac{1}{r^2}\frac{\partial(r^2 T^{rr})}{\partial r} - \frac{2p}{r} = -\varphi_1\, T^{0r},$$

$$T^{00} = w\gamma^2 - p, \qquad T^{0r} = w\gamma^2\beta, \qquad T^{rr} = w\gamma^2\beta^2 + p,$$

where T is the energy-momentum tensor, γ the Lorentz factor, $w=e+p$ the enthalpy density, p the pressure, e the energy density. All quantities are measured in the fluid comoving frame. In the following analysis we use the ultrarelativistic approximation of these equations, obtained by expanding velocity up to the third contributing order in γ^{-2} and the equation of state up to the first order.

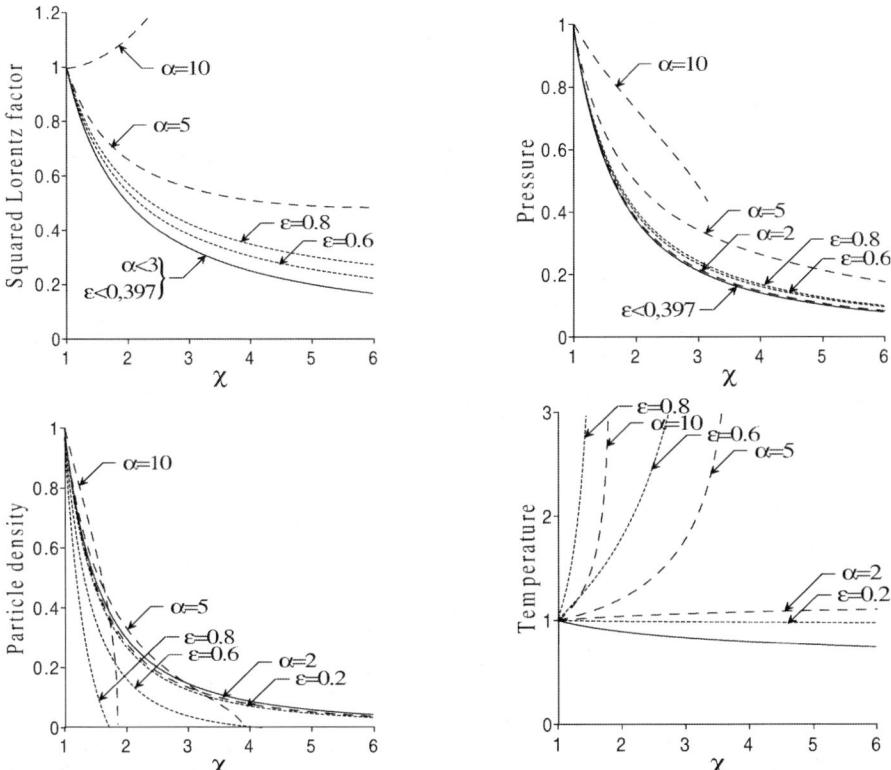

Figure 1. The distributions of the Lorentz factor, pressure and particle density of the shocked gas as functions of self-similar variable χ. All quantities are normalized to unity at the shock front where $\chi = 1$. Long-dashed lines correspond to the losses uniformly distributed in the downstream medium and decreasing with time as $t^{-\cdot}$. Short-dashed lines are for the case of localized losses with $\delta = \eta = 0$. The solid line — solution of Blandford and McKee (1976).

Because of the lack of space, here we consider only equal losses for the energy and momentum ($\varphi_0 = \varphi_1 = \varphi$).

In the case of localized losses we treat them as discontinuities of the energy, momentum and particle number fluxes at the shock front, which are characterized by three parameters ε, δ, η equal to the fractions of corresponding fluxes lost at the shock front in the front comoving frame. We obtain the following expressions for the pressure p_2, number density n_2 and the Lorentz factor γ_2 immediately behind the shock front:

$$p_2 = \frac{2}{3}\xi_1\Gamma^2 w_1, \quad \gamma_2^2 = \frac{1}{2}\xi_2\Gamma^2, \quad n_2 = 2\sqrt{2}\xi_3\Gamma n_1, \quad \text{where } \xi_1 = \frac{3}{2}\frac{(1-\delta)\xi_2}{\xi_2^2 - \xi_2 + 1},$$

$$\xi_2 = 1 + \frac{3(1 - (1-\varepsilon)/(1-\delta))}{1 + \sqrt{4 - 3(1-\varepsilon)^2/(1-\delta)^2}}, \quad \xi_3 = \frac{(1-\eta)}{\sqrt{2}}\frac{\sqrt{\xi_2}}{1-\xi_2}.$$

They differ from the Taub adiabat only by numerical factors ξ_i, which become unity in the absence of losses. Here $w_1 = n_1 m_p c^2$ is the enthalpy density and n_1 the number density of the external gas.

To find a self-similar solution we assume $\Gamma^2 = t^{-m}$ and introduce the similarity variable $\chi = \frac{t-r}{t-R}$, where r is distance from the center and R the current radius of the shock front. We find self-similar solutions in the case of localized losses and in the case of $\varphi = \frac{\alpha}{t}$, $\alpha = const$, but they also exist for non-uniform distributions of losses if $\alpha = \alpha(\chi)$. The velocity, pressure and particle density are found in terms of variables Γ, χ.

From the energy balance equation we find that the power law index m is in the range $3 \leq m \leq 6$. There are two regions in the parameter space where the solutions are qualitatively different. In the case of small losses, $\alpha < 3$ or $\varepsilon + \frac{2-\sqrt{109}}{14}\delta < 1 + \frac{2-\sqrt{109}}{14}$, the index m rises from 3 to 6 as the losses increase. The pressure, Lorentz factor and number density of the downstream are proportional to powers of the similarity variable whose indices are different from those in the solution of Blandford and McKee (1976). On the contrary, large losses lead to the universal deceleration law of the shock: m is equal to 6. The problem is fully integrable but solutions can not be written as explicit. In the high-loss solutions there appears an expanding spherical cavity bounded by the contact discontinuity and the temperature at its edge tends to infinity.

The solutions obtained are presented in Fig.1.

Conclusion

We have analyzed the dynamics of relativistic shock wave with losses due to escape of neutral particles from plasma flow. Both for localized and for distributed losses there are self-similar solutions, which are different from those found previously for lossless case. We find that increasing of the losses change the dynamics of the shock deceleration qualitatively. In the case of small losses, the role of ejected material asymptotically vanishes and the Lorentz factor of the shock decreases as t^{-m} with m varying from 1,5 (no losses) to 3. In the opposite case of large losses, the shock decelerates in accordance with universal law $\Gamma \sim t^{-3}$ and the energy content in the ejecta constitutes a significant fraction of the total energy budget. Also, in the presence of large losses, the temperature and the Lorentz factor of the fluid behind the shock can be non-monotonic functions of distance from the shock.

References

Baring, M. G.; Harding, A. K., 1995, Adv. Sp. Res. 15(5), p. 153-156

Blandford R. D., and McKee C. F., 1976, Phys. Fluids **19**, 1130.

Derishev, E. V.; Aharonian, F. A.; Kocharovsky, V. V.; Kocharovsky, Vl. V., 2003, Phys. Rev. D 68, 043003

Meszaros P., 2002, Ann. Rev. Astron. Astrophys. 40, p. 137-169
Piran, T., 2004, Rev. Mod. Phys., in press, astro-ph/0405503
Taub, A. H., 1948, Phys. Rev. 74, p. 328

THE DYNAMICS OF THE CARINA SPIRAL ARM

Smadar Naoz and Nir J. Shaviv
Racah Institute of Physics, Hebrew University of Jerusalem, Jerusalem 91904, Israel

Abstract   We map and study the dynamics of the Milky Way's spiral arms in our vicinity, using the birth place of open clusters. Since these objects are located in the galactic plane and are born primarily in the arms, they are an excellent tool for analyzing the dynamics of the spiral structure. Using their birth place, we find evidence for multiple spiral patterns with different speeds. The Carina arm appears now to be a superposition of both. The results are important for proper modeling of cosmic ray diffusion in our galactic vicinity.

Introduction

There is no doubt that the Milky Way is a spiral armed galaxy. Yet, our edge-on view and dust obscuration through the galactic disk precludes us from mapping the spiral arm structure directly. It is for this reason that it difficult to unequivocally determine the spiral arm nature of our galaxy. In fact, different claims for different spiral arm geometries and dynamics are abundant in the literature. Understanding the spiral arm dynamics is, however, an important problem in astronomy. Besides the inherent interest in knowing our galactic geography, the solution to this problem is relevant for the understanding of various aspects of our galaxy. In particular, cosmic ray acceleration takes place in the supernovae confined primarily to the galactic spiral arms. Thus, mapping the structure and dynamics of spiral arms is paramount to the prediction of cosmic ray propagation in the Milky Way.

The standard understanding today is that spiral structure can be regarded as a density wave, which revolves in the galactic disk. Lin and Shu, 1964 & 1966 were the first to apply wave mechanics techniques, obtain the *density wave theory* and deduce the properties of the density wave solutions. We work under the infrastructure set by the theory.

Spiral Structure Analysis

In this section, we will present the methods we used to map and study the dynamics of the Milky Way's spiral arms in our vicinity. In our analysis, we

M. M. Shapiro et al. (eds.), Neutrinos and Explosive Events in the Universe, 207–210.

used the Dias et al., 2002, and Loktin et al., 1997, databases of open clusters. These include the galactic coordinates and age estimates for open clusters at typical distances of up to a few kpc. We used Olling and Merrifield, 1998, for the rotation curves.

We work in cylindrical galactic-centered coordinates, r, ϕ and z. The solar system's position in the Milky Way has not been ascertained yet, it ranges between 7.1 kpc and 8.5 kpc. Similarly, the rotational velocities range between 184 and 200 $km\ sec^{-1}$ for a solar system at 7.1 kpc and $220 - 240\ km\ sec^{-1}$ for 8.5 kpc. We carried out all our analyses while assuming five possible locations and velocities, and working in a frame of reference rotating with the spiral pattern speed, Ω_P. In this frame, the differential velocity of each cluster is $\Delta\Omega = \Omega(r) - \Omega_P$. Furthermore, we consider only relatively young clusters, with ages $t \leq 50$ million years. This implies that we preform our analysis only on clusters which did not have time to move too far from the arm.

The most detailed $v - l$ maps of the Milky Way are the Dame et al., 2001, observations of molecular gas using CO as a tracer. In this data set, we can clearly identify two different arms (with a $\sim \pi/2$ phase separation, i.e., probably part of an $m = 4$ set): The Carina arm which is located inwards to the solar galactic radius $R_\odot$, and the Perseus arm, externally located relative to $R_\odot$. The solar system is located in the Orion arm. We perform our analysis on the *Sagittarius-Carina* arm (Naoz and Shaviv 2004), which has the richest cluster data.

We work in the frame of reference of the arms. In this frame of reference, the birth location is given by $\phi_{birth} = \phi - (\Omega - \Omega_P) \cdot t$, where t is the age of the clusters. $\Omega(r)$ was taken from Olling and Merrifield, 1998. We assume a logarithmic arm profile given by

$$r = e^{a\phi.\ \dots\ +b}. \tag{1}$$

Here a is related to the pitch angle i through the relation $i = \tan^{-1}(a)$. In other words, a describes the slope of the spiral arm in the $\ln r$ vs. ϕ_{birth} plot, while b is the interaction point with the $\ln r$ axis. We find a best fit for three parameters: Ω_P, the inclination angle i and the intersection point with the ϕ-axis. For each Ω_P we find the best fit, according to a weight function, $\mathcal{W}$, with which we evaluate the goodness of the fit. The definition we chose is

$$\mathcal{W}(a,b) = -\sum_{i=1}^{N} \max\left([1 - \frac{\Delta_i}{d}], 0 \right) \tag{2}$$

$$nonumber \Delta_i = |\ln r_i - (a\phi_{birth,i} + b)|, \tag{3}$$

where N is the number of data (i.e., cluster) points, and d is a scaling factor which satisfies the condition $d \geq \Delta_i$ for $i \in \{1, N\}$, such that the argument in the summation will be positive. The condition $\Delta_i \leq d = 0.05$ enables us to disregard data points which do not lie within ∼ 700 *pc*. We used a "brute force" method in which we simply go over the parameter space with a fine grid. This is possible only because we have 3 dimensions in the parameter space. Since we have neither good knowledge of the statistical properties of the weight function (eq. 3), nor of the statistical or physical variance in the cluster data itself, we used the *Bootstrap Method* (e.g., Press et al., 1986) to estimate the confidence levels around the obtained minima. This method allows us to evaluate the errors of the obtained results using the data itself, hence the name bootstrap.

We now proceed to apply the above methods to the Carina arm. We begin by calculating the best fit for the spiral arms as a function of the assumed Ω_p. We plot the minimized weight function $\mathcal{W}$ as function of Ω_p and find a bimodal pattern (see fig. 1). This can be interpreted as *two different of sets of spiral arms* in the Milky Way.

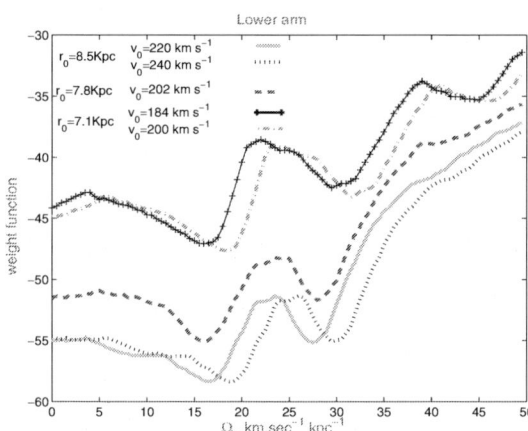

Figure 1. Weight function vs. the pattern speed, Ω_P, for the lower (Carina arm). The five graphs represent different assumed $R_\odot$ and $v_\odot$. Each graph is obtained by minimizing the weight function while fixing the value of Ω_p. The two minima in $\mathcal{W}$ demonstrate that there are probably two sets of spiral arms. An apparent feature is that pictures where the sun is placed at larger radii yields overall better fits. This strongly suggests that $R_\odot > 8\ kpc$. This however is still not substantiated and therefore requires further analysis.

In fig. 1, we also find that the more dominating minimum is of the lower pattern speed. The actual results for the fits to a logarithmic arm (eq. 1) are shown in fig. 2(a,b), while the results for second (higher) pattern speed are shown in fig. 2(c,d).

Summary

We have analyzed the Loktin et al., 1997, and the Dias et al., 2002, open cluster data in order to derive the pattern speed of the Sagittarius-Carina spiral arm. In our analysis, we obtained two arm sets that coexist in the vicinity of the Carina arm. In the "velocity spectrum" diagram, the lower Ω_P is more dominant, with lower values of the weight function $\mathcal{W}$. We find $\Omega_{P,\mathrm{Carina},1} = 16.5^{+1.2}_{-1.4 sys} \pm 1.1_{stat}\ km\ sec^{-1}\ kpc^{-1}$. For the higher pattern speed in the Carina arm, we find $\Omega_{P,\mathrm{Carina},2} = 29.8^{+0.6}_{-1.3 sys} \pm 1.3_{stat}\ km\ sec^{-1}\ kpc^{-1}$.

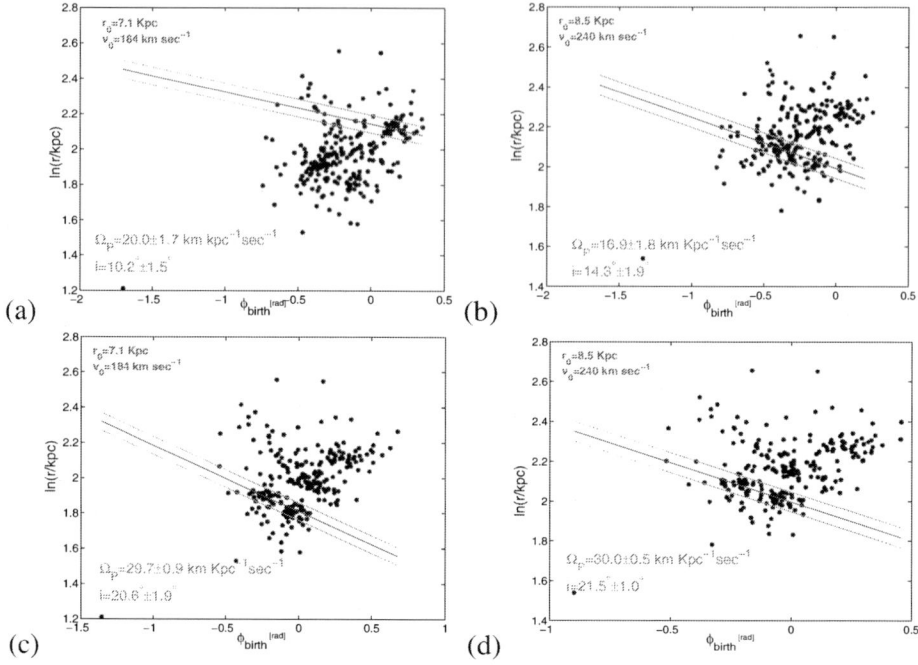

Figure 2.   (a),(b) The first set of arms for two limiting values of the solar system's location in the galaxy. (c), (d) The second set of arms. The lines depict the best fitting logarithmic arms, while the thiner lines are 700 *pc* strips delineating the clusters which enter the spiral arm weight function.

References

Dame, T. M., Hartmann, D., and Thaddeus, P. (2001). The Milky Way in Molecular Clouds: A New Complete CO Survey. *ApJ*, 547:792–813.

Dias, W. S., Alessi, B. S., Moitinho, A., and Lepine, J. R. D. (2002). Optically visible open clusters catalog (Dias, 2002). *VizieR Online Data Catalog*, 7229.

Lin, C. C. and Shu, F. H. (1964). On the Spiral Structure of Disk Galaxies. *Ap. J.*, 140:646.

Lin, C. C. and Shu, F. H. (1966). On the Spiral Structure of Disk Galaxies, II. Outline of a Theory of Density Waves. *Proceedings of the National Academy of Science*, 55:229–234.

Loktin, A. V., Matkin, N. V., and Gerasimenko, T. P. (1997). Catalogue of open cluster parameters from UBV-data. (Loktin, 1994). *VizieR Online Data Catalog*, 5096.

Naoz, S. and Shaviv, N. J. (2004). Open cluster birth place analysis and the Milky Way Spiral arm dynam ics. *Submitted to Astrophys. J.*

Olling, R. P. and Merrifield, M. R. (1998). Refining the Oort and Galactic constants. *MNRAS*, 297:943–952.

Press, W. H., Flannery, B. P., and Teukolsky, S. A. (1986). *Numerical recipes. The art of scientific computing.* Cambridge: University Press, 1986.

MAGNETIC FIELD GENERATION AND ELECTRON ACCELERATION IN COLLISIONLESS SHOCKS

Christian B. Hededal[1], Jacob Trier Frederiksen[2], Troels Haugboelle[1], Aake Nordlund[1]

[1]*Niels Bohr Institute for Astronomy, Physics, and Geophysics, Juliane Maries Vej 30, 2100 Koebenhavn Oe , Denmark*

[2]*Stockholm Observatory, Roslagstullbacken 21, 106 91 Stockholm, Sweden*

Abstract   Using a three dimensional relativistic particle–in–cell code we have performed numerical experiments of plasma shells colliding at relativistic velocities. Such scenarios are found in many astrophysical objects e.g. the relativistic outflow from gamma ray bursts, active galactic nuclei jets and supernova remnants. We show how a Weibel like two–stream instability is capable of generating small–scale magnetic filaments with strength up to percents of equipartition. Such field topology is ideal for the generation of jitter radiation as opposed to synchrotron radiation. We also explain how the field generating mechanism involves acceleration of electrons to power law distributions ($N(\gamma) \propto \gamma^{-p}$) through a non–Fermi acceleration mechanism. The results add to our understanding of collisionless shocks.

Keywords: Collisionless shock waves, particle acceleration, gamma ray bursts, plasma instabilities

Introduction

Many astrophysical objects emit non–thermal radiation when expelled plasma interacts with the surrounding media. These objects include gamma ray bursts (GRBs) and their afterglows, the jets from active galactic nuclei, jets from quasars and supernova remnants. The non–thermal radiation is believed to be emitted in strong collisionless shocks. Despite the vast prevalence and wide astrophysical applicability, collisionless shocks remain poorly understood. The observed non–thermal emission suggests that strong particle acceleration and magnetic field generation takes place in these shocks. It was suggested by Medvedev and Loeb, 1999 that a Weibel–like two–stream instability is able to generate a strong magnetic field in the shock transition region.

M. M. Shapiro et al. (eds.), Neutrinos and Explosive Events in the Universe, 211–215.

Recently, due to the increase in computer power, this has been verified with particle–in–cell (PIC) simulations (Frederiksen et al., 2003, Frederiksen et al., 2004, Hededal et al., 2004, Medvedev et al., 2004, Nishikawa et al., 2003, Nishikawa et al., 2004, Silva et al., 2003).

Still, todays computer power does not allow us to fully resolve the shock transition region. We can, however, explain the non-linear kinetic plasma dynamics that generate the electromagnetic fields needed for transmission of momentum between the colliding plasma populations. Here, we report on 3D PIC simulations of the shock formation in the counter–streaming region of two colliding plasma shells.

Numerical Experiments

We have performed numerical experiments using a three dimensional relativistic kinetic electromagnetic particle–in–cell code The code works from first principles by solving the Lorentz force equation for the particles and the Maxwell's equations for the electromagnetic fields.

We let two electron-proton plasma populations (with a density difference of a factor of three) collide in the reference frame of the denser population. In this frame we continuously inject the less dense population with a bulk Lorentz factor, $\Gamma = 15$ in the z-direction. The computational box consists of $125 \times 125 \times 2000$ gridzones or $37 \times 37 \times 600\Delta_e^3$ where Δ_e is the electron skin depth c/ω_e. Using 16 particles pr. cell this adds up to almost 10^9 particles.

Both populations have a rest frame temperature corresponding to a thermal velocity $v_{th} = 0.01c$. In order to be able to resolve both electron and ion dynamics the ion-electron mass ratio is $M_i/m_e = 16$. This is clearly a strong assumption but it allow us qualitatively to understand how particles with different mass affects the Weibel instability. In these experiments both populations are initially unmagnetized.

Results

When the simulation runs and the plasma populations stream through each other, we observe how the Weibel instability collects particles into current filaments (Medvedev and Loeb, 1999). First the electrons go through the instability and then further downstream the electrons thermalize to one single population and the heavier ions goes through the instability.

The ion filaments are more robust than the preceding electron filaments since the thermalized electron will Debye shield the ion filaments. Even further downstream the current filaments acts as 2D macro particles in the transverse plane and are themselves collected into larger filaments as explained by Frederiksen et al., 2004 and Medvedev et al., 2004. This behavior can be seen in fig. 1.

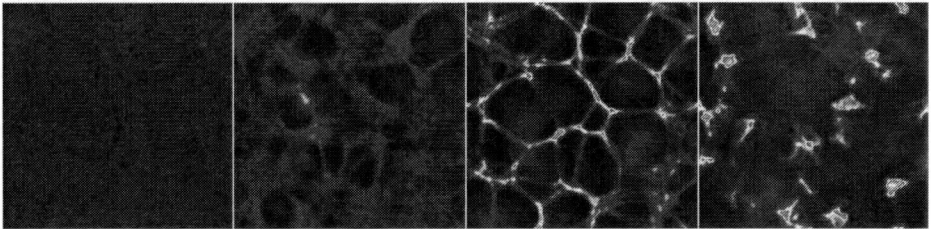

Figure 1.   This figure shows the generation of ion current filaments. Here we see the jet head on. The four slices show the ion current density at different depths of the shock at a fixed time. The different depths are z={60, 100, 120, 160} electron skin depths.

Surrounding the current filaments is generated a strong magnetic field with $\epsilon_B \sim 0.01$. Here, ϵ_B is the field generation efficiency i.e. the amount of total injected kinetic energy that goes into magnetic field energy. It is important to realize that ϵ_B is not a simple parameter since it varies strongly along the flow-direction down through the shock.

Inside the Debye sphere, strong electron acceleration takes place. The electrical field that surrounds the ion current channel accelerates the electrons toward the filaments where they are deflected on the induced magnetic field. The scenario is depicted in fig. 2. It have previously been shown that the ion filaments are generated in a self-similar coalescence process (Medvedev et al., 2004) which implies that a spatial Fourier decomposition exhibits power law behavior. As a result, the electrons are accelerated to a power law distribution function (fig. 3) as shown by Hededal et al., 2004.

The electrons are trapped in the potential field of the ion channels and are repeatedly accelerated and decelerated. This means that energy losses due to escape are small, and that the electrons remain trapped long enough to have time to loose their energy via a combination of bremsstrahlung and synchrotron or jitter radiation. The fact that the electrons cannot escape the ion filaments without being decelerated also implies that they are are not available for recursive acceleration as suggested in Fermi acceleration (Hededal et al., 2004).

Summary

We have performed numerical experiments of relativistic collisionless plasma shocks using a self-consistent three dimensional particle–in–cell code. We find that the Weibel instability is capable of generating turbulent magnetic fields with a strength up to percents of equipartition. The magnetic field is induced around ion current filaments. These filaments also accelerates electrons to power law distributions. The suggested acceleration scenario does not rule out ion Fermi acceleration but might overcome some of the problems pointed

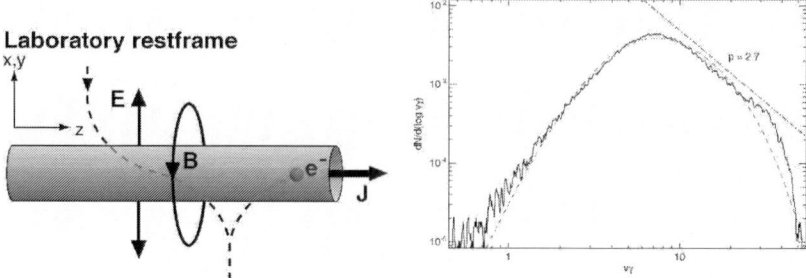

Figure 2.   An ion current channel surrounded by an electric – and a magnetic field.   Electrons in the vicinity of
the current channels are thus subject to
a Lorentz force with both an electric and
magnetic component, effectively accelerating the electrons.

Figure 3.   The normalized electron four
velocity distribution function downstream
of the shock.   The dot–dashed line is
a power law fit to the non–thermal high
energy tail, while the dashed curve is a
Lorentz boosted thermal electron population.

out by Baring and Braby, 2004 regarding the apparent contradiction between
standard Fermi acceleration of electrons and spectral observations of GRBs.

The microphysics in the field generation and particle acceleration described
here is clearly beyond the reach of the magneto hydrodynamic approximation.
A parameter study utilizing a PIC code working from first principles is necessary to fully understand the interdependence between the relative bulk Lorentz
factors of the colliding plasma shells, the power law index of the non–thermal
electron population, ϵ_B and in a broader sense the detailed evolution and structure in collisionless shocks.

References

Baring, M.G. and Braby, M.L. (2004). *ApJ*, 613:460–476.

Frederiksen, J. T., Hededal, C. B., Haugboelle, T., and & Nordlund, Aa (2003). *Proceedings of
the 2002 Niels Bohr Summer Institute, ArXiv Astrophysics e-prints*, astro-ph/0303360.

Frederiksen, J. T., Hededal, C. B., Haugboelle, T., and Nordlund, Aa. (2004). *ApJL*, 608:L13–
L16.

Hededal, C. B., Haugboelle, T., Frederiksen, J. T., and & Nordlund, Aa (2004). *Submitted to
ApJL*, astro-ph/0408558.

Medvedev, M. V., Fiore, M., Fonseca, R., Silva, L., and Mori, W. (2004). *submitted to ApJ*,
astro-ph/0409382.

Medvedev, M. V. and Loeb, A. (1999). *ApJ*, 526:697–706.

Nishikawa, K.-I., Hardee, P., Richardson, G., Preece, R., Sol, H., and Fishman, G. J. (2003).
ApJ, 595:555–563.

Nishikawa, K.-I., Hardee, P., Richardson, G., Preece, R., Sol, H., and Fishman, G. J. (2004). *Submitted to ApJ*, astro-ph/0409702.

Silva, L. O., Fonseca, R. A., Tonge, J. W., Dawson, J. M., Mori, W. B., and Medvedev, M. V. (2003). *ApJL*, 596:L121–L124.

III

HIGH ENERGY GAMMA RAY AND NEUTRINO ASTRONOMY

HIGH ENERGY NEUTRINO ASTRONOMY AND UNDERWATER DETECTORS

Emilio Migneco
Dept. of Physics and Astronomy,
University of Catania and
Laboratori Nazionali del Sud - INFN
Via S. Sofia 62 I-95123 Catania, Italy
migneco@lns.infn.it

Giorgio Riccobene
Laboratori Nazionali del Sud - INFN
riccobene@lns.infn.it

Abstract   Neutrinos are considered promising probes for high energy astrophysics. Many indications suggest, indeed, that cosmic objects where acceleration of charged particles takes place, e.g. GRBs and AGNs, are the sources of the detected UHE-CRs. Accelerated hadrons, interacting with ambient gas or radiation, can produce HE neutrinos. Differently from charged particles and gamma rays having $E_\gamma >$TeV, neutrinos can reach the Earth from far cosmic accelerators, traveling in straight line, therefore carrying direct information on the source. Theoretical models indicate that a detection area of $\simeq 1$ km$^{\cdot}$ is required for the measurement of HE cosmic ν fluxes. The underwater/ice Cherenkov technique is widely considered the most promising experimental approach to build high energy neutrino detectors. The first generation of underwater/ice neutrino telescope, BAIKAL and AMANDA, despite their limited size have already set first constraints on TeV neutrino production astrophysical models. The quest for the construction of km$^{\cdot}$ size detectors have already started: in the South Pole the ICECUBE neutrino telescope is under construction; the ANTARES, NEMO and NESTOR Collaborations are working towards the installation of a neutrino telescope in the Mediterranean Sea.

Keywords: high energy neutrino astronomy, underwater telescopes

M. M. Shapiro et al. (eds.), Neutrinos and Explosive Events in the Universe, 219–241.
© 2005 *Springer. Printed in the Netherlands.*

Introduction to high energy neutrino astronomy

MeV neutrino astrophysics

Solar and Supenovae physics have indicated the peculiar role of neutrinos in the investigation of burning processes in stars, opening the way to the so called *low energy neutrino astrophysics*. The first measurement of neutrino fluxes produced in the Sun by nuclear fusion chains, was presented by Davis's group in 1968 [1]. Davis experimental setup, the very first underground detector for astrophysical neutrinos, was a tank containing 615 tonnes of C_2Cl_4, a common liquid cleaner, located at 1500 m depth in Homestake Gold Mine (USA). Neutrino interactions were identified measuring the number of radioactive ^{37}Ar nuclei produced in the detector through the reaction $\nu+^{37}\text{Cl}\rightarrow^{37}\text{Ar}+e^-$ ($E_{threshold}$ =0.81 MeV). The results was an experimental upper limit of 3 Solar Neutrino Units (1 SNU = 1 capture per second per 10^{36} target atoms) well below the theoretical expectations based on the Solar Standard Model [2]. The solar neutrinos flux deficit was related to the problem of neutrino oscillation, which has been verified only in recent days by kton scale underground detectors studying MeV neutrinos from the Sun (Superkamiokande [3] and SNO [4]), MeV ν produced at nuclear reactors (Kamland [5]) and atmospheric neutrinos in the GeV energy range (Superkamiokande, Macro [6]).

The birth of experimental neutrino astronomy is, however, commonly associated with the detection of anti neutrinos from SN 1987A by IMB (8 events [8]) and Kamiokande (12 events [10]) on March 10th 1987. The time coincidence between neutrinos and photons was a clear signature of the event. Kamiokande [9], a tank of 2140 tonnes of hyper-pure water surrounded by 1100 large photocatode PMTs built in Kamioka mine (Japan), permitted also the identification of the direction of SN1987A neutrinos. The detector, in fact, reconstructs the Čerenkov light front produced by neutrino-induced electrons, through the reaction $\overline{\nu_e} + p \rightarrow n + e^-$: an excess of $E_e > 10$ MeV electrons was measured in the direction of the Large Magellanic Cloud.

These results opened the way to neutrino astronomy and also to the use of cosmic ν to study fundamental particle physics in phase-space regions not accessible to accelerators based experiments.

High energy cosmic rays

Cosmic rays (CR), whose first studies date back to the beginning of XX century, are still a puzzling subject for physicists. Up to date measurements show that CR flux extends over 10 orders of magnitude in energy, up to 3×10^{20} eV, and over 28 orders of magnitude in flux, down to few particles per 100 km^2 per century (see figure 1).

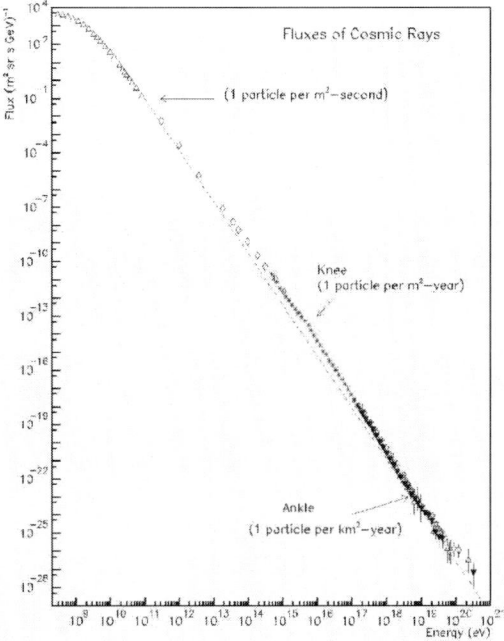

Figure 1. All particle cosmic ray spectrum from [11].

While the lower energy region of CR spectrum (E_{CR} <GeV) is well explained by solar activity, at higher energy there is not direct evidence of connection with sources. Cosmic Ray flux is mostly composed by charged particles and Galactic magnetic field ($B \sim 3\ \mu$G) randomises their arrival direction at the Earth. This implies that the reconstruction of charged CR direction is not affordable up to $E \simeq 10^{20}$ eV. On the other hand neutrons have a decay length which is too short to reach the Earth from the originating source (roughly 10 kpc at $E_\nu \simeq 10^{18}$ eV).

The bulk of CR spectrum is understood in terms of the Fermi acceleration mechanism [12], whose theoretical description was revised in a more effective version by Bell [13]. It takes place in sources where plasma (e^+e^- or/and pe) contained by strong magnetic fields, is driven by strong shock waves. The spectrum of Fermi accelerated particles follows an $E^{-(2\div 2.2)}$ power law and the maximum energy that a particle can reach is a function of confinement time within the shock:

$$E_{max} \approx \beta_{shock}\ Z\ B\ R, \tag{1}$$

where Z is the nucleus atomic number, $\beta_{shock} \times c$ is the shock wave velocity, B and R are the source magnetic field and the source linear extension respectively. Plugging in 1, B in units of μG and L in kpc, finds E_{max} in units of 10^{18} eV. From the same relation one can deduce that Galactic sources ($B_{Galaxy} \simeq 3\mu G$, $R \simeq 100$ pc) cannot accelerate protons to extremely high energies.

The experimental CR spectrum shows a peculiar behaviour: at $E > 10^{14.5}$ eV the CR flux changes both its spectral index (from $\alpha \simeq 2.7$ to $\alpha \simeq 3.0$) and its composition, from proton-dominated to nuclei-dominated: the higher is energy, the heavier are nuclei. This region is called the *knee*. At $E > 10^{18.5}$ eV (the *ankle* region) the CR spectrum features change again. Above this energy the CR flux similar to the pre-knee region: the spectral index 2.7 and flux is proton-dominated. The standard paradigm is that CR flux below the ankle is originated by galactic sources and the change in chemical composition is attributed to the escape of HE protons from the the Milky Way[1]. Since Galactic sources cannot accelerate particles to extremely high energies (as shown in equation 1), the detection of cosmic protons with energies up to $E > 10^{19}$ eV suggests the presence of extragalactic sources in which the Fermi acceleration mechanism takes place.

According to equation 1 there are only few classes of cosmic objects capable to accelerate protons at $E >$EeV, among these Gamma Ray Bursters (GRB) [16] and powerful Active Galactic Nuclei (AGN) [17] are the most favourable candidates. These sources, the most luminous bursting ($L_{GRB} \simeq 10^{53}$ erg/sec) and steady ($L_{AGN} \simeq 10^{46}$ erg/s) objects in the Universe, are typically located at cosmological distance. Protons having $E > 10^{20}$ eV possibly accelerated in AGNs and GRBs, could be good astrophysical probes being only slightly bent by cosmic magnetic fields but, at these energies, they are absorbed in the Universe through the interaction with Cosmic Microwave Background Radiation (CMBR), the so called GZK effect [18]. The cross section of the process $p + \gamma \rightarrow \Delta^{+}$ is $\sigma_{p\gamma} \sim 100$ μbarn and the average CMBR density is $n_{CMBR} \sim 400$ cm^{-3}, therefore the absorption length of UHE protons in the Universe is roughly:

$$L_{p,CMBR} \simeq (\sigma_{p\gamma} \cdot n_{CMBR})^{-1} < 50 \text{ Mpc}. \qquad (2)$$

High energy (TeV) gamma rays are also produced in the discussed sources. In fact, protons accelerated via Fermi mechanism can interact, within the source or in its vicinity, with gas clouds or inter-stellar medium (pp) and with ambient radiation ($p\gamma$) producing pions; neutral pions then decay into high energy γs. HE gamma rays can also be generated in a different, purely electromagnetic, way (electron Bremstrahalung and Inverse Compton Scattering), as supposed in the case of close (≤ 100 Mpc) AGNs Mkn-421 and Mkn-501 [19]. However the detection of $E_\gamma \simeq 10$ TeV rays from point-like sources is limited to few tens Mpc, due to pair production interaction of HE gamma rays with diffuse infrared and microwave cosmic background.

In this scenario only high energy neutrinos, produced in cosmic sources by charged pion decay, offer the possibility to directly observe TeV÷PeV radiation emitted by far cosmic objects and to disentangle between the occurrence of purely electromagnetic or hadronic processes. Thus, the observation of cosmic HE neutrinos can probe hadronic processes and hopefully extend our horizon in the far Universe.

High energy neutrinos

Hadronic source models predict that, in cosmic objects, high energy pions are produced (directly or through Δ resonance) by pp or $p\gamma$ interactions (photomeson production). While neutral pions decay into gamma rays, neutrinos are generated through decay chains of charged pions. If muons cooling time in the source is larger than their decay time, high energy electron neutrinos are also produced:

$$
p + p, \gamma \rightarrow N +
\begin{array}{l}
\pi_0 \rightarrow \gamma + \gamma \\[1em]
\left.\begin{array}{l} \\ \pi \\ \end{array}\right] \\[1em]
\pi^{\pm} \rightarrow \nu_\mu + \mu \\[1em]
\left.\begin{array}{l} \\ \end{array}\right] \\
\mu \rightarrow e + \nu_\mu + \nu_e.
\end{array}
\tag{3}
$$

In a first guess, the source ν spectrum is expected to follow the primary protons one (E^{-2}), the fraction of proton energy that goes into pions is roughly 0.2 and $E_\nu \simeq 0.05 E_p$. The reaction chains 3 also predict that ν_μ and ν_e are produced in a ratio of 2:1. Taking into account $\nu_\mu \leftrightarrow \nu_\tau$ oscillations and assuming present experimental values of Δm^2 and $\sin^2 \theta$ [20] equipartition between the three leptonic flavours is expected at the Earth.

A number of astrophysical high energy neutrino sources have been suggested as neutrino candidates. Indeed, $p\gamma$ interactions are expected to occur in several astrophysical environments.

In Supernova remnants (SNRs) protons, accelerated through Fermi mechanism, can interact with gas in dense SN shells, producing both neutral and charged pions [31]. Decay of neutral pion originates gamma rays (see 3). The observation of $\simeq$ 10 TeV gamma rays from Supernova Remnant RXJ1713.7-3946 claimed by the CANGAROO air Čerenkov telescope Collaboration [21], seems to validate this hypothesis, but the result is still under discussion[2]. The neutrino flux, produced by charged pion decay, is expected to be similar to the hadronic high energy gamma one.

Recent results from HESS air Cherenkov telescope array Collaboration [22] concern the observation of TeV gamma ray from the Sgr A* region (the Galactic centre). Preliminary results show an E^{-2} gamma spectrum that could be originated by hadronic interactions, an interpretation which also implies the production of intense high energy neutrino fluxes [23]. Photomeson $p\gamma$ interactions can occur in astrophysical environments that show dense low energy photons fields: microquasar (μQSO) jets [24], AGNs (blazars and BL Lacs [15, 37]) and in GRBs [25] are examples. AGNs and GRBs are particularly relevant since they are the candidate sources of UHECR. Starting from this hypothesis Waxman and Bahcall set an upper bound (the so called WB limit) to high energy neutrino fluxes that can reach the Earth. The limit is obtained assuming that the energy density injection rate of $10^{19} \div 10^{21}$ eV CRs is $\simeq 10^{44.5}$ erg Mpc^{-3}year^{-1}, assuming that particles are accelerated at their sources with E^{-2} spectrum, and posing, conservatively, that pions carry a fraction of proton energy $f_\pi < 1$ and neutrinos carry about 1/4 of this amount. The resulting upper limit for diffuse neutrino flux is:

$$E_\nu^2 \Phi_\nu \simeq \frac{c}{4\pi} \frac{f_\pi}{4} E_p^2 \left(\frac{dN_p}{dE_p dt}\right) t_{Hubble} \simeq 10^{-7.5} \text{GeV} /(\text{cm}^2 \text{ s sr}) \qquad (4)$$

This limit has been discussed and recalculated by Mannheim, Protheroe, Rachen [26] and, again, by Waxman and Mannheim [27], taking into account the effect of propagation of CR and of cosmological evolution of the source distribution. However equation 4 limit sets a strong reference value for the discussions on the dimensions of future neutrino telescopes.

Astrophysical objects that do not contribute to UHECR spectrum are not constrained by the WB limit. In *optically thick* sources (for which the optical depth is $\tau_{p\gamma} \equiv R_{source} \cdot (\sigma_{p\gamma} n_\gamma) \gg 1$) all nucleons interact while neutrinos can escape giving rise to a ν flux not constrained by relation 4. Galactic objects also are not included in the WB constraints, μQSO jets [28] and plerions [29] are examples: in both cases the maximum energy for Fermi protons is well below 10^{19} eV. The advantage of Galactic sources observation is due to the fact that though they are much fainter than extragalactic sources ($L_{\mu QSO} \simeq 10^{33}$ erg/s) they are close to the Earth and can therefore produce observable neutrino fluxes. In μQSO, for instance, photomeson interaction of PeV protons on ambient X synchrotron radiation could produce directional neutrino fluxes at Earth [28]. Waxman-Bahcall limit does not apply also to a different kind of processes, known as *top-down*, which foresee the production of high energy CR, gammas and neutrinos by the decay or annihilation of particles with mass $M_X > 10^{21}$ eV, relics of the primordial Universe such as Topological Defects or GUT scale WIMPS (for an exhaustive review see [30]).

Neutrino telescopes

As shown in the previous section, light and neutral neutrinos are optimal probes for high energy astronomy, i.e. for the identification of astrophysical sources of UHE particles.To fulfill this task neutrino detectors must be design to optimise reconstruction of particle direction and energy, thus they are commonly referred as *neutrino telescopes* (for a clear review see [31]).

HE neutrinos are detected indirectly following weak Charged Current (CC) interaction with nucleons in matter and the production of a charged lepton in the exit channel of the reaction. The low νN cross section ($\sigma_{\nu N} \simeq 10^{-35}$ cm^2 at $\simeq 1$ TeV) and the expected astrophysical ν fluxes intensities require a ν interaction target greater than 1 GTon. Markov and Zheleznykh proposed the use of natural water (lake or seawater or polar ice) to detect neutrinos [32] using the optical Čerenkov technique to track the charged lepton outgoing the νN interaction. Underwater neutrino telescopes are, then, large arrays of optical sensors (typically photomultipliers tubes of about 10" diameter) which permit charged leptons tracking in water by timing the Čerenkov wavefront emitted by the particle (see figure 2).

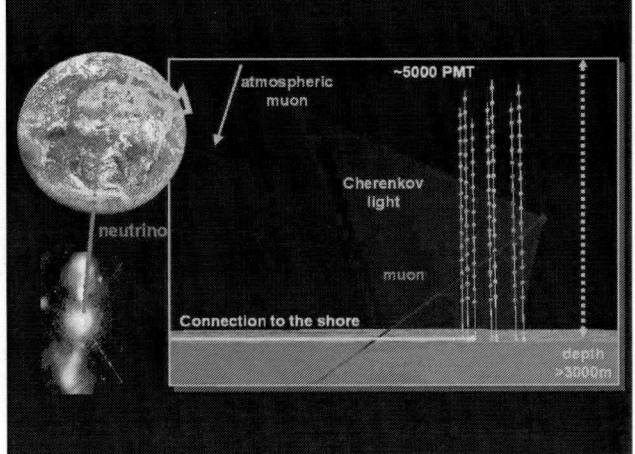

Figure 2. Detection principle of an underwater neutrino telescope. Astrophysical neutrinos can reach the Earth and interact in water or in rocks generating an upgoing muon. An array of $\simeq$ 5000 optical detectors tracks Čerenkov photons generated along the muon track. A water shielding ≥ 3000 m is effective to reduce the atmospheric μ background, allowing the reconstruction of upgoing muon tracks.

When an upward going particle is reconstructed this is a signature of neutrino event, being the atmospheric upgoing muon background completely filtered by the Earth. Seawater has a threefold use: huge (and inexpensive) neutrino target, Čerenkov light radiator and shielding for cosmic muon background.

Neutrino detection

Neutrino weak interaction cross section is dominated at $E_\nu > 100$ GeV by charged current Deep Inelastic Scattering (DIS) [33]:

$$\nu_l + N \to l + X \qquad (5)$$

where l is the lepton flavor ($l = e, \mu, \tau$), N represents the hit nucleon, X the outgoing hadron(s). The $\nu + N$ CC cross section increases linearly with neutrino energy up to $\simeq 5$ TeV energy, above this value its slope changes to $E^{0.4}$, as shown in figure 3. This leads to two implications: a) the number of detectable neutrino-induced events increase with energy; b) the absorption length of neutrinos in the Earth

$$L_{\nu N} \simeq (\sigma_{\nu N}^{total} < \rho Earth >)^{-1} \qquad (6)$$

is, for $E \geq 100$ TeV, comparable to the Earth diameter.

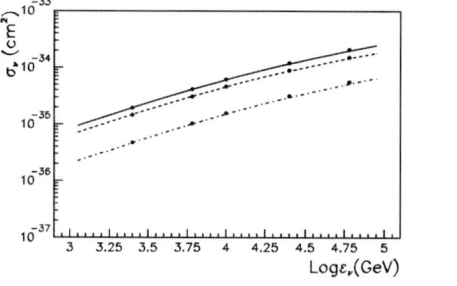 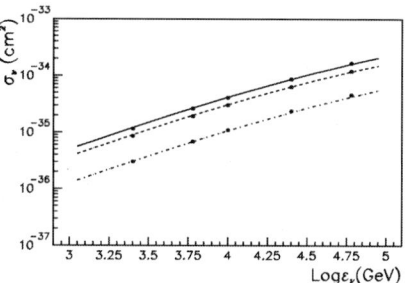

Figure 3. Left Panel - Total (solid line) cross section for νN scattering in matter, Charged Current (dashed) and Neutral Current (dash-dotted) cross sections are also indicated. Right Panel - Same plot for $\bar{\nu} N$ scattering . For details see reference [33]

Underwater/ice neutrino telescopes are expected to identify lepton flavour by reconstructing event topology due to different propagation of e, μ and τ in water. Among different flavours, muon detection is favoured. Muons take, in average $50 \div 60\%$ of neutrino energy and μ-range in water is, at $E \simeq$ TeV, of the order of kilometres (see figure 4). Muon neutrino detection allows neutrino astronomy: the angle between the outgoing muon and the interacting neutrino decreases as a function of neutrino energy (see figure 4). At high energy muon track is therefore almost co-linear to the neutrino one and allows pointing back to the ν cosmic source.

In order to calculate the number of detectable events expected by cosmic neutrinos it is important to introduce the quantity $P_{\nu\mu}$, which is the probability

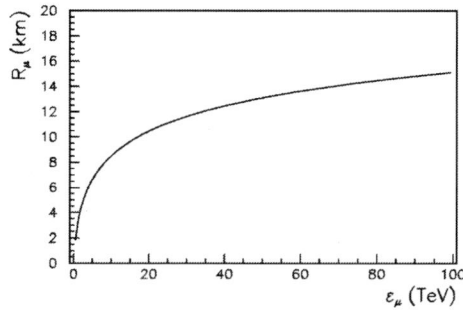
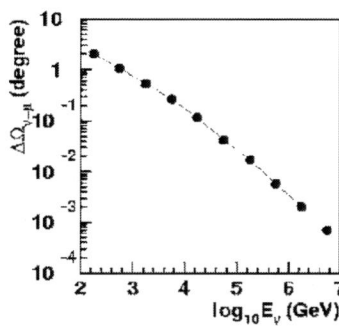

Figure 4. Left Panel - Average muon range in water as a function of muon energy. Right Panel - Median of the distribution $\Delta\Omega_{\nu-\nu_\mu}$ (muon exit angle in space with respect to the ν_μ direction) as a function of neutrino energy.

to convert neutrinos into detectable muons. $P_{\nu\mu}$ is a function of the neutrino interaction cross section and of the average muon range $R(E_\mu, E_\mu^{min})$:

$$P_{\mu\nu}(E_\nu) = N_A \int_{E_\mu^{min}}^{E_\nu} dE\mu \frac{d\sigma_{\nu N}^{CC}}{dE\mu} R(E_\mu), \qquad (7)$$

where N_A is the Avogadro number and E_μ^{min} is the minimum detectable muon energy, or detector threshold. For a given detector the value of $P_{\mu\nu}$ has to be calculated via simulations [31]. As a rule of thumb $P_{\mu\nu} \simeq 1.3 \times 10^{-6}$ for TeV neutrinos ($E_\mu^{min} = 1$ GeV) and increases with energy[3] as $E^{0.8}$. Since, in the interesting energy range, the muon range is of the order of kilometres, and even larger, detectable muons can be originated far from the detector horizon. The parameter usually quoted to describe detector performances is the detector *effective area* A_{eff} for muons, i.e. the surface intersecting the neutrino-induced muon flux folded with the detection efficiency for muons. The rate of events produced by a neutrino flux $\Phi_\nu(E_\nu, \vartheta)$ per unit of detector effective area, is then expressed by:

$$\frac{N_\mu(E_\mu^{min}, \vartheta)}{A_{eff} \, T} = \int_{E_\mu^{min}}^{E_\nu} dE_\nu \Phi_\nu(E_\nu, \vartheta) P_{\nu\mu} e^{-\frac{Z \cdot \vartheta \cdot}{L_{\nu N} \cdot E_\nu \cdot}}, \qquad (8)$$

being $L_{\nu N}$ the neutrino absorption length in the Earth and $Z(\vartheta)$ the Earth column depth.

Plugging the WB limit flux (see formula 4) into equation 8 and integrating over solid angle, one gets a rate of about 10^2 upgoing events per year for a 1 km^2 effective area detector with $E_\mu \simeq 1$ TeV threshold. This number set the scale of dimension for astrophysical neutrino detectors.

Besides these isotropically distributed events, a number of point-like sources can produce cluster of ν events. Bright AGN blazars could produce intense ν fluxes. Atoyan and Dermer estimated a high energy neutrino flux as high as $E_\nu^{-2}\Phi_\nu \simeq 10^{-10}$ erg cm^{-2}s^{-1} for 3C273 [37]), which can produce $\simeq 10$ detectable muons per year in a km^3 telescope. Close and/or intense GRBs, like the well known GRB-030329, are also expected to be detected as point sources (recent papers foresee some muon events per burst [34] in a km^2). In the GRB case time and direction coincidence between νs and MeV γs will permit effective atmospheric ν background rejection (see next sections), enhancing the detector discovery capabilities. Bednarek calculated a neutrino flux from the Crab SNR of $E_\nu^{-2}\Phi_\nu \simeq 10^{-10}$ erg cm^{-2}s^{-1}, this flux could produce few muon events per year in a km^2 detector [35]. A larger number of events, up to some tens or hundreds, is expected from several Galactic microquasars; SS433 and GX339-4, in particular, are expected to emit ν fluxes of the order 10^{-9} erg cm^{-2}s^{-1} [28] .

Underwater Cherenkov technique

The underwater Čerenkov technique allows the tracking of charged relativistic particles. In water (whose refractive index for blue light is, $n \simeq 1.35$) Čerenkov photons are emitted along particle track at $\vartheta_{\check{C}} \simeq 42^o$. The time sequence of photons hits on PMTs is correlated by the causality relation (see figure 5):

$$c(t_j - t_0) = l_j + d_j \tan(\vartheta_{\check{C}}). \qquad (9)$$

The space-time pattern of Čerenkov wavefront can be reconstructed during off-line analysis fitting relation 9. The reconstructed muon direction will be affected by indetermination on PMTs position (due to underwater position monitoring) and on hit time (PMT transit time spread, detector timing calibration,...).

Particle energy loss via Čerenkov radiation is only a negligible fraction of the total one and the number of Čerenkov photons emitted by a charged relativistic particle in water is roughly 300 per cm of track[4]. Simulations show that an underwater detector having an instrumented volume of about 1 km^3 equipped with $\simeq$5000 optical modules can achieve an affective area of $\simeq 1$ km^2 and an angular resolution of $\simeq 0.1°$ for $E_\mu > 10$ TeV muons [36].

Water(ice) optical properties determine the detector granularity (i.e. the PMT density). As shown in figure 6, water is transparent only to a narrow range of wavelengths ($350 \leq \lambda \leq 550$ nm). For blue light, such as Čerenkov radiation, the absorption length of clear ocean is $L_a \simeq 70$ m. This number roughly sets the spacing distance between PMTs, thus $\simeq 5000$ PMTs could fill up a volume of one km^3.

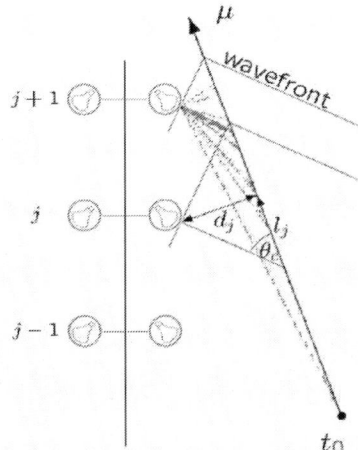

Figure 5. The lepton Čerenkov wavefront is reconstructed using the information on photon hit and PMT positions (see equation 9).

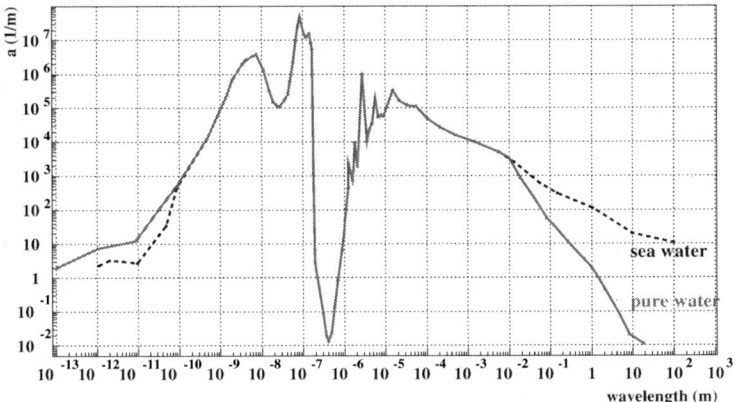

Figure 6. Light absorption spectrum as a function of wavelength for pure water (solid line) and seawater (dashed line). The absorption coefficient $a(\lambda)$ is defined by the law $I(x, \lambda) = I \cdot (x, \lambda)e^{-a \cdot \lambda \cdot x}$ [38]. The light absorption length is then defined as $a(\lambda) \equiv 1/L_a(\lambda)$.

The estimate of the performances of the detector requires detailed Montecarlo simulations that have to take into account the detector layout and the physical characteristics of the Čerenkov radiator which surrounds the detector: light refraction index, light absorption and scattering coefficients, in seawater (or ice). These quantities must be accurately measured *in situ* [38].

Background

Neutrino detectors must identify faint astrophysical neutrino fluxes among a diffuse atmospheric background. The cosmic muon flux, which at sea surface is about 10 orders of magnitude higher than the number of neutrino-induced upgoing muons, strongly decreases below sea surface as a function of depth and as a function of zenith angle: it falls to zero near the horizon and below. This is the reason why astrophysical neutrino signals are searched among upward-going muons. At 3000 m depth, an underwater neutrino telescope is hit by cosmic muon flux still about 10^6 times higher than the upgoing atmospheric neutrino signal, therefore accurate reconstruction procedures are needed to avoid the mis-reconstruction of downgoing tracks as *fake* upgoing.

In the case of the background produced by atmospheric neutrinos, energy cuts and statistical arguments can be used to discriminate these events from astrophysical ones during data analysis. In fact, atmospheric ν flux is expected to produce diffuse events with a known spectral index ($\alpha \simeq -3.7$ at $E_\nu >$ 10 TeV) while neutrino fluxes coming from astrophysical point sources are expected to follow an E^{-2} and to be concentrated within a narrow angular region, in the direction of their source, whose dimension is essentially given by the detector angular resolution.

Another background source is the optical noise in seawater. This background is due to the presence of bioluminescent organisms and radioactive isotopes. Radioactive elements in water (mainly ^{40}K) originate electrons above the Čerenkov threshold. ^{40}K decay produces an uncorrelated background on PMTs that has been measured to be about 20÷30 kHz for 10" PMT (at 0.5 single photoelectron -s.p.e.- threshold) [39]. These signals must be eliminated by the event trigger and reconstruction algorithms. Optical noise is also due to bioluminescent organisms living in deep water. These organisms (from small bacteria to fishes) produce long lasting ($\simeq 10^{-3}$ s) bursts of light that saturate close PMTs for the period of emission. In oceanic deep seawater (as measured at 3000 m depth in the Ionian Sea Plateau) bioluminescent signals are rare (few per hours) and do not affect the average optical noise rate on PMTs. On the contrary, in biologically active waters, bioluminescence signals may produce an intense background noise up to several hundreds kHz (on 10" PMTs, 0.5 s.p.e) [39]. This high rates strongly worsen the telescope track reconstruction capabilities and, in the worst case, could not be sustainable by data-rate transmission.

Future cubic kilometre arrays

As shown in previous sections, the expected number of detectable high energy astrophysical neutrino events is of the order of $10 \div 100$ per km^2 per year [31, 25] and only detectors with an effective area (A_{eff}) of 1 km^2 scale could allow the identification of their sources. Starting from the second half of '90s, two small scale neutrino telescopes AMANDA [45] ($A_{eff} \simeq 0.1$ km^2) and BAIKAL [43] ($A_{eff} \simeq 10^4$ m^2) have demonstrated the possibility to use underwater/ice Čerenkov technique to track $E_\nu > 100$ GeV neutrinos, measuring the atmospheric neutrino spectrum at high energies. This success opened the way to the construction of km^3 underwater neutrino telescopes: ICECUBE [47], that will extend the AMANDA detectors at South Pole, is going to be deployed starting from austral summer 2004-2005, to be completed within 2010. In the Northern Hemisphere three collaboration ANTARES, NEMO and NESTOR, which are building demonstrator detectors and prototypes, aim at the construction of the km^3 detector in the Mediterranean Sea.

The running neutrino telescopes: Baikal-NT and AMANDA

After the pioneering work carried out by the DUMAND collaboration offshore Hawaii Island [41, 42], Baikal was the first collaboration which installed an underwater neutrino telescope and, after more than ten years of operation, it is still the only neutrino telescope located in the Northern Hemisphere. The BAIKAL NT-200 is an array of 200 PMTs, moored between 1000 and 1100 m depth in lake Baikal (Russia) [43]. The deployment and recovery operations are carried out during winter, when a thick ice cap (about 1 meter) is formed over the lake. BAIKAL is an high granularity detector with a threshold $E_\mu \simeq$ 10 GeV and an estimated effective detection area $\leq 10^5$ m^2 for TeV muons. The limited depth and the qualities of lake water (light transmission length of $15 \div 20$ m, high sedimentation and bio-fouling rate, optical background due to bioluminescence) limit the detector performances as a neutrino telescope. The collaboration has presented data analysis of ν_e events, which set an upper limit for diffuse astrophysical neutrino fluxes $E_\nu^2 \Phi_\nu < 4 \times 10^{-7}$ cm^{-2} sec^{-1} sr^{-1} GeV (at 90% CL, not including systematic uncertainties) [44].

AMANDA (for a detailed discussion on AMANDA and ICECUBE see A. Silvestri lecture in this School [46]) is currently the largest neutrino telescope installed. In the present stage, named AMANDA II, the detector consists of 677 optical modules (OM) pressure resistant glass vessel hosting downward oriented PMTs and readout electronics. OMs are arranged in 19 vertical strings, deployed in holes drilled in ice between 1.3 and 2.4 km depth. Vertical spacing between OMs is $10 \div 20$ m, horizontal spacing between strings is $30 \div 50$ m. The ice optical properties have been mapped as a function of depth: at detector installation depth the average light ($\lambda = 400$ nm) absorption length is $L_a \simeq 100$ m (in the ocean $L_a \simeq 70$ m), the effective light scattering length is $L_b \simeq 20$ m ($L_b > 100$ m in the ocean). This makes AMANDA a good calorimeter for astrophysical events, with a resolution 0.4 in log(E) for muons and 0.15 in log(E) for electron cascades. The detector angular resolution is between $1.5°$ and $3.5°$ for muons and $\simeq 30°$ for cascades. AMANDA data have permitted to measure for the first time the upgoing atmospheric neutrino spectrum in the energy range from few TeV to 300 TeV. For diffuse astrophysical muon neutrino fluxes ($100 < E_\nu < 300$ TeV) a 90% upper limit $E_\nu^2 \Phi_{\nu_\mu} < 2.6 \times 10^{-7}$ GeV cm^{-2} s^{-1} sr^{-1} was set. For point sources the detector reached a sensitivity $E_\nu^2 \Phi_{\nu_\mu} \simeq 7 \times 10^{-8}$ GeV cm^{-2} s^{-1} calculated, over 807 days live time (years 2000-2003). Among a sample of 3369 neutrino candidates (3438 expected atmospheric ν) no significant excess due to astrophysical point sources has been detected.

ICECUBE

The ICECUBE telescope will be the natural extension of AMANDA to the km^3 size. When completed (expected in 2010) it will consist of 4800 PMT displaced in 80 strings. All the PMTs will be downward looking and simulations show that an average spacing of 125 m between PMTs is a good compromise between the two requirements of angular resolution $\leq 1°$ ($E_\nu > 1$ TeV) and effective area $\simeq 1$ km^2. Simulations run by the ICECUBE collaboration show that in three years of live time the detector will reach a sensitivity $E^2 \Phi_\nu = 4 \times 10^{-9}$ GeV cm^{-2} s^{-1} sr^{-1} for a diffuse E^{-2} neutrino spectrum [47]. It is also worthwhile to mention that the under-ice detectors are not affected by radioactive and biological optical noise. This makes them suitable for the search of low energy neutrino fluxes from Galactic SuperNova explosions.

The Mediterranean km^3

The contemporary observation of the full sky with at least two neutrino telescopes in opposite Earth Hemispheres is an important issue for the study of transient phenomena. Moreover ν events detection from the Northern Hemisphere is required to observe the Galactic Centre region (not seen by ICE-CUBE), already observed by HESS as an intense TeV gamma source. In the Norther Hemisphere a favourable region is offered by the Mediterranean Sea, where several abyssal sites (> 3000 m) close to the coast are present and where it is possible to install the detector near scientific and industrial infrastructures.

An underwater detector offers, compared to ICECUBE, the possibility to be recovered, maintained and/or reconfigured; detector installation at depth ≥ 3500 m will reduce atmospheric muon background by a factor ≥ 5 with respect to 2000 m depth. The long light scattering length (L_b) of the Mediterranean abyssal seawater preserve the Čerenkov photons directionality and will permit excellent pointing accuracy (order of $0.1°$ for 10 TeV muons). On the other hand the light absorption length in water (L_a) is shorter than in ice, then it reduces the photon collection efficiency for a single PMT. Differently from deep polar ice, the sea is a biologically active environment where organisms produce background light (bioluminescence).The selection of a marine site with optimal oceanographic and optical parameters is, therefore, a major task for the Mediterranean collaborations involved in the km^3 project (see figure 7).

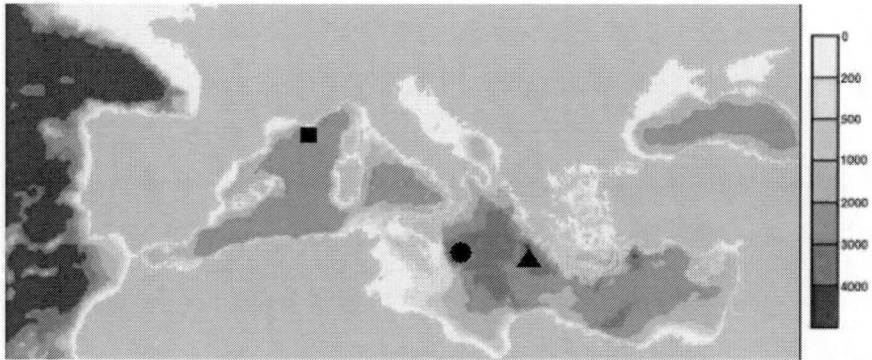

Figure 7. Abyssal sites candidate to host the Mediterranean km* detector: ANTARES-site at 2400 m depth near Toulon (square), NEMO-site at 3500 m depth near Capo Passero (circle), NESTOR-site at 3800 m depth near Pylos (triangle)

Demonstrator detectors: NESTOR and ANTARES. NESTOR [48], the first collaboration that operated in the Mediterranean Sea , proposes to deploy a modular detector at 3800 m depth in the Ionian Sea, near the Peloponnese coast (Greece). Each module is a semi-rigid structure (*the NESTOR tower*), 360 m high and 32 m in diameter, equipped with $\simeq$ 170 PMTs looking both in upward and downward directions. After a long R&D period, during March 2003 NESTOR has successfully deployed 12 PMTs at 3800 m depth acquiring, on-shore, underwater optical noise and cosmic muon signals (745 events reconstructed) for about 1 month. In the next future the collaboration aims at the deployment of the first tower with $\times 10^4$ m^2 effective area for E_μ >10 TeV muons.

ANTARES (see L. Sulak presentation in this School [42]) will be a *demonstrator* neutrino telescope with an effective area of 0.1 km^2 for astrophysical ν [49]. It will be located in a marine site near Toulon (France), at 2400 m depth. Between December 2002 and March 2003 the collaboration has deployed a junction box and two prototypal lines equipped with oceanographic instruments and optical modules hosting 10" PMTs. Instruments were connected to the shore through electro-optical cables. The lines have been recovered for maintenance on 2003. Data recorded by optical modules in the prototypal line shown an unexpectedly high optical background ranging from 60 to several hundreds kHz, well above the one produced by ^{40}K decay, then probably due to bioluminescence. ANTARES will be a high granularity detector consisting of 12 strings, each equipped with 25 equidistant stories made of 3 PMTs (total

75), placed at an average distance of 60 m. The PMT are 45° downward oriented, in order to avoid their obscuration by sediments and bio-fouling. Two new prototype lines will be deployed in the first months of 2005 and the whole detector installation is scheduled to be completed in 2007 [42, 50].

Research and development for the km³: NEMO. The construction of km³ scale neutrino telescopes requires detailed preliminary studies: the choice of the underwater installation site must be carefully investigated to optimise detector performance; the readout electronics must have a very low power consumption; the data transmission system must allow data flow transmission, as high as $\simeq 100$ Gbps, to shore; the mechanical design must allow easy detector deployment and recovery operations, moreover the deployed structures must be reliable over more than 10 years; the position monitoring system has to determine the position of OM within $\simeq 10$ cm accuracy. In order to propose feasible and reliable solutions for the km³ installation the NEMO (NEutrino Mediterranean Observatory) Collaboration has been conducting an intense R&D activity on all the above subjects since 1998 [53].

NEMO has intensively studied the oceanographic and optical properties in several deep sea (depth ≥ 3000 m) sites close the Italian coast. Results indicate that a large region located 80 km SE of Capo Passero (Sicily) is excellent for the installation of the km³ detector. The bathymetric profile of the region is extremely flat over hundreds km², with an average depth of $\simeq 3500$ m. Deep sea currents are, in the average, as low as 3 cm s^{-1}, and never stronger than 15 cm s^{-1}. Seawater oceanographic parameters (temperature and salinity) and inherent optical properties (light absorption and attenuation) were measured as a function of depth using a set-up based on the AC9, a commercial transmissometer. Figure 8 shows that at shallow depth, water properties change as a function of season, on the contrary at large depth (> 1500) the water column has constant characteristics (within the experimental data fluctuations). The average value of blue ($\lambda = 440$ nm) light absorption length is $L_a = 66 \pm 5$ m[5]. The same device measured $L_a(440\text{nm}) \simeq 48$ m in Toulon site [39] and $L_a(488\text{nm}) = 27.9 \pm 0.9$ m in Baikal lake [40].

The optical background noise was also measured at 3000 m depth in Capo Passero. Data collected in Spring 2002 and 2003, for several months, show that optical background induces on 10" PMTs (0.5 s.p.e.) a constant rate of $20 \div 30$ kHz (compatible with the one expected from ^{40}K decay), with negligible contribution of bioluminescence bursts. These results were confirmed by biological analysis that show, at depth> 2500, extremely small concentration of dissolved bioluminescent organisms [39].

The detector design must be optimised in order to get an effective area of $\simeq 1$ km² and a pointing accuracy $< 0.1°$ for 10 TeV muons, an energy resolution of the order of some tens percent in $\log(E)$ and an energy threshold close

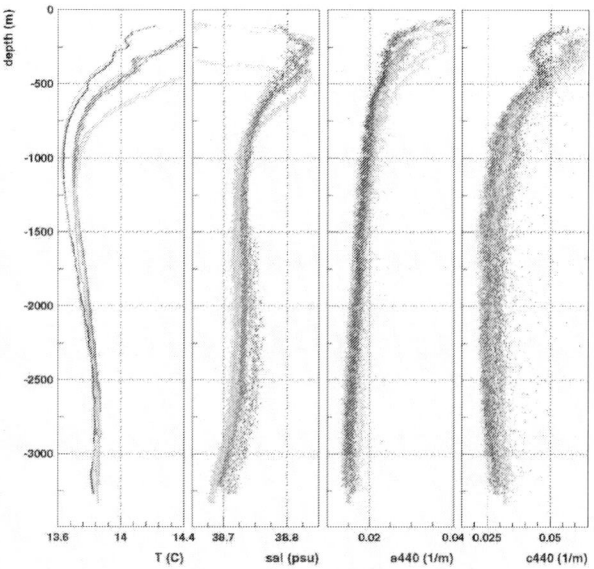

Figure 8. Profiles of temperature (T), salinity (S), absorption coefficient (a) and attenuation coefficient (c) and at 440 nm, measured in the Capo Passero site. The 13 superimposed profiles refer to five different campaigns: December 1999 (red dots, 2 profiles), March 2002 (green, 4 profiles), May 2002 (yellow, 2 profiles), August 2002 (orange, 3 profiles) and July 2003 (light blue, 2 profiles).

to 100 GeV. NEMO proposes an innovative structure to host OMs: the NEMO *tower*. It is designed to deploy, during a single operation, a large number of PMTs ($\geq$ 60) arranged in a 3-dimensional shape in order to locally permit event trigger and track reconstruction [51]. The structure is mechanically flexible, being composed by a sequence of $16 \div 20$ stories hosting OMs, interlinked by a net of syntectic fiber ropes (see figure 9). Each storey is $15 \div 20$ m long and hosts two optical modules (one downward looking and one looking horizontally) at each storey end[6]. Before the deployment the tower is assembled in a compacted structure allowing easy transportation and mooring (figure 9). When anchored on the seabed, the structure is released and it reaches its operational tower shape, pulled up by a buoy that tensions a lattice of ropes. The vertical inter-spacing between stories is $\simeq$ 40 m and an additional spacing of 150 m is added at the base of the tower, between the anchor and the lowermost storey to allow for a sufficient water volume below the detector. The operational height of the tower is therefore $H = 150 + (N_{storey} - 1) \times 40$ m.

In order to compare the physics performances of a modular detector made of *NEMO towers* (5832 PMTs, 10" diameter, arranged in 81 towers forming a 9×9 square matrix, with inter-tower distance 140 m and storey length 20 m)

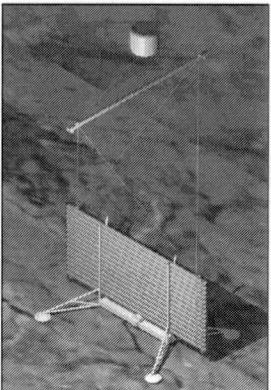

Figure 9.   Pictorial view of the NEMO tower. Before the deployment the tower is assembled in a transportable structure together with its anchor and the releasing system. After the deployment a large buoy pulls up the whole tower. The stories are kept in tension by means of buoys and synthetic fiber ropes. In the present project each tower holds 64 PMT displaced in 16 stories.

versus a homogeneous lattice detector (5600 PMTs arranged in 400 strings, 60 m horizontal and vertical distance between PMTs) software simulations were carried out by means of software packages developed by the ANTARES Collaboration [52]. Results of simulations, performed using the environmental characteristics similar to the ones of Capo Passero site show that the two detectors have similar values of A_{eff} and angular resolution $\simeq 0.1°$ for >10 TeV muons (see figure 10) [36]. The homogeneous lattice has, anyhow, a design which requires long and expensive sea operations due to the very large number of strings.

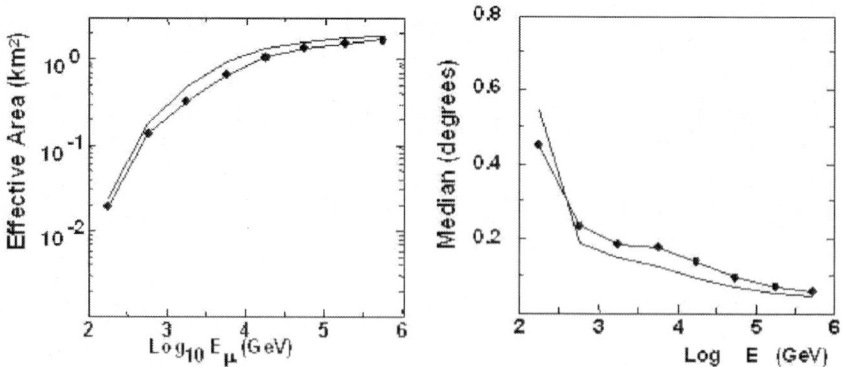

Figure 10.   Effective area and angular resolution of a NEMO "9 × 9" detector (solid line with diamonds) and of a "homogeneous lattice" detector(solid line), see text and reference [36] for details.

Technological challenges for the km³. In recent years many innovations have been applied to underwater technology: DWDM (dense wavelength division multiplexing) is permitting a large increase in speed and bandwidth of the optical fibres data transmission; newly developed materials (synthetic fibres, alloys, plastics, ...) can improve long term underwater reliability of complex deployed structures; deep sea ($\simeq$ 3000 m depth) operations with remotely operated vehicles (ROV) or autonomous underwater vehicles (AUV) have been developed and standardised. The design of the Mediterranean km³ will directly profit of these ultimate technologies.

Concerning the km³ data transmission system, it is considered a wise approach to transmit all PMT signals, acquired at low threshold level, to shore. In such a way all filtering and event reconstruction procedures can be applied by on-shore data acquisition systems. This approach results into a very high data transmission rate (order of 100 Gbps) that has to be transmitted $50 \div 100$ km away to the shore. The above considerations impose the use of a fibre optics transmission system based on DWDM telecommunication technique and composed by mainly passive optical components to reduce power consumption and increase reliability [53]. The NEMO-km³ architecture is designed to collect data hierarchically: each storey (hosting 4 OMs) transmits data to a tower main module using an assigned wavelength (or *colour*). Each tower module collects data from the $N_{stories}$ (there are then $N_{stories}$ wavelengths channels) and sends all data, multiplexed in one single fibre, to a collector. This collector can receive data from a number of towers, multiplexes them and sends them to the shore. A mirror system on shore de-multiplexes the signals, recovering data from each storey, then from each very single OM. The data flow system from shore to the OMs is identical, but is performed using different wavelengths.

Fast and low power consumption PMT read-out electronics is designed using of VLSI-ASIC boards. The LIRA chip, developed by NEMO, is capable to sample PMT signals at 200 MHz sampling frequency with 10 bits dynamic range, moreover trigger thresholds can be remotely controlled on shore. The LIRA maximum dissipation is <200 mW.

Materials that have to be used in the km³ detector construction must resist to an extremely hostile environment where pressure ($\simeq$300 bar at 3000 m depth), together with salinity, reduce the lifetime of most metals and alloys used in surface and shallow water applications. NEMO is designing and realising underwater vessels (junction boxes and other submarine containers) decoupling the two problems: an outer vessel, made of glass epoxy, resistant to marine corrosion is filled with not-corrosive oil. The inner pressure resistant vessel, made of common steal and pressure resistant, will host fragile electronic components. Another crucial point for deep sea detectors installation is feasibility and reliability of underwater connections. NEMO is studying the possibility to operate underwater connections at 3500 m depth using custom-designed ROV and/or AUV.

In order to test technical solution for the km^3 installation, the INFN installed a deep sea test site at 2000 m depth, 25 km E offshore the port of Catania (Sicily) supported by the infrastructure of LNS, Laboratori Nazionali del Sud. An underwater station consisting of a basic NEMO module (junction boxes, towers and data acquisition system) will be installed within 2007. The NEMO test site is also available for oceanographic research such as the GEOSTAR-INGV (Istituto Nazionale di Geofisica e Vulcanologia) project, the Italian station of ESONET (European Seismic Observatories Network) [54].

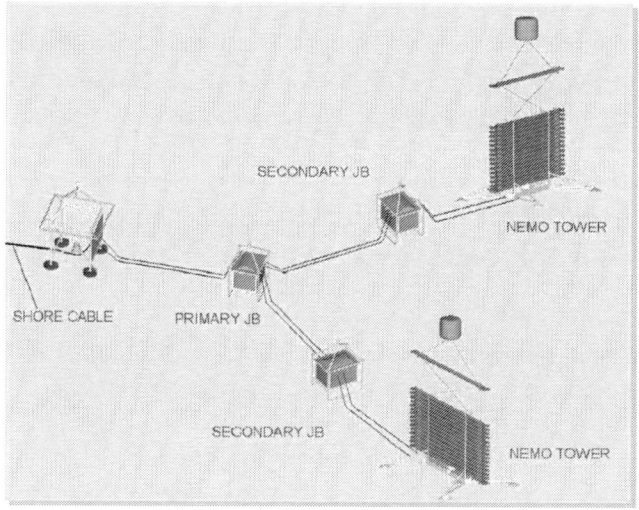

Figure 11. The NEMO test site will be installed at 2000 m depth, 25 km offshore Catania. It will host a complete basi NEMO module composed of one primary JB, two secondary JBs and two towers (see text).

Conclusions

The forthcoming km^3 neutrino telescopes are *discovery* detectors that could widen the knowledge of the Universe. These detectors have high potential to solve questions as the detection of UHECR sources, the investigation of hadronic processes in astrophysical environments or massive dark matter. Strong scientific motivations suggest the construction of two km^3 scale detectors in the Northern and the Southern Hemisphere.

In the South Pole the ICECUBE detector is extending AMANDA to the km^3 size and it is planned to be completed within year 2010, being probably the fist km^3 telescope running. In the Mediterranean Sea ANTARES is going to install a 0.1 km^2 neutrino telescope demonstrator, NESTOR has deployed

12 PMTs in deep water, aiming at a 2×10^4 m^2 tower and NEMO is installing a technological demonstrator for km^3 detector and proposed Capo Passero as an optimal installation site. The research and development activities conducted by the three Mediterranean collaborations can represent a valuable experience in the construction of the underwater km^3 detector, which will be the result of their efforts. A step in this direction is represented by their common application to the EU of a proposal (the KM3-NET) for a design study of the km^3 Mediterranean Telescope.

Notes

^{1}. The gyroradius of $E > 10^{\bullet\bullet}$ eV protons is larger than Galaxy thickness ($\simeq 50$ pc), moreover due to the effect of galactic magnetic field fluctuations, protons leave our Galaxy even at smaller energies. At the same energies heavier nuclei, with larger Z, are confined.

$^\bullet$. The analysis shown that gamma spectrum was compatible only with a hadronic origin(π. decay)

$^\bullet$. For $E_\nu \gg E_\mu^{min}$ the $P_{\mu\nu}$ shape is independent on E_μ^{min}

$^\bullet$. Indetermination on muon energy is large, indeed, since the energy loss in Čerenkov light is very small and since only a fraction of the muon track is sampled.

$^\bullet$. The NEMO Collaboration measured also the value of blue light attenuation length ($L_c^{-\bullet} = L_a^{-\bullet} + L_b^{-\bullet}$) $L_c = 35 \pm 5$ m. Several authors usually quote the water effective scattering length, which is defined as $L_b^{eff} = L_b/ < \cos \vartheta >$, where $< \cos \vartheta >$ is the average cosine of the light scattering distribution; in Capo Passero $L_b^{eff} \gg 100$ m.

$^\bullet$. The final configuration of the tower (number and length of storeys, number of optical modules per storey, distance between the storeys) has to be optimized following the results of numerical simulations. However, the modular structure of the tower will permit the modification of these parameters to experimental requirements.

References

[1] R. Davis et al. *Physical Review Letters*, 20:1205, 1968.

[2] J. Bahcall *Physical Review Letters*, 12:300, 1964.

[3] Y. Fukuda et al. *Nuclear Instruments and Methods*, A501:418, 2003. SuperKamiokande web page http://www-sk.icrru.u-tokyo.ac.jp

[4] Q.R.Ahmed et al. *Physical Review Letters*, 12:300, 1964. SNO web page http://www.sno.phy.queensu.ca

[5] K. Eguchi et al. *Physical Review Letters*, 90:021802, 2003. KAMLAND webpage http://wwwawa.tohoku.ac.jp/html/KamLAND/index.html

[6] M. Ambrosio et al. *Physics Letters*, B566(2-3):35, 2003.

[7] R. Becker-Szendy et al. *Nuclear Instruments and Methods*, A(324):363, 1993.

[8] R.M. Bionta et al. *Physical Review Letters*, (58):1494, 1987.

[9] K. Hirata et al. *Physical Review Letters*, (63):16, 1989.

[10] K. Hirata et al. *Physical Review Letters*, (58):1490, 1987.

[11] J.W. Cronin, T.k. Gaisser and S.P. Swordy Cosmic Rays at the Energy Frontier *Scientific American*, January 1997.

[12] E. Fermi *Physical Review*, 75:1169, 1949.

[13] A.R. Bell *Monthly Not. R. Astr. Soc.*, 182:147, 1978. for a description see also M.S. Longair *High Energy Astrophysics*, Vol 1. and 2. Cambridge Univerity Press, 1994.

[14] V.L. Ginzburg and S.I. Syrovatskii *Soviet Physics Uspekhi*, 3:504, 1961.

[15] K. Mannheim *Science*, 279:684, 1998.

[16] E. Waxmann and J. Bahcall, *Phisical Review Letters*, 78:2292, 1997.

[17] P.L. Biermann and P.A. Strittmatter *Astrophysical Journal*, 322:643, 1997.

[18] G.T. Zatzepin and V.A. Kuzmin *Soviet Physics JETP Letters*, 4:78, 1966 and K. Greisen *Physical Review Letters*, 16:748, 1966

[19] L. Costamante and G. Ghisellini *Astronomy and Astrophysics*, 56:384, 2002.

[20] Y. Ashie et al. *Phisical Review Letters*, 93:101801, 2004.

[21] R. Enamoto et al. *Nature*, 416:823, 2002.

[22] F. Aharonian et al. *astro-ph/0408145*, 2004.

[23] F. Aharonian and A.Neronov *astro-ph/0408303*, 2004.

[24] A. Levinson and E. Waxman *Phisical Review Letters*, 87:171101, 2001.

[25] E. Waxman and J. Bahcall *Physical Review*, D64:023001, 1999.

[26] K. Mannheim, R.J. Protheroe and J.P Rachen *Physical Review* , D63:023003, 2001.

[27] E. Waxman and K. Mannheim *Europhysics News*, 32(6), 2001.

[28] C. Distefano et al. *Astrophysical Journal*, B78:2292, 2001.

[29] D. Guetta et al. *Astroparticle Physics*, 19:403, 2003.

[30] P. Battacharije et G. Sigl *Physics Reports*, 327:109, 2000.

[31] T.K. Gaisser, F. Halzen and T. Stanev *Physics Reports*, D258:173,1995.

[32] M.A. Markov and I.M. Zheleznykh *Nuclear Physics*, 27:385, 1961.

[33] R. Gandhi et al., *Physical Review*, D58:93009, 1998.

[34] S. Razzaque, P. Meszaros, E. Waxman. *Physical Review*, D69:023001, 2004.

[35] W. Bednarek *Astronomy and Astrophysics*, D407:1, 2003.

[36] P. Sapienza for the NEMO Collaboration. Status of Simulations in NEMO. In *VLVνT Workshop Proc.*, pages 114–118, Amsterdam, 2003. Editor E. de Wolf, NIKHEF, Amsterdam, The Netherlands. Proceedings available at http://www.vlvnt.nl/proceedings

[37] A. Atoyan and C. Dermer *Physical Review Letters*, 87:221102, 2001. and A. Atoyan and C. Dermer, *New Astron. Review*, 48:381, 2004.

[38] C.D. Mobley. *Light and Water: radiative transfer in natural waters* Academic Press, 1994.

[39] G. Riccobene for the NEMO Collaboration Overview over Mediterranean Optical Properties. In *VLVνT Workshop Proc.*, pages 93–96, Amsterdam, 2003. Proceedings available at http://www.vlvnt.nl/proceedings

[40] V. Balkanov et al. *Nuclear Instrumnets and Methods*, A498:231, 2003.

[41] A. Roberts et al. *Review of Modern Physics*, 64:259, 1992. DUMAND webpage http://www.phys.hawaii.edu/dmnd/dumand.html

[42] L. Sulak. *These Proceedings*

[43] I. Belolaptikov et al. *Astroparticle Physics*, 7:263, 1997.

[44] R. Wischnewski for the BAIKAL Collaboration. In *28th International Cosmic Ray Conference Proc.*, Tsukuba, Japan, 2003.

[45] E. Andres et al. *Nature*, 40:441, 2001. AMANDA web page http://amanda.uci.edu

[46] see A. Silvestri for the AMANDA Collaboration. In *these Proceedings*

[47] J. Ahrens et al. *Astroparticle Physics*, 20:507, 2004. ICECUBE web page http://icecube.wisc.edu

[48] NESTOR web page http://icecube.wisc.edu

[49] *ANTARES: A Deep Sea Telescoper for High Energy Neutrinos.* available at ANTARES web page http://antares.in2p3.fr

[50] E.V. Korolkova for the ANTARES Collaboration. *The status of the ANTARES experiment* In *Proceedings of the CRIS 2004*, Catania, Italy, 2004. Proceedings available at `http://www.ct.infn.it/ cris`.

[51] M.S. Musumeci for the NEMO Collaboration Mechanical Structures for a Deep Seawater Neutrino Detector. In *VLVνT Workshop Proc.*, pages 153–156, Amsterdam, 2003. Proceedings available at `http://www.vlvnt.nl/proceedings`

[52] D.J.L Bailey, PhD Thesis, University of Oxford, UK, 2003.

[53] E. Migneco for the NEMO Collaboration. *NEMO: Status of the project* In *Proceedings of the CRIS 2004*, Catania, Italy, 2004. Proceedings available at `http://www.ct.infn.it/čris`. NEMO web page `http://nemoweb.lns.infn.it`

[54] See GEOSTAR-ESONET web page at `http://www.ingrm.it/GEOSTAR/esonet.htm`

RECENT RESULTS FROM THE MILAGRO GAMMA RAY OBSERVATORY

J. A. Goodman
Department of Physics
University of Maryland, College Park, Md, USA 20742
goodman@umdgrb.umd.edu

for the Milagro Collaboration

Abstract   The Milagro gamma-ray observatory utilizes a large water Cherenkov detector to observe TeV extensive air showers produced by high energy particles impacting the Earth's atmosphere. Milagro is different from Atmospheric Cherenkov Telescopes that are used to study TeV gamma-rays in that it views a wide field (2 steradian over-head sky) and it continuously operates (>90% live time). These factors give Milagro the potential for discovery of new sources with unknown positions and times, such as gamma-ray bursts, flaring AGN, and observation of extended sources like the Galactic plane or large supernova remnants. The Milagro detector consists of a 4800 m² pond instrumented with 723 8" PMTs which detect Cherenkov light produced by secondary air-shower particles. A sparse array of 175 4000l water tanks surrounding the central pond detector has recently been added which will extend the physical area of Milagro to 40,000 m² and substantially increase the sensitivity of the instrument. Based on three years of operation,consitutes Milagro has established its sensitivity through the detection of the Crab plerion and active galaxy Markarian 421. A summary of the recent results from the Milagro collaboration is presented with a focus on the first observation of the galactic plane in the TeV range and evidence for two newly observed TeV sources: diffuse emission from the Cygnus Region, and evidence for an extended TeV hot spot near the EGRET unidentified 3EG J0520+2556.

Keywords: Milagro, Gamma-ray astronomy

Introduction

Milagro is a unique TeV gamma-ray observatory capable of continuously monitoring the overhead sky. The directions of gamma-rays impacting the earth's atmosphere are reconstructed through the detection of air-shower particles that reach the earth. The shower particles are detected with a 60m × 80m

M. M. Shapiro et al. (eds.), Neutrinos and Explosive Events in the Universe, 243–254.
© 2005 *Springer. Printed in the Netherlands.*

pond of purified water instrumented with an array of photomultiplier tubes arranged into two layers, the air-shower layer and the muon layer. The pond is located in the center of a sparse 200m x 200m array of 175 outrigger tanks. The air-shower layer is situated at a depth of 1.4m below the surface of the pond. The shallow depth allows the accurate measurement of shower particle arrival times used in direction reconstruction. The angular resolution for gamma-ray induced air-showers is $\approx 0.75°$ without the outriggers and $\approx 0.45°$ with the outriggers. The muon layer is located at about 6m below the surface. (See Figure 1). The greater depth (17 radiation lengths) is utilized to detect the presence of penetrating muons and hadrons. Simple cuts have been developed to distinguish between gamma-ray induced and hadron induced air-showers using the pattern of hits in the bottom layer. The median energy of gamma rays detected from a crab-like spectrum is ≈ 3 TeV.

The Milagro Gamma Ray Observatory has been operational since July 2000. The detector functions both during the day and night and observes the entire overhead sky (≈ 2 steradians). Cosmic ray events are collected and logged at a rate of roughly 1500Hz. Nearly all of the events collected are due to air-showers induced by cosmic ray protons and nuclei. These charged particles are useless for astronomy because the interstellar magnetic fields bend charged particle trajectories decoupling their momentum from their direction of origin. Gamma-ray sources are identified as statistically significant excesses on the nearly uniform cosmic ray background.

The Milagro detector's large field of view and continuous duty cycle make it an ideal instrument for the discovery of previously unknown sources. Recent publications cover topics including detection of the Crab Nebula[1], limits on TeV emission from GRB [2] and a TeV all-sky survey of the northern celestial hemisphere[3]. Recently we have presented papers on the detection of diffuse TeV emission from the Galactic plane[4], limits on TeV emission from satellite detected GRB[5], a study of nearby AGN[6] and limits on relic neutralino annihilation derived from TeV flux limits from the sun[7]. The focus of this paper is the search for extended sources of TeV gamma rays with the Milagro detector.

Milagro's wide field makes it well suited to observe extended sources. Although Milagro has a relatively poor angular resolution compared to Atmospheric Cherenkov Telescopes (ACTs), this disadvantage is eliminated when studying sources with angular extent $\geq$ Milagro's angular resolution. At least two categories of diffuse very high-energy gamma-ray sources are known to exist at GeV[8] and TeV energies: Supernova remnants (SNR)[9] and the Galactic plane. In the analysis presented here, the northern sky ($0° < \delta < 60°$) is searched for TeV sources with angular extent up to $\approx 5°$ by re-binning the Milagro data to optimize the existing all-sky search methods for detection of diffuse gamma-ray emission.

Figure 1. Aerial view of the Milagro pond (top) with outriggers marked as red circles. The bottom picture shows a view of the inside of the Milagro detector. The two layers of PMTS at depths of 1.4m and 5.5m are clearly visible.

Data Analysis

The Milagro collaboration has developed a standard set of cuts that this analysis utilizes. The cuts were optimized using studies of the detector simulation and confirmed through observation of the Crab Nebula [1] and Markarian 421[6]. Events with greater than or equal to 20 PMT hits utilized by the shower angle fitter ($NFIT \geq 20$) and the "compactness" parameter (C) greater than 2.5 are retained. The $NFIT$ cut preserves about 80% of the data, only removing events with obviously poor angle fits. About 8% of the background data pass the compactness cut. This cut has been determined from simulations to have an efficiency of $\approx 50\%$ for gamma rays. (See Figure 2). Application of this cut increases the sensitivity of Milagro by a factor of $\approx 1.6 (= \frac{0.5}{\sqrt{0.08}})$.

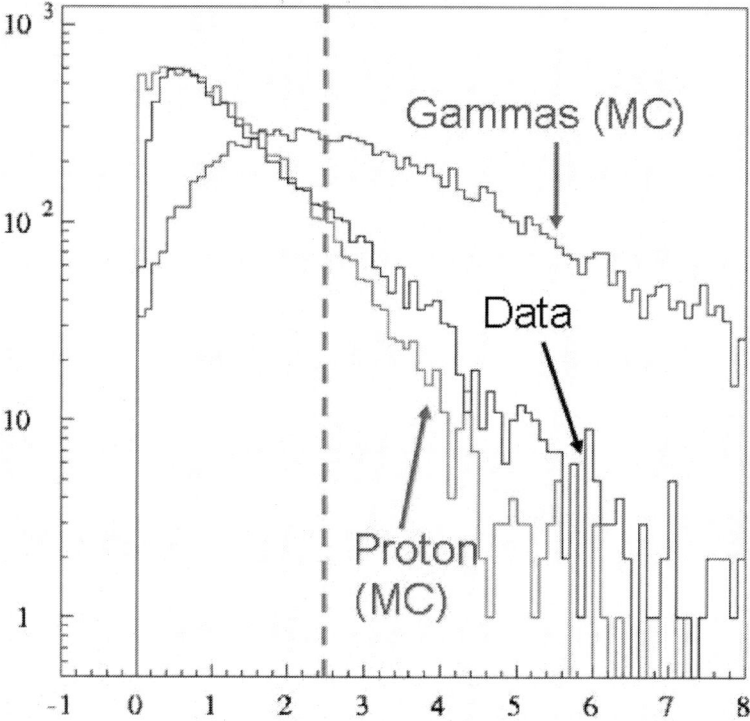

Figure 2. The distribution of the compactness parameter for Monte Carlo protons and gammas as well as data (which is mostly protons). $\sim 8\%$ of the background data pass the compactness cut of 2.5 (shown above), while simulations show that$\sim 50\%$ of gamma rays are retained.

The excess at each position in the celestial sky is computed by counting the number events from that sky position and subtracting the estimated background. For a given point the background is computed from data collected at the same local detector coordinates (θ, ϕ), but at a different time, so that the celestial angles of the background event sample do not overlap with the source position under consideration. The method of Li and Ma[10] is used to compute the final probability of the observed excess or deficit.

The optimal square bin size for detection of a point gamma-ray source with Milagro is $2.1°$ on a side corresponding to an angular resolution of $\approx 0.75°$. To search for diffuse gamma-ray sources with angular extent greater than or equal to the angular resolution, the standard Milagro sky maps are searched using a range of bin sizes from 2.1° to 5.9° in steps of 0.2 °. In total, 20 separate searches are performed on the same maps. The results, however, are highly correlated. For each bin size, points with probability >5 σ are recorded for further study. Monte Carlo simulations of the map analysis process are used to compute the post-trials probability for each source candidate.

Data collected between 17 August 2000 and 5 May 2004 are used in this analysis. In total, 1305 live days are included.

Results

In addition to the high significance detection of the Crab and Mrk421, the analysis also identifies two additional excesses that substantially distinguish themselves from background plus a 5σ detection of the galactic plane . The most significant source candidate has P=5.9σ and is located at RA=78.8° and δ=26.0° (l=178.8,b=-7.3) and was identified with a 2.9° bin size. This candidate is spatially coincident with EGRET unidentified source 3EG J0520+2556. The second most improbable point has P=5.5σ, is located at RA=308° and δ=42° (l=81°,b=1°) and was identified with a 5.9° bin. This source candidate lies in the Galactic plane and is spatially coincident with the Cygnus Arm.

The Galactic Plane

Diffuse emission from the galactic plane is the dominant source in the gamma-ray sky [11]. The origin of very-high-energy diffuse emission is conventionally understood to be due to the decay of π^0's produced by the interaction of cosmic-ray hadrons with interstellar matter. The flux of gamma-rays measured by EGRET below 1 GeV fits models well, while between 1 and 40 GeV models predict fluxes 60% less than that measured. It has been suggested [12] that the enhanced emission at high energies is due to inverse-Compton scattering from cosmic-ray electrons. Candidate sources for very-high-energy electrons are supernova remnants, where X-ray observations [13] of synchrotron radiation indicate the presence of electrons above 30 TeV. If cosmic-ray electron cooling and not hadronic interactions are the dominant source of diffuse gamma-ray emission from the galactic plane then, the flux might be an order of magnitude higher than previously thought at TeV energies. For this reason, ACTs have been spending an increasing amount of time searching for TeV gamma-rays from the galactic plane. Milagro, because of its large aperture, high duty cycle, and good background rejection (for a diffuse source the background rejection capability of Milagro is comparable to that of a single mirror ACT such as Whipple), is the world's most sensitive instrument for the detection of diffuse emission near 1 TeV.

Using 36 months of Milagro data beginning in July of 2000 we looked at the inner (20°-100°) and outer (140°-220°) regions of the galaxy and applied our standard gamma-hadron cut to the data. We also corrected for the diurnal variation of the zenith angle distribution of events (due to changes in the atmosphere from night to day). Due to the small signal-to-background ratio (1:10,000) extreme care must be taken in understanding systematic effects. One possibility is that there is a large scale anisotropy in the cosmic rays that is causing some (or all) of the observed signal. We measured this effect by excluding the galactic plane and fitting the latitudinal distribution to a flat line. The outer galaxy shows no significant excess. However, there is a 5σ excess in the inner galaxy. In Figure 3 we show the significance map of the galaxy. The region of the sky visible from Milagro can easily be seen. The region of the inner galaxy shows an enhancement along and just north of the galactic equator. This region is where EGRET showed its strongest enhancement in the 100 MeV energy range. It is also what is known as the Cygnus region. In Figure 4 we show the longitude and latitude profiles of the galaxy. Once again a the enhancement of the inner galaxy can be seen. The 5σ excess is seen by summing the entire inner galaxy with a $\pm5^o$ latitude band as was suggested by the EGRET results. This constitutes the first detection of the galactic plane at TeV energies [4].

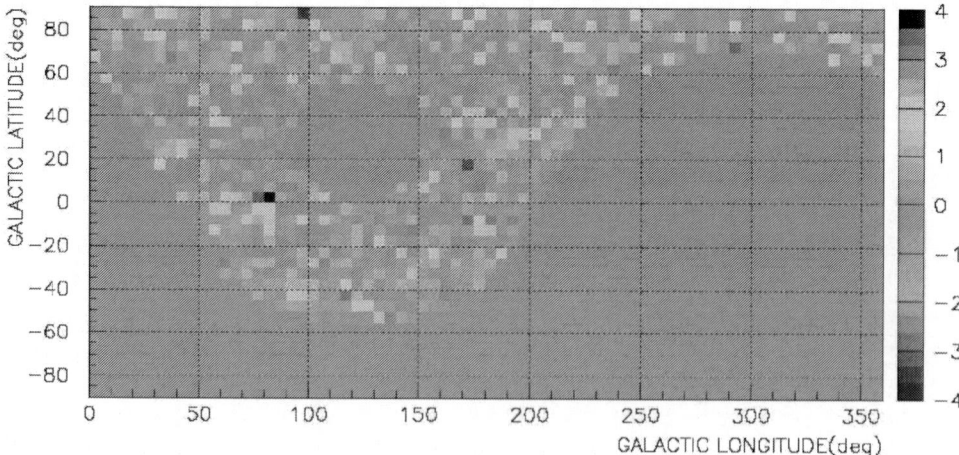

Figure 3.   Significance map in galactic coordinates.

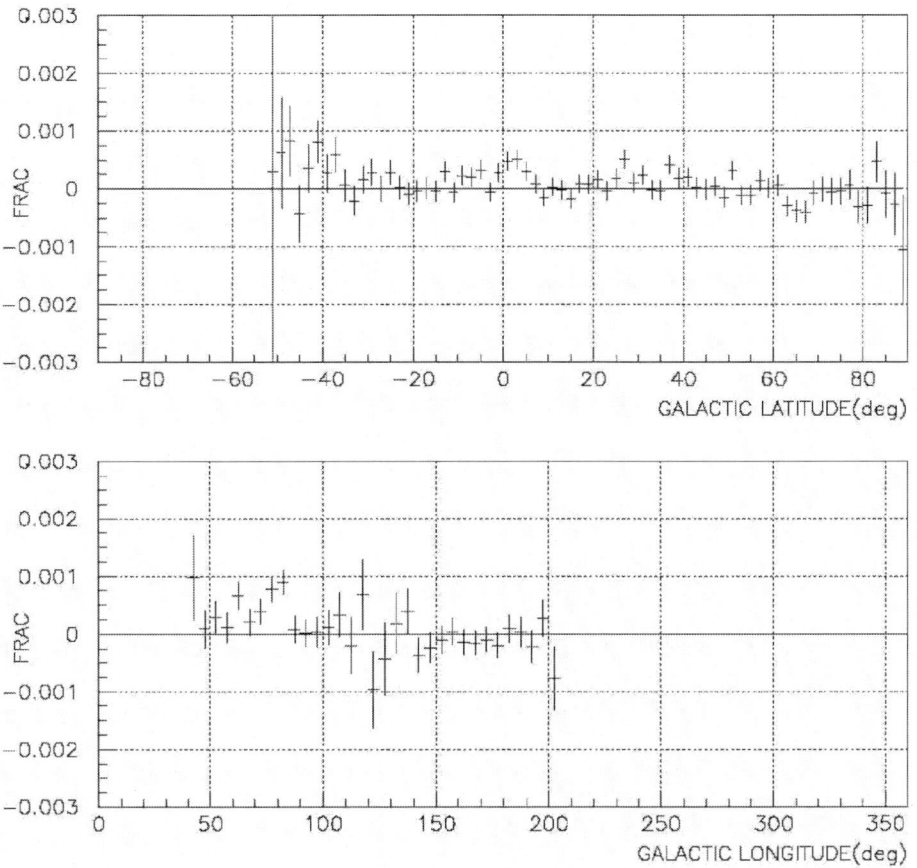

Figure 4. Latitude and Longitude for the inner galaxy as seen from Milagro.

TeV0520+2556

The first point-source candidate is spatially coincident with the EGRET unidentified source 3EG J0520+2556. Monte Carlo simulations indicate that the probability of observing an excess this significant at any point in the sky at any bin size is 0.8%. Figure 5 shows a significance map in the region of the Crab, showing this source. This candidate was first reported by Milagro in 2002 [14] as a possible TeV gamma-ray source. In the initial Milagro all-sky search, a larger more conservative bin size of 3.0 ° was used. With this bin size, this source candidate was identified as a "hot spot" worth investigating further. Greater understanding of the performance of the Milagro detector lead to a reduction in the standard bin size for gamma-ray point sources searches from 3.0° to 2.1°. The reduction in bin size diminished the observed significance, however, the point remained one of the most significant in the sky and is one of only 11 points with significance greater than 4 σ in the published Milagro sky survey[3]. As this source candidate was originally identified in 2002, it is reasonable to break the data into 2 subsets, events used in the initial all-sky search and events collected since. The data set used in the analysis reported in 2002 contained data collected between 17 August 2000 and 9 December 2001 (465 days of detector on time). Data collected since the initial report and used in this analysis include 840 days of detector on time from 10 December 2001 to 5 May 2004. In the 465 day sample, a 4.4 σ excess is observed. At the exact position of the maximum in the 465 day sample, a 3.7 σ excess is observed in the 840 day sample, and a local maximum is found a few tenths of a degree away with a significance of 4.4 σ. Figure 6 shows the significance maps for the two data subsets.

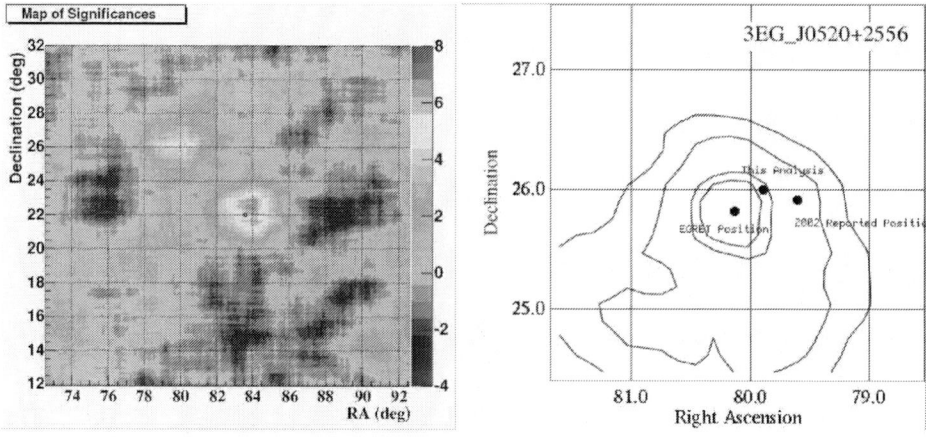

Figure 5. Significance map of the region containing the Crab nebula (RA=84.6°, δ=22.0°) and the highly improbable excess coincident with the EGRET unidentified source 3EG J0520+2556 (RA=78.8°, δ=26.0°). Also shown is the EGRET significance contour for 3EG J0520+2556.

To estimate the spatial extent of the excess associated with this source candidate the excess was fit to a Gaussian convoluted with Milagro's point spread function. A four parameter fit was performed where the width (σ), position (RA, δ) and amplitude were allowed to vary. The best estimate of the source position and width are RA=78.8°±0.4°, $\delta = 26.0°±0.4°$, and $\sigma = (0.8 \pm 0.4)°$. The excess is inconsistent with a point source hypothesis at the $\approx 2\sigma$ level. The flux of this candidate is 85% of the Crab. The excess was also found to grow steadily over time. There is no evidence of episodic emission.

The 2002 announcement this source candidate coincident with 3EG J0520 + 2556 lead to follow-up observations[15] by the Whipple collaboration which found no evidence of a gamma-ray point source at the position reported by the

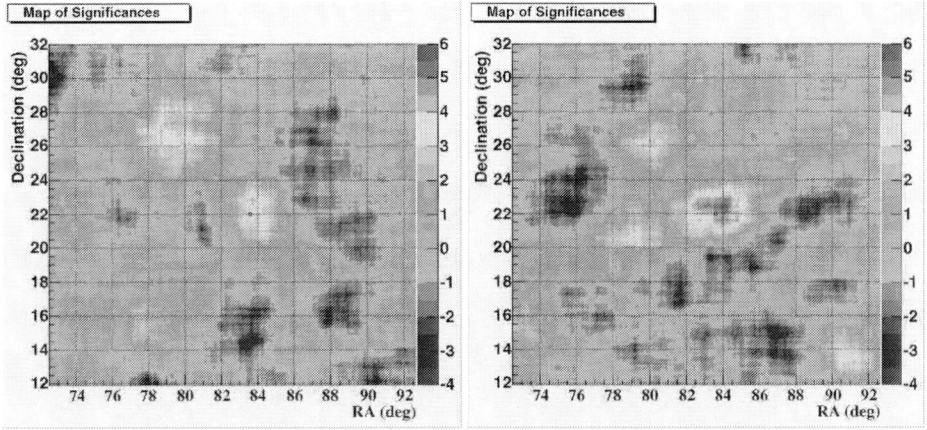

Figure 6. Significance map for the Crab and Milagro "Hot Spot". The panel on the left shows the data set utilized in the initial all-sky survey. The panel on the right shows that data collected after the hot spot was reported.

Milagro Collaboration and at the position of 3EG J0520+2556. The flux upper limits set by Whipple rule out the hypothesis that the Milagro excess is due to a gamma-ray point source, but Whipple did not publish a limit on the diffuse gamma ray flux. The sensitivity of the Whipple telescope to extended sources is substantially lower than to point sources, due to its outstanding angular resolution, so it is likely that if this source is diffuse, with an angular extensions greater than $\approx 0.5^o$, it would not be detectable by Whipple with the limited telescope time dedicated to it.

Cygnus Arm

The second extended gamma-ray source candidate is found to be coincident with the Cygnus Arm. The probability of observing an excess this significant at any point in the sky at any bin size is 2.0%. Figure 7 shows a significance map of the region of the sky containing the Cygnus Arm. The Cygnus Arm is a spiral arm within our Galaxy that extends radially away from observers on the earth. This is a known dense region of gas and dust and is observed by EGRET to have diffuse GeV gamma-ray emission comparable to the Galactic bulge, making it the brightest region of GeV gamma rays in the northern sky[11]. Similar to the GeV and TeV [4] emission from the broader Galactic plane, VHE emission from the Cygnus region is believed to be dominated by interactions of cosmic ray hadrons with interstellar gas and dust. The 5.5 σ excess observed at this location is broad and inconsistent with a single point source hypothesis. The excess has a flux approximately twice the Crab. Despite the seemingly

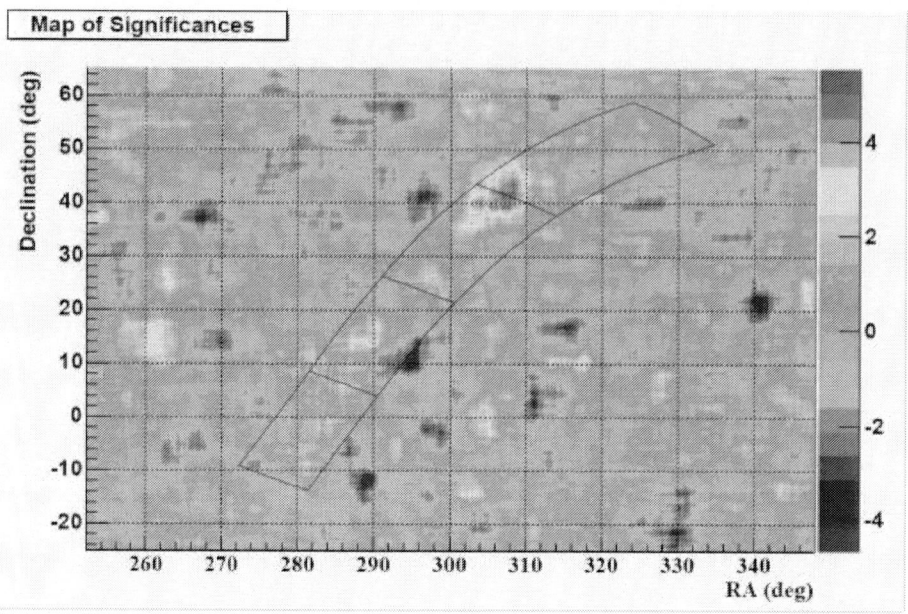

Figure 7.   Significance Map showing the Galactic plane as it transverses the mid northern declinations.   The excess in the vicinity of the Cygnus region is clearly visible at RA=308.°,δ=42.°. The Galactic plane from l=20° to l=100° and b=±5° is superimposed on the plot. This map was made using a 5.9° binsize.

large flux, this known gamma-ray source is a difficult target for ACTs because of its size. No ACT has a camera large enough to image the entire source region. The excess is observed to grow steadily over time. There is no evidence of episodic emission.

It should be noted that while this constitutes the hottest spot in the galactic plane, even with this region excluded the galactic plane excess reported above still remains significant.

Conclusion

Presented here is evidence for the first detection of TeV emission from the galactic plan and two previously unknown sources of TeV gamma rays, one coincident with 3EG J0520+2556, and another coincident with the Cygnus Arm. The new sources are found to be diffuse. The two source candidates have probabilities of 5.9σ and 5.5σ respectively. When all the trials of the all-sky search with a range of 20 bin sizes are counted the probabilities of observing these excesses are reduced to 0.8% and 2.0% respectively. The source candidate coincident with 3EG J0520+2556 was reported as a "hot spot" by the Milagro collaboration in 2002. Data collected since the initial report show a 3.7σ excess. The Cygnus Arm is the brightest source of GeV gamma rays in the northern sky. The observed location and shape of the Milagro excess coincide with the 100 MeV observations. While this source contributes to the inner galactic plane excess, this excess remains significant even with this new source removed.

The Milagro experiment has recently been expanded with the addition of an array of outrigger tanks. Inclusion of the outriggers in the real time reconstruction will improve the significance of the sensitivity of Milagro by both through improvement in the angular resolution ($0.75^o \rightarrow 0.45^o$) and by the improving gamma/hadron separation.

Acknowledgments

The Milagro Collaboration consists of: R. Atkins, W. Benbow, D. Berley, E. Blaufuss, D. G. Coyne, T. DeYoung, B. L. Dingus, D. E. Dorfan, R. W. Ellsworth, L. Fleysher, R. Fleysher, M. M. Gonzalez, J. A. Goodman, T. J. Haines, E. Hays, C. M. Hoffman, L. A. Kelley, C. P. Lansdell, J. T. Linnemann, J. E. McEnery, R. S. Miller, A. I. Mincer, M. F. Morales, P. Nemethy, D. Noyes, J. M. Ryan, F. W. Samuelson, P. M. Saz Parkinson, A. Shoup, G. Sinnis, A. J. Smith, G. W. Sullivan, D. A. Williams, M. E. Wilson, X. W. Xu and G. B. Yodh

Many other people helped bring Milagro to fruition. In particular, we acknowledge the efforts of Scott DeLay, Neil Thompson and Michael Schneider. This work has been supported by the National Science Foundation (under grants PHY-0075326, -0096256, -0097315, -0206656, -0245143, -0245234, -0302000, and ATM-0002744) the US Department of Energy (Office of High-Energy Physics and Office of Nuclear Physics), Los Alamos National Laboratory, the University of California, and the Institute of Geophysics and Planetary Physics. .

References

[1] Atkins, R. et al. 2003b, *ApJ* 595, 803.

[2] Atkins, R. et al. 2004, *ApJ* 604, 25.

[3] Atkins, R. et al. 2004, *ApJ* 608, 680.

[4] "Discovery of Diffuse Gamma Ray Emission from the Galactic Plane using the Milagro Detector," Heidelberg Workshop 2004.

[5] "Search for VHE emission from GRB with Milagro," Proceedings of International Symposium on High Energy Gamma-ray Astronomy, Heidelberg, Germany, July 26-30 (2004)

[6] "Studies of Nearby Blazars with Milagro," Proceedings of International Symposium on High Energy Gamma-ray Astronomy, Heidelberg, Germany, July 26-30 (2004)

[7] "Search for Relic Neutralinos with Milagro," Proceedings of International Symposium on High Energy Gamma-ray Astronomy, Heidelberg, Germany, July 26-30 (2004)

[8] Hartman, R.C. et al 1999, *ApJS* 123, 79.

[9] Green D.A., 2004, 'A Catalogue of Galactic Supernova Remnants (2004 January version)', Mullard Radio Astronomy Observatory, Cavendish Laboratory, Cambridge, United Kingdom

[10] Li T.P. and Ma Y.Q., 1983, ApJ 272, 317.

[11] Hunter, S.D. et al 1997, *ApJ* 603, 355.

[12] Porter, T.A. and Protheroe, R.J., 1997, J Phys G, 23, 1765

[13] Koyama, M., et al., 1995, Nature, 378, 255.

[14] Sinnis C. et al. 2002, Proceedings of the American Physical Scociety and the High Energy Astrophysics AAS Meeting.

[15] Falcone A. et al. 2003, astro-ph/0305575.

PHYSICS POTENTIAL AND FIRST RESULTS OF THE MAGIC TELESCOPE

Ester Aliu[1], Roger Firpo[1], Florian Goebel[2], Daniel Mazin[2], Satoko Mizobuchi[2,3], Raquel de los Reyes[4], Javier Rico[1], Nuria Sidro[1] and Nadia Tonello[2] for the MAGIC Collaboration

[1]*Institut de Fisica d'Altes Energies, 08193 Bellaterra (Barcelona), Spain.*

[2]*Max-Planck-Institut fuer Physik, Foehringer Ring 6, 80805 Muenchen, Germany.*

[3]*Department of Physics, Ehime University, 3 Bunkyo-cho, Matsuyama 790-8577, Japan*

[4]*Dept. de Fisica Atomica, nuclear y molecular, UCM, Ciudad Universitaria s/n, Madrid, Spain.*

Abstract   The Major Atmospheric Gamma ray Imaging Cherenkov (MAGIC) telescope is a new generation imaging air Cherenkov telescope designed for gamma ray astronomy in the energy range between 30 GeV and several TeV. Data taking in commissioning phase started in October 2003. A short overview of the physics potential and a description of the telescope is followed by more detailed studies on possible measurements of pulsars, active galactic nuclei, microquasars and gamma ray bursts. Finally, first results obtained during the commissioning phase are presented.

Keywords: Imaging Air Cherenkov Telescopes, Gamma Ray Astronomy

Introduction

The gamma ray energy region between 30 GeV and 300 GeV is essentially unexplored due to the small collection efficiency of satellite experiments and the high energy threshold (< 300 GeV) of old generation ground based telescopes. In order to close this gap the MAGIC telescope (see Figure 1) is aiming at the lowest energy threshold among the new generation Imaging Air Cherenkov Telescopes (IACTs). With a 17 m diameter high reflectivity mirror dish and a high quantum efficiency (QE) camera MAGIC is designed for an energy threshold as low as 30 GeV.

A compelling motivation for the search of gamma ray sources in the energy range between 30 GeV and 300 GeV is the apparent discrepancy between the well populated sky-map of sources below 10 GeV observed by EGRET (more than half of them still unidentified) and the handful of sources discovered by

M. M. Shapiro et al. (eds.), Neutrinos and Explosive Events in the Universe, 255–267.

Figure 1 The MAGIC telescope.

IACTs above 300 GeV. MAGIC observations in this energy gap should allow to study the mechanisms which cut off the spectra of several EGRET sources. Due to a higher flux sensitivity and a better angular resolution compared to EGRET it may also help identify many unidentified EGRET sources.

The physics program of MAGIC covers several galactic and extragalactic types of sources such as supernova remnants (SNR), pulsars, microquasars and active galactic nuclei (AGNs). High energy gamma rays traveling cosmological distances are absorbed via interactions with the extragalactic background light which limits observation with old generation Cherenkov telescopes to $z \leq 0.1$. The low energy threshold of MAGIC allows to extend the observation of extragalactic sources up to $z \approx 1$ and beyond. A unique feature of MAGIC is the possibility to observe gamma ray bursts (GRBs) only 30 seconds after an alert provided by satellite detectors. Furthermore, MAGIC may contribute to measurements of fundamental physics such as searches of cold dark matter and quantum gravity effects.

Description of the Telescope

MAGIC incorporates many technological innovations in order to fulfill the requirements imposed by the physics goals. The telescope is installed at 2200 m above see level on the Canary island La Palma (Spain). The 17 m diameter tessellated reflector dish consists of 964 0.5×0.5 m^2 diamond milled aluminum mirrors. The parabolic reflector shape conserves the time structure of the Cherenkov pulses to increase the signal-to-noise ratio with respect to the night sky background.

Great care has been taken to reduce the overall weight of the telescope to allow for fast repositioning required for fast reactions to GRB alerts. The telescope frame is made of light but stiff carbon fiber tubes and weighs less than 20 tons. The maximum repositioning time for a complete turnaround of the telescope has been measured to be 22 seconds, well below the design specification of 30 seconds.

The 3.5° field of view (FOV) camera is equipped with 576 high QE photomultipliers (PMTs). The inner area is composed of 396 0.1° FOV PMTs surrounded by 180 0.2° FOV PMTs. The QE is enhanced up to 30% using a diffuse scattering, wavelength shifting coating which also extends the sensitivity to the ultra-violet (UV) range.

The analog PMT signals are transfered via 162 m optical fibers using Vertical Cavity Surface Emitting Lasers (VCSELs). In the electronics room the signals are split. One branch is sent to a software adjustable threshold discriminator which generates a digital signal for the trigger system. In the other branch the pulses are stretched to 6 ns and again split into a high and low gain channel. The high gain signal is amplified by a factor 10 while the low gain signal is delayed by 55 ns before it is merged again with the high gain signal using a fast GaAs analog switch. The combined signal is then digitized by a 8 bit 300 MHz Flash ADC.

The trigger decision is generated by a 2-level system using the digital trigger signals generated by the 325 innermost camera pixels. The level-1 trigger is a fast next neighbor coincidence logic. The second level performs more complex topological filter algorithms.

MAGIC physics potential

In the following, we summarize the prospects for the detection of several kinds of gamma ray sources.

Active Galactic Nuclei. AGNs constitute the main population of high energy (HE) and very high energy (VHE) emitting sources found so far, thus being the main targets to be observed with MAGIC. The MAGIC Telescope has been designed to achieve an energy threshold of about 30 GeV and a high flux sensitivity, to highly increase the AGN catalog in the VHE range. Here we compile, based on EGRET data, a list of AGNs that we expect will be detected by MAGIC. The 3rd EGRET catalog [1] and the detections of the former generation of IACTs [2] are the natural references for MAGIC. EGRET was sensitive to gamma rays of energy up to 10 GeV and the threshold for Cherenkov telescopes was about 500 GeV. Most of the detected sources were AGNs (Blazars). In spite of their substantially better sensitivity, IACTs only detected 8 blazars while the EGRET catalog lists 66 AGNs. Since MAGIC is sensitive to the energies from 30 GeV to few TeV, it is the perfect instrument to investigate why EGRET AGNs seem to disappear at TeV energies.

In order to produce a list of observable AGNs, the 3rd EGRET catalog sources are considered. The energy spectrum for these sources is extrapolated to MAGIC energies assuming a simple power law. The effect of the infra-red

Table 1.   AGN observation prospects with MAGIC.

	source name	type	z	estimated flux (ph cm^{-2} s^{-1})	obs. time (hours)
Mrk 421	3EG_J1104+3809	BL Lac	0.031	8.55 $\times 10^{-11}$	3.2
OD+160	3EG_J0237+1635	BL Lac	0.94	6.41 $\times 10^{-11}$	4.3
3C 279	3EG_J1255-0549	FSRQ	0.538	6.07 $\times 10^{-11}$	4.6
W Comae	3EG_J1222+2841	BL Lac	0.102	6.03 $\times 10^{-11}$	4.6
CTA026	3EG_J0340-0201	FSRQ	0.852	1.67 $\times 10^{-11}$	23.8
4C+29.45	3EG_J1200+2847	FSRQ	0.729	2.23 $\times 10^{-11}$	27.5
3C 66A	3EG_J0222+4253	BL Lac	0.444	34.3 $\times 10^{-11}$	0.8
3C454.3	3EG_J2254+1601	FSRQ	0.859	25.3 $\times 10^{-11}$	1.1
OJ+287	3EG_J0853+1941	BL Lac	0.306	15.8 $\times 10^{-11}$	1.8
4C+15.54	3EG_J1605+1553	BL Lac	0.357	15.5 $\times 10^{-11}$	1.8
4C+38.41	3EG_J1635+3813	FSRQ	1.814	8.96 $\times 10^{-11}$	3.1
	3EG_J1224+2118	FSRQ	0.435	6.95 $\times 10^{-11}$	4.0
4C+15.05	3EG_J0204+1458	FSRQ	0.405	6.49 $\times 10^{-11}$	4.3
	3EG_J2359+2041	FSRQ	1.066	5.70 $\times 10^{-11}$	4.9
	3EG_J0958+6533	BL Lac	0.368	4.80 $\times 10^{-11}$	5.8
	3EG_J0952+5501	FSRQ	0.901	4.16 $\times 10^{-11}$	6.7
NRAO530	3EG_J1733-1313	FSRQ	0.902	2.48 $\times 10^{-11}$	11.2
	3EG_J0450+1105	FSRQ	1.207	2.54 $\times 10^{-11}$	16.7
	3EG_J1512-0849	FSRQ	0.361	1.56 $\times 10^{-11}$	18.5
	3EG_J0828+0508	BL Lac	0.18	1.96 $\times 10^{-11}$	23.4
NRAO190	3EG_J0442-0033	FSRQ	0.844	1.65 $\times 10^{-11}$	25.9
PKS 2155-304	3EG_J2158-3023	BL Lac	0.116	0.93 $\times 10^{-11}$	29.8
3C 273	3EG_J1229+0210	FSRQ	0.158	1.55 $\times 10^{-11}$	33.3
	3EG_J0530+1323	FSRQ	2.06	1.83 $\times 10^{-11}$	34.3
	3EG_J1614+3424	FSRQ	1.401	1.95 $\times 10^{-11}$	34.8
CTA102	3EG_J2232+1147	FSRQ	1.037	1.74 $\times 10^{-11}$	36.3
	3EG_J0422-0102	FSRQ	0.915	1.19 $\times 10^{-11}$	48.9
	...	...	...	...	...
BL Lacertae	3EG_J2202+4217	BL Lac	0.069	1.43 $\times 10^{-11}$	58.4
	3EG_J1738+5203	FSRQ	1.375	0.81 $\times 10^{-11}$	139

(IR) background absorption is taken into account and a cutoff energy of 5 GeV is simulated if the spectral index is below 2. The resulting flux is then used to compute the needed observation time to detect the source with 5σ significance, using a nominal sensitivity of 10^{-11} ph cm^{-2} s^{-1} at 30 GeV. A total of 28 AGN have been found to be detectable with MAGIC in less than 50 hours of observation (see Table 1). This number may increase if we observe sources in their high flux state.

Pulsars.   Since their discovery pulsars have been observed in a wide range of wavelengths. Several theoretical models of emission have been developed to explain these data. The EGRET detector of the CGRO satellite measured pulsed emission from several high energy pulsars up to energies of about 20 GeV. The energy band between 20-300 GeV is at present unexplored.

Due to physical phenomena occurring in the pulsar magnetosphere, the pulsed spectrum at high energies shows a cutoff at a specific value [3], experimentally constrained to the range of 5-200 GeV, model dependent. Its detection is of relevant importance to discriminate between different pulsed emission models of pulsars. To simulate the behavior of the telescope at the pulsar's cutoff (5-200 GeV) for gamma sources and cosmic ray background, we have to compute the effective area for gammas and protons through Monte Carlo (MC) simulations. At these energies we have to take into account the effect of the Earth's magnetic field on the extensive atmospheric shower (EAS) development. The particles of the shower are deflected by the magnetic field component perpendicular to their trajectory. This effect results in a decrease in the number of collected photons and therefore the telescope sensitivity. For MAGIC location, the maximum effect takes place between North and South directions at 50° of zenith angle.

We have assumed a reflector of 17m of diameter (MAGIC telescope) and trigger threshold ~200 photons within 1.3° of the pointing direction. The decrease in the effective area is shown in Figure 2 for different zenith angles.

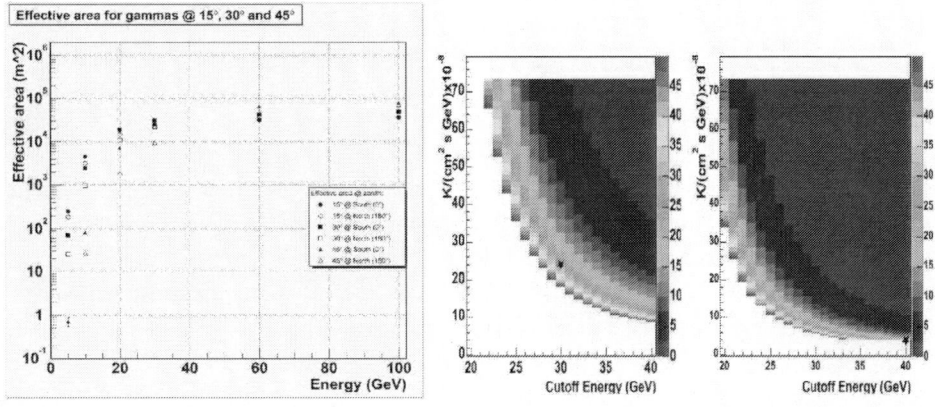

Figure 2. MAGIC effective area for gammas (E = 5-100 GeV) for North and South directions at zenith angles 15°, 30° and 45°.

Figure 3. Observation times ranges for Crab and PSR1951+32. K is the flux at 1GeV and $E.$ is the spectral cutoff. The star corresponds to the pulsar position.

The effect gives rise to a decrease of 50% in the MAGIC effective area for gammas at 5 GeV while there is hardly any effect in the background. We can calculate the observation times needed to detect the six high-confidence pulsars detected by EGRET at energies greater than 1 GeV (Figure 2) and any other possible pulsar spectrum (Figure 3). For this we have used the spectra fitted from the EGRET measurements [3] the proton flux from cosmic rays in the VHE range [4].

The observation time needed to obtain a given significance depends on the gamma and background rates and the quality of the analysis. The latter improves through timing analysis [5], up to a factor 3 in the case of Crab data.

Table 2. Rates for the EGRET pulsars (forth column) and observation time (fifth column) to detect them with 5σ with a proton rate $R = 70$ Hz for $\phi = 0^{\circ}$ / $\phi = 180^{\circ}$ (South/North) directions. PSR1706-44, PSR1055-52 and Vela are not visible to MAGIC. The first two columns show the culmination zenith angle and the energy cutoff of the pulsar spectra.

Pulsar	θ_{culm} ($^{\circ}$)	$E_{\cdot}$ (GeV)	R(Hz)	t_{obs} (hour)
Crab	7	30	0.25/0.10	30/180
PSR1951+32	4	40	0.30/0.14	20/100
Geminga	11	5	0./0.	∞/∞

Microquasars. Microquasars are a selected class of Xray binaries that produces relativistic radio-jets [6]. The origin of the jets is related to the matter accreted by the compact object, a neutron star or a black hole, from the companion star.

The word *microquasar* was proposed for the first time in 1992 to refer to this new class of galactic sources, not only because of the observed morphological similarities with the distant quasars, but also because of the physical similarities: the system itself behaves as a scaled-down version of quasars.

If the compact object is a black hole (BH) some parameters scale with its mass. BHs in quasars have about 10^8 solar masses and may result when huge quantities of gas collapse into the central region of a new born galaxy, whereas BHs in microquasars contain only a few times the mass of the sun ($\sim 10 M_{\odot}$), and may be the remains of a star after a supernova explosion. Since the mass of the microquasar is about 7 orders of magnitude smaller than that of a quasar, phenomena taking place in time scales of years in quasars can be studied in time scales of minutes in microquasars.

Following a leptonic model, gamma rays may arise from inverse Compton (IC) processes when the relativistic electrons of the jet interact with external photons from the massive companion star. If the jets are hadronic, the gamma rays may come from the decay of neutral pions in the interaction between protons from the jet and ions from the wind of the companion star.

High-mass microquasars have been proposed as the counterparts of variable EGRET sources on the galactic plane, like AX J1639.0-4642, LS 5039 or LSI +61 303. Figure 4 shows the integrated flux of the source LSI +61 303 when extrapolating it from the EGRET data ($1 - 10^4$MeV). Also an inhomogeneous model of microquasar leptonic jet to explore the gamma ray emission have

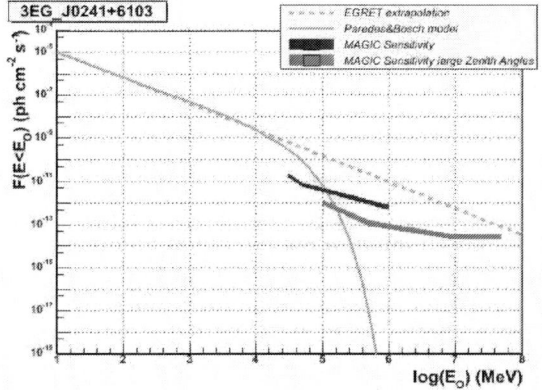

Figure 4 Microquasar candidate LSI+61 303.

been developed [7] and its spectrum is compared to the MAGIC sensitivity curve. For a very conservative MAGIC energy threshold about 100GeV, the observation time for a 5σ detection will be less than 50 hours. Two possible microquasar candidates as Cygnus X3 and GRS 1915+105 are currently being observed by MAGIC. The source GRS 1915+105 was observed at May 2004, when also a flare was seen by ASM (2-10 keV).

From the theoretical point of view, microquasars might emit at hundreds of GeV and beyond. The new generation of Cherenkov imaging telescopes like MAGIC will help clarify this point, improving the understanding of microquasars and hence of quasars.

Gamma Ray Bursts. GRBs are the most energetic and intense phenomena in the Universe which occur at cosmological distances. Due to technical limitations, the prompt gamma-rays at GeV energies have not been well studied. For ground based IACTs, the absorption process due to interaction with IR background photons is the most significant problem, since the cut-off in the gamma-ray energy spectrum is expected at about 30-100 GeV for the GRBs at $z = 1$ or more distance. Nevertheless, it is interesting to observe GRBs with a few tens GeV prompt gamma rays. In internal - external model [8], a few tens GeV prompt gamma-ray emissions are expected from accelerated electrons in the internal shocks. Photon emissions are explained with synchrotron process and inverse Compton process. According to some theoretical models [9; 10], two bumps are expected in the energy spectrum, the first bump is at a few tens keV and the second one is at a few tens GeV. The entire shape of the spectrum depends on the magnetic field strength and surrounding medium density. As of the fundamental physics, one of the interesting topics is a test of the quantum gravity and the special relativity using the high energy gamma rays from GRBs. Quantum gravity theory predicts the tiny time delay in the propagation of high energy photons. For several tens of GeV energy and the cosmolog-

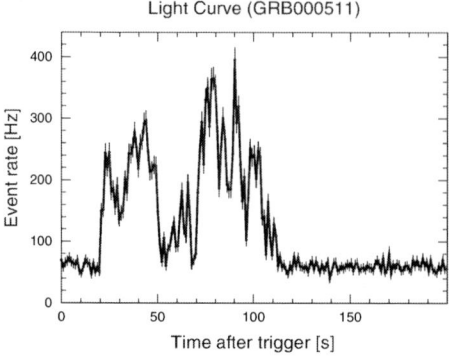

Figure 5 Expected light curve for GRB000511 in a simulated observation by MAGIC. We calculated the expected event rate of gamma-ray from 20 to 117 s (end of duration time of BATSE) at 10-40 GeV. A 60 Hz background rate is added.

ical distance, this time delay becomes an order of a few seconds and can be detectable provided we have enough statistics in the detection of GRBs.

MAGIC is a unique imaging Cherenkov telescope which allows us to observe the prompt high energy gamma rays, because of its low threshold energy and the fast slewing time. For example, after detecting the GRB alert from the satellite such as SWIFT [11], MAGIC can be pointed to a GRB within 22 seconds. Even at 10 GeV, MAGIC has an effective area of 10^5 m^2.

Using the information in BATSE GRB current catalog of 2704 GRBs [12; 13] and the spectral parameters in a sub-sample of 95 bright GRBs [14; 15], we have estimated the detectability of GRBs by MAGIC. For each GRB, we simulated the possible light curves observed by MAGIC as follows. The flux at 300keV is used for the normalization factor of the energy spectrum. The spectrum is extrapolated to sub-TeV energies with a spectral index of -2.2. We sampled the observation starting time randomly between 15 and 30 s which corresponds to the time delay from the GRB alert by SWIFT.

In Figure 5 we show a typical example of expected light curve (this event is generated from the light curve observed by BATSE for GRB000511). In this case, the observation starting time is 20 s after the trigger time of satellite detector. The background rate from cosmic ray protons is estimated to be about 60Hz after the minimum event selection using the directional information.

From this simulation, we estimated the detectability of GRBs by MAGIC. We found that GeV gamma-rays are detectable for 24% GRBs by MAGIC with reasonable statistical significance ($\geq 5\sigma$). If we assume 14% of the duty cycle of MAGIC observation time and 15% of its sky coverage, we expect that MAGIC has a chance to observe 4.2 GRBs per year and to detect significant signals for 1 GRB per year.

The analysis chain for the MAGIC telescope

The MAGIC Cherenkov telescope, using the imaging technique, collects the Cherenkov light produced by atmospheric showers. The parameters of the shower image allow us to apply methods to distinguish the shower images produced by γ-rays and by hadrons, that are background for us.

Calibration of the signal.   The digitized signal of each PMT must be converted from FADC counts to number of photons that produced the signal. For this purpose, during telescope operation, so-called calibration runs are taken, in which the telescope is triggered by light pulses emitted by LEDs of different wavelengths. The light intensity is uniform over the entire camera. The integral of the signal produced by a pixel from a calibration pulse is proportional to the measured charge. The charge can be converted from FADC counts to number of photons that hit the PMT with different methods: F-factor, muons or a blind pixel, currently under installation. The results presented below are based on the F-factor calibration. We convert the signal charge in number of photo-electrons (phel) by means of: $N_{phel} = \frac{Q}{\sigma_Q} F^2$ where F is a quantity measured in the laboratory and representing the intrinsic noise of the PMTs (F = 1.15), Q is the mean charge for a pixel and σ_Q^2 is the variance of the charge distribution. The charge and the variance are calculated from calibration runs analysis, after pedestal subtraction. The final conversion from phel to photons is done using the QE determined after measurements in laboratory.

Image cleaning and image parameter calculation.   Image cleaning is needed to recognize those pixels that contain the physical information about the shower and reject those whose signal is due to night sky light fluctuations. The parameterization of the image is done in terms of moments of the light distribution, such as LENGTH, WIDTH, SIZE (number of photons in the image), called Hillas parameters [16]. We calculate the distance (DIST) of the center of gravity of the image to the assumed position of the source and the orientation of the image with respect to the direction to the source position. The latter parameter is an angle called ALPHA. With the image shape parameters we can reconstruct the parameters of the original shower (arrival direction, energy) and perform the rejection of the background.

γ/hadron separation methods.   The technique to distinguish images generated by γ-rays from those generated by hadrons is based on their different image parameter distributions. The optimization can be done studying MC events. This method requires an extremely good description of the data by the MC. A MC independent method that compares observations of the γ-ray source (ON) and OFF-source, i.e. with no sources in the FOV will be used in the following. We apply selection criteria based on the image parameters. The different image parameter dependencies are taken into account in the so-called dynamical cuts, where the cuts on WIDTH and LENGTH depend quadratically on SIZE (Supercuts [17]). In the so-called Scaled cuts [18] approach, WIDTH and LENGTH are scaled to the mean values for OFF source in bins of SIZE.

The strongest discrimination power belongs to the parameter ALPHA whose typical distribution after cuts of other parameters is shown in Figure 6. From this plot, the number of detected γ events is calculated, since those events appear as a peak in the distribution at low ALPHA: the γ shower images are expected to point to the source position, while background images are homogeneously distributed (i.e. the ALPHA distribution is flat). The number of excess events and their significance are calculated after all cuts, subtracting the OFF data ALPHA distribution from the ON data ALPHA distribution [19].

Source position evaluation.   The correct calculation of the parameters that depend on the source position, like DIST and ALPHA, is critical for a correct analysis. We correct for the bending of the structure applying a model [20] which is tuned using star images on the camera plane. When the image of a star is focused on a pixel, the anode current of that PMT is higher with respect to the mean. We can compare the star map with catalogs and, by triangulation, determine the sky coordinates on the camera plane. Independently, we use the so-called "false source method": the image parameters that depend on the source position are calculated assuming different source positions across the FOV of the camera. For each position in the FOV, the ALPHA plot of the ON and OFF data sets allow us to calculate the number of excess events and the significance. The position with the highest significance corresponds to the most likely source position.

Outlook and further analysis steps.   We have presented the analysis method used during the commissioning phase of the MAGIC telescope. We are developing and testing new analysis techniques to analyze data using image parameters. A new method based on fit of the whole camera information called Model Analysis, is currently under development. We are working to complete the analysis chain with programs to calculate gamma ray fluxes as function of time (light curve) and unfold energy spectra.

First detections with MAGIC

The MAGIC telescope is in commissioning phase until October 2004. For that reason, two established TeV sources such as the Crab Nebula and the blazar Mrk421, have been observed extensively, in order to assess the telescope performance such as the angular resolution, the sensitivity and the energy threshold. This report, which presents an analysis based in a restricted sub-sample of Crab Nebula and Mrk421 data, is concerned with the first observations and detections by the MAGIC telescope.

Crab Nebula.   The Crab Nebula was first established as a TeV gamma-ray source by the Whipple telescope in 1989 [21] and for the moment being, is the best firmly established TeV steady emitter in the northern sky. MAGIC will not use the Crab Nebula only as a standard candle but can also expect interesting physics from it as the detection of the Crab pulsar and the IC peak [22; 23] once the 30 GeV energy threshold has been reached.

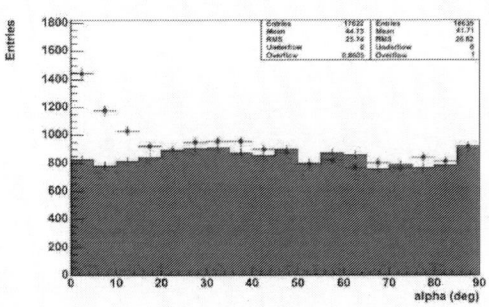
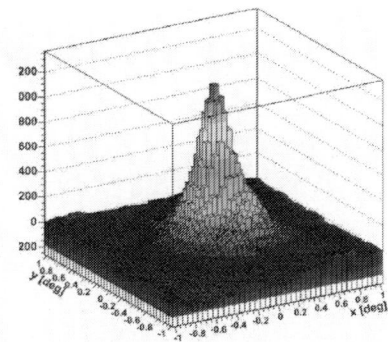

Figure 6.   ALPHA distribution of Crab Nebula events after selection cuts. ON (dots) and OFF (histogram) samples are normalized above 20°.

Figure 7.   Sky plot of γ-ray excess in degrees around the Mrk421 position.

In March 2004, Crab was observed for a total of 3.7 hours ON-source and 4.2 hours OFF-source over a zenith angle range between 25° and 50°. The ALPHA distributions for ON and OFF data samples after selection cuts for SIZE greater than 2000 photons are shown in Figure 6. A clear signal is observed for lower values of ALPHA with an excess of 1302 events and a significance of 15.1σ.

Mrk421.   The first extragalactic source found to emit gamma-rays in the TeV range was Mrk421 [24]. This blazar has also exhibited the fastest flux variations in the TeV range [25]. Mrk421 has been observed with the MAGIC telescope because it has been in a flaring state during the first months of 2004.

Here we present an analysis of the observations performed in February 2004 for a total of 1.6 hours ON source and 0.8 hours OFF source over a zenith angle range between 25° and 50°. A standard analysis has been applied, giving a number of excesses of 1594 events (25σ). The performance of the source position evaluation was evaluated by performing a "false source" method, which yields the sky plot shown in Figure 7.

Conclusions and Outlook.   The established TeV sources Crab Nebula and Mrk421 have been clearly detected by the MAGIC telescope during its commissioning period. These detections have been reached using a short data sample and a conservative imaging analysis at relatively high energies (above 100-200 GeV).

Light curve and spectra analysis are underway and soon will be presented as well as new methods for gamma/hadron separation. Calibration and analysis of the remaining Crab and Mrk421 data and other sources taken during these first months of operation is currently in progress. The analysis of low energy events, which is one of the aims of the MAGIC telescope is ongoing.

References

[1] R.C. Hartman *et al.*, ApJS **123** (1999) 79.

[2] D. Horan and T. C. Weekes, New Astron. Rev. **48** (2004) 527.

[3] O. C. de Jager *et al*, ApJ **457** (1996) 253.

[4] B. Wiebel-Sooth, *Measurement of the all particle energy spectrum and chemical composition of cosmic rays with the HEGRA detector*, PhD thesis, Universitt Wupertal, 1998.

[5] E. de Ona-Wilhelmi, *The Optimization of the MAGIC Telescope for Pulsar Observations*, PhD Thesis, Universidad Complutense de Madrid, 2004.

[6] I. F. Mirabel and L. F. Rodriguez, Ann. Rev. Astron. Astrophys **37** (1999) 409.

[7] V. Bosch-Ramon and J. M. Paredes, astro-ph/0401260, 2004.

[8] P. Mészáros and M. J. Rees, APJ **476** (1997) 232.

[9] A. Pe'er and E. Waxman, astro-ph/0311252, 2003

[10] A. Pe'er and E. Waxman, astro-ph/0407084, 2004

[11] http://swift.gsfc.nasa.gov/docs/swift/about_swift/

[12] http://www.batse.msfc.nasa.gov/batse/grb/catalog/current/

[13] W. S. Paciesas *et al.*, APJS **122** (1999) 465.

[14] R. D. Preece *et al.*, APJS **126** (2000) 19.

[15] R. D. Preece *et al.*, APJ **496** (1998) 849.

[16] A. M. Hillas, *Cherenkov light images of EAS produced by primary gamma*, in NASA, Goddard Space Flight Center 19th International Cosmic Ray Conference, Vol 3 (1985) 445.

[17] P. T. Reynolds *et al.*, ApJ **404** (1993) 206.

[18] F.A. Aharonian *et al.*, ApJ **539** (2000) 317.

[19] T. Li and Y. Ma, ApJ **272** (1983) 314.

[20] P. Wallace, "TPoint", http://:/www.tpsoft.demon.co.uk/

[21] T. C. Weekes *et al.*, ApJ **342** (1989) 379.

[22] A. M. Hillas *et al.* ApJ **503** (1998) 744.

[23] O. C. de Jager and A. K. Harding, ApJ **396** (1992) 161.
[24] M. Punch *et al.*, Nature **358** (1992) 477.
[25] J. A. Gaidos *et al.*, Nature **383** (1996) 319.

FIRST RESULTS FROM THE H.E.S.S. CHERENKOV TELESCOPE SYSTEM IN NAMIBIA

Martin Raue for the H.E.S.S. Collaboration

Institut f"ur Experimentalphysik, University of Hamburg
Luruper Chaussee 149, 22761 Hamburg, Germany

martin.raue@desy.de

Abstract   H.E.S.S. (High Energy Stereoscopic System) is a new system of four imaging atmospheric Cherenkov telescopes for gamma ray astronomy optimized for the 100 GeV to 10 TeV energy range located in Namibia on 1835 m a.s.l.. Each telescope is equipped with a 107 m˙ mirror and a high resolution camera build out of 960 photomultipliers. This article reports on first results on the Crab Nebula, the supernova remnant SN 1006, the galactic center and the active galactic nucleus (AGN) PKS 2155-304.

Keywords: Gamma Astronomy, Cherenkov Telescope

Detector

The H.E.S.S. experiment consists of a system of four atmospheric Cherenkov telescopes located in Namiba ($23°$ 16' S, $16°$ 30' E, 1835 m a.s.l.). The telescopes are alt-az mounted Davies-Cotton reflectors with a focal length of 15 m. The total mirror area of each telescope is $107\,m^2$. The telescopes are equipped with a high resolution camera made out of 960 photomultipliers (PMT), resulting in a field of view of 5 degrees with a pixel size of $0.16°$. The telescopes work in coincidence mode to allow for stereoscopic reconstruction of the shower parameters. With this observation mode an angular resolution of the shower direction of $0.1°$ per event and an energy threshold at zenith of about 100 GeV is achieved. The first H.E.S.S. telescope started data taking in June 2002. Since December 2003 all four telescopes are in operation. (Hofmann, 2003)

First Results

Crab nebula.   The Crab Nebula, first detected in the TeV regime by the Whipple experiment in 1989 (Weekes, 1989), is the standard candle of the

269

M. M. Shapiro et al. (eds.), Neutrinos and Explosive Events in the Universe, 269–273.

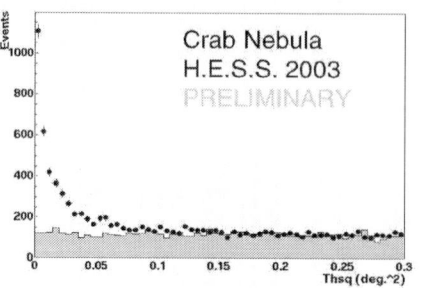 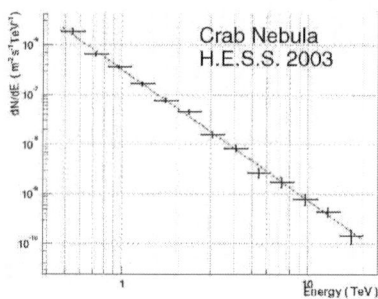

Figure 1. Left: Detection plot of the Crab Nebula from the H.E.S.S. 2003 data set. Shown is the number of detected events vs. the squared angular distance to the source position for the signal region and a control background region (grey shaded). Right: Differential energy spectrum. The dashed line is a power law fit to the data, which yields a spectral index of $\Gamma = 2.63 \pm 0.04$ and an integral flux above 1 TeV of $\Phi = (1.98 \pm 0.07) \cdot 10^{-7} \, m^{-2} \, s^{-1}$.

TeV-γ astronomy. It's high and steady flux makes it an ideal target for calibration and performance studies. Due to it's northern position the Crab Nebula culminates at a rather large zenith angle of about 45 ° at the H.E.S.S. site. The Crab Nebula was observed in 2002 with a single telescope and in 2003 with a two and three telescope setup. The results presented here are derived from the 2003 three telescope dataset. The calibration and analysis methods are described elsewhere (Aharonian et al., 2004b). Fig. 1 left shows the detection plot of number of events vs. squared angular distance from the target position for the signal and a control background region (grey shaded). A clear excess of over 50σ was found, which results in a rate of about 11 γ/min. The differential energy distribution is shown in Fig. 1 right. It is well described by a power law $d\phi/dE = \phi_0 \cdot (E/1 \text{ TeV})^{-\Gamma}$ with spectral index of $\Gamma = 2.63 \pm 0.04$ and an integrated flux above 1 TeV of $\phi(> 1 \text{ TeV}) = (1.98 \pm 0.07) \cdot 10^{-7} m^{-2} s^{-1}$, which is in good agreement with other measurements (Aharonian et al., 2000, Hillas et al., 1998, Tavernet and The CAT Collaboration, 1998).

SN 1006. SN 1006 is a shell type supernova remnant, which are prime candidates as production sites for galactic cosmic rays. SN1006 was reported as a gamma ray emitter by the CANGAROO collaboration in 1998 (Tanimori et al., 1998). H.E.S.S. observations have been performed with two telescopes during the commission phase in 2003. No significant excess was found in the field of view in 14h of lifetime. Taking the significance of 0.25 σ found at the hotspot position reported by the CANGAROO experiment, an upper limit (99% confidence level, following Helene, 1983) of 1.6% of the Crab Flux or 8 % of the flux reported by the CANGAROO experiment was derived.

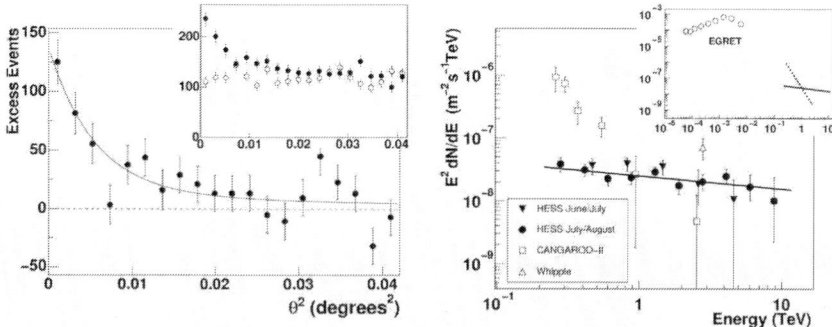

Figure 2. Left: Signal from the direction of the galactic center as measured by H.E.S.S.. The solid line marks the expectation for a point source. Right: Spectral energy distribution for the galactic center as derived by H.E.S.S. in comparison with measurements by CANGAROO (Tsuchiya et al., 2004) and Whipple (Kosack et al., 2004).

Galactic Center. The galactic center region shows a high concentration of potential TeV-γ ray sources like supernova remnants, the supermasive black hole Sgr A* or dark matter annihilation. Non thermal radiation has been detected in the radio, X-ray and GeV regime. Recently a detection in TeV has been reported by the CANGAROO collaboration (Tsuchiya et al., 2004) (and a marginal detection by Whipple (Kosack et al., 2004)). H.E.S.S. observations where carried out in 2003 with a two telescope setup. Clear detections with 9.7 and 6.4 σ were made (Fig. 2). A power law fit to the spectrum yields a rather hard spectral index of $\Gamma = 2.2 \pm 0.1$, which is not compatible with the spectrum reported by CANGAROO (Fig. 2 right). The measurements have not been carried out simultaneously, which could imply variability of the source. But in the H.E.S.S. dataset no significant indications for variability have been found. The position of the excess has been resolved on the arcmin scale to be compatible with Sgr A* ($(14 \pm 30)''$ in b and $(12 \pm 30)''$ in l from Sgr A*). (Aharonian et al., 2004c)

PKS 2155-304. PKS 2155-305 is a high-frequency peaked BL Lac object at a redshift of z= 0.117. It is very bright at all wavelength and shows strong variability. The first TeV-γ detection was made by the Mark 6 telescope in 1999 (Chadwick et al., 1999), but it was not detected in the follwing years. H.E.S.S. observations were carried out in 2002 with a single telescope and in 2003 with two and three telescopes (see Aharonian et al., 2004a). PKS 2155 was detected with a total significance of more than 45 σ. This detection was the first confirmation of PKS 2155 and establishes it as a source of TeV-γ-rays.

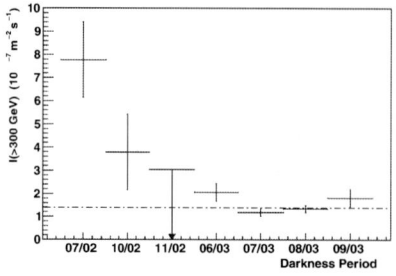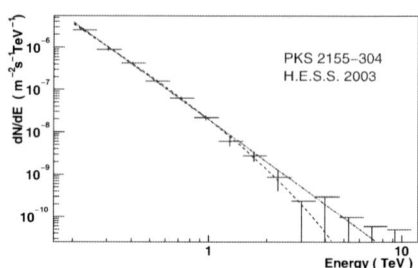

Figure 3. Left: Integrated flux above 300 GeV versus darkness period for PKS 2155 as measured by H.E.S.S.. Variability is clearly detected. Right: Time averaged differential energy distribution of PKS 2155-304 for the 2003 dataset. The spectrum is well described by a power law with a spectral index of $\Gamma = 3.32 \pm 0.06$.

Strong variability was seen on month and day level (Fig. 3 left). A time averaged spectrum is shown in Fig. 3 right. It is well described by a power law with spectral index $\Gamma = 3.32 \pm 0.06$. A power law with an exponential cut-off does not give a significant better fit. For TeV sources at this redshift a significant absorbtion from interaction with the diffuse extragalactic background radiation is expected. This, together with a discussion of different models, will be the subject of an upcoming paper. A new multiwavelength campaign is on the way in 2004 to provide more data with higher quality.

Summary

Already during the commission phase (with two or three out of the four telescopes in operation) shows the H.E.S.S. Cherenkov telescope system its great potential. Several detections have been made: The Crab Nebula, the standard candle of the TeV astronomy, has been detected with high significance. A signal from the galactic center was detected compatible with the position of Sgr A*. The derived energy spectrum is rather hard and differs from what was reported by earlier detections. The blazar PKS 2155 has been confirmed as a source of TeV-γ rays. Its high redshift (z= 0.117) makes it an ideal candidate to study TeV-γ absorbtion by the diffuse extragalactic background radiation. The supernova remnant SN 1006 has been observed for $\sim$14h and no significant access was found. An upper limit of 8% of the reported flux from the CANGAROO experiment was derived. The full H.E.S.S. detector is in operation since the December 2003 and we await more and exciting results.

References

Aharonian, F. A., Akhperjanian, A.-G., Aye, K., et al. (2004a). H.E.S.S. observations of PKS 2155-304. *Astronomy and Astrophysics*. submitted

Aharonian, F. A., Akhperjanian, A.-G., Aye, K., et al. (2004b). H.E.S.S. observations of the Crab Nebula. in preparation.

Aharonian, F. A., Akhperjanian, A.-G., Aye, K., et al. (2004c). Very high energy gamma rays from the direction of Sagittarius A*. *Astronomy and Astrophysics*, 425:L13.

Aharonian, F. A., Akhperjanian, A. G., Barrio, J. A., et al. (2000). The Energy Spectrum of TeV Gamma Rays from the Crab Nebula as measured by the HEGRA System of imaging air Cerenkov Telescopes. *The Astrophysical Journal*, 539:317.

Chadwick, P. M., Lyons, K., McComb, T. J. L., et al. (1999). Very high energy gamma rays from PKS 2155-304. *The Astrophysical Journal*, 513:161.

Helene, O. (1983). Upper limit of peak area. *Nuclear Instruments and Methods*, 212:319.

Hillas, A. M., Akerlof, C. W., Biller, S. D., et al. (1998). The Spectrum of TeV Gamma Rays from the Crab Nebula. *The Astrophysical Journal*, 503:744.

Hofmann, W. (2003). Status of the H.E.S.S. Project. In *28th International Cosmic Ray Conference*, volume 5, page 2811.

Kosack, K., Badran, H. M., and Bond, I. H. (2004). TeV Gamma-Ray Observations of the Galactic Center. *The Astrophysical Journal*, 608:L97–L100.

Tanimori, T., Hayami, Y., Kamei, S., et al. (1998). Discovery of TeV gamma rays from SN 1006: Further evidence for the supernova remnant origin of cosmic rays. *The Astrophysical Journal*, 497:L25.

Tavernet, J. P. and The CAT Collaboration (1998). Measurement of the Gamma-ray Spectrum of the Crab Nebula Above 250 GEV with the CAT Cherenkov Telescope. In *Abstracts of the 19th Texas Symposium on Relativistic Astrophysics and Cosmology, held in Paris, France, Dec. 14-18, 1998. Eds.: J. Paul, T. Montmerle, and E. Aubourg (CEA Saclay)*.

Tsuchiya, K., Enomoto, R., and Ksenofontov, L. T. (2004). Detection of Sub-TeV Gamma Rays from the Galactic Center Direction by CANGAROO-II. *The Astrophysical Journal*, 606:L115–L118.

Weekes, T. C. (1989). Observation of TeV gamma rays from the Crab Nebula using the atmospheric Cherenkov imaging technique. *The Astrophysical Journal*, 342:379.

THE AMANDA NEUTRINO TELESCOPE

Andrea Silvestri[†] for the AMANDA Collaboration
[†]*Department of Physics and Astronomy, University of California, Irvine, CA 92697, USA*
silvestri@HEP.ps.uci.edu

Abstract   We present new results from the Antarctic Muon And Neutrino Detector Array (AMANDA), located at the South Pole in Antarctica. AMANDA-II, commissioned in 2000, is a multipurpose high energy neutrino telescope with a broad physics and astrophysics scope. We summarize the results from searches for a variety of sources of ultra-high energy neutrinos: TeV-PeV diffuse sources by measuring either muon tracks or cascades, neutrinos in excess of PeV by searching for muons traveling in the down-going direction, point sources, neutrinos originating from GRBs, and dark matter in the center of the Earth or Sun.

Keywords: Neutrino Detector, Neutrino Telescopes, Neutrino Astronomy, Antarctic Ice, Ultra High Energy Neutrinos, AMANDA

Introduction

AMANDA is the first neutrino telescope constructed in transparent ice, and deployed between 1500 m and 2000 m beneath the surface of the ice at the geographic South Pole in Antarctica. It is designed to search for neutrinos that originate in the most violent phenomena in the observable universe. Galactic objects like Supernova Remnants (SNR) and extragalactic objects such as Active Galactic Nuclei (AGN) are expected to be the powerful engines accelerating protons and nuclei to the highest energies, which eventually interact to generate neutrinos. AMANDA has searched for point sources in the entire northern sky. It has also searched for a diffuse flux of neutrinos of cosmic origin. With the energy threshold of a few tens of GeV, the detector has measured atmospheric neutrinos, and searched for neutrinos correlated with gamma ray bursts (GRB), and high energy neutrinos from dark matter annihilation in the center of the Earth or Sun. The best sensitivity of the detector is achieved for neutrinos with E_ν between 10^3-10^6 GeV. However, the very quiet dark noise background in the Antarctic ice enables the detection of MeV neutrinos from a supernova (SN) within our Galaxy.

M. M. Shapiro et al. (eds.), Neutrinos and Explosive Events in the Universe, 275–285.
© 2005 *Springer. Printed in the Netherlands.*

Science Goals

Neutrinos are the only high energy particles able to propagate undeflected and unattenuated from the furthest reaches of the Universe. Extragalactic UHE γ-ray astronomy falters for energies greater than a few tens of TeV due to interactions with infrared and Cosmic Microwave Background photons. The information carried by the neutrino messengers from distant, unexplored regions of the universe may help to unravel longstanding mysteries associated with the origin of the highest energy cosmic rays.

Galactic supernova remnants have been suggested as one source of cosmic rays, and recent observations of high energy γ-rays from SNR (Aharonian et al., 2004) are compatible with this idea. If the γ-rays are produced by collisions between energetic hadrons and ambient material and fields in the the SNR, then the sources emit neutrinos with a flux that is correlated with the γ-ray flux (Stanev, 2002). Several models of the physical mechanisms that drive AGN emission suggest concurrent high energy neutrino emission from the core (Stecker et al., 1991, Stecker and Salamon, 1996) or from jets (Mannheim, 1995, Protheroe, 1996a) of blazars at fluxes that are detectable in AMANDA. Several models generically predict high energy charged particle and neutrino emission, but the diffuse ν-flux predictions by these models are constrained by the observed cosmic ray fluxes (Waxman and Bahcall, 1999). Therefore, the upper bound on the diffuse neutrino flux from possible sources of high energy cosmic rays provide an important target for detector sensitivity.

AMANDA

The AMANDA-II neutrino telescope (Andres et al., 2001, Barwick et al., 2002) is an array of 677 Optical Modules (OM) arranged in 19 strings. An OM is a pressure sphere housing of about 35 cm in diameter enclosing the PMT with optical coupling gel for continuity of the index of refraction, and dedicated electronics. For the detector, Hamamatsu 14-stage PMTs have been selected, which are operated at a gain of $\sim 10^9$ and experience a dark noise background in the ice between 300-1500 Hz. From 1997 to 1999, data were taken with the inner configuration consisting of 302 OMs arranged in 10 strings (AMANDA-B10). AMANDA will be integrated into IceCube (Ahrens et al., 2004c) in 2005.

Atmospheric μ and ν Background.   Down-going atmospheric muons are the major contribution to the background. The muon tracks are reconstructed with a maximum likelihood approach which models the arrival times and amplitudes of Cherenkov photons recorded by the photomultiplier tubes (PMTs) (Ahrens et al., 2004a), achieving an angular resolution of 1.5°-2.5°. Up-going atmospheric neutrinos are another source of background, and can be separated from presumed signal events only by their softer energy spectra and characteristic angular distribution. Atmospheric muons and neutrinos provide a convenient "beam" to test detector performance. Above a few PeV, the Earth becomes opaque to neutrinos, and only those moving down or horizontally can reach the detector. Such events need to be separated from large muon bundles from down-going atmospheric air shower events. In addition, the mass composition of high-energy CRs at energies above 10^{15}eV can be measured (Gaisser, 2004) in combination with the air shower array SPASE-2 at the South Pole (Ahrens et al., 2004b).

Physical Properties of Antarctic Ice

The physical and optical properties of the Antarctic ice provide fundamental information for a correct description of experimental data. Detailed analyses have been developed to study the ice at the South Pole as a function of depth and wavelength (Woschnagg et al., 2004). Results of these analyses show that the scattering coefficient of the light at a wavelength of 300-600 nm for a depth up to 1400 m is strongly effected by bubbles randomly distributed in the younger shallower ice layers (see Fig. 1). Pressure increases with depth, reducing the

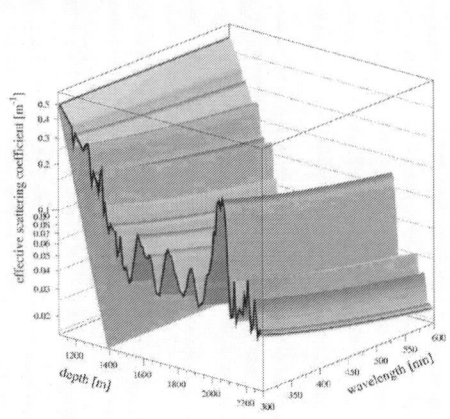

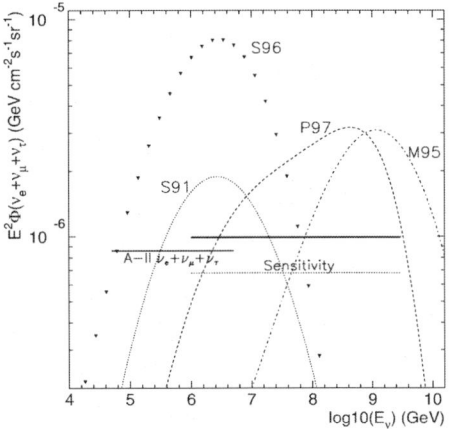

Figure 1. Scattering coefficient as a function of depth and wavelength.

Figure 2. AMANDA-II limits for the diffuse flux of all ν-flavors calculated for an E^{-2} spectrum. For model predictions shown, see (Ackermann et al., 2004a, Ackermann et al., 2004b)

bubble contribution drastically to a level where trace dust concentration begins to dominate. The effects of the dust layers can be observed from 1800 to 2300 m. The scattering length is fairly independent of wavelength at shallow depths but shows a power law dependency due to dust at depths below 1350 m. These results can be summarized with the average values of the absorption length

$\langle\lambda_{abs}\rangle\sim 110$ m and the scattering length $\langle\lambda_{sct}\rangle\sim 20$ m, both measured at 400 nm, where the PMTs exhibit the highest quantum efficiency.

Science Results

To perform unbiased analyses, a collaboration-wide policy of blindness was established, where cut selections are optimized on a fraction of data or on time-scrambled data set. We present upper confidence limits for null results following the treatment described in (Feldman and Cousins, 1998) and incorporate systematic uncertainties into the calculation of confidence intervals according to (Conrad et al., 2003). The contributions to systematic uncertainties is predominantly due to variations of the optical properties of the ice, the absolute sensitivity of the OM, the neutrino cross section and the muon propagation. The combined systematic uncertainty is typically 30%, although the value varies slightly with the analysis method.

Search for Diffuse Neutrino Flux.   The following three sections present different methods to search for diffuse flux of neutrinos. Fig.2 summarizes the results of the two search methods sensitive to all ν-flavors. The experimental limits assume 1:1:1 ratio of neutrino flavors at the Earth due to oscillation. The dotted and dashed lines represent a sample of model predictions (Stecker and Salamon, 1996, Protheroe, 1996a, Szabo and Protheroe, 1992, Stecker et al., 1991, Nellen and Biermann, 1993, Protheroe, 1996b), adjusted for oscillation if necessary, which have been excluded by these analyses. The model M95 (Mannheim, 1995), dash-dotted line, is not quite excluded. The leftmost (rightmost) solid line is the AMANDA limit obtained from the cascade (UHE) analysis. The length of the experimental lines corresponds to the energy interval that contains 90% of the signal for a spectrum proportional to E^{-2}.

(i) Neutrino-Induced Cascades.   The spatial topology from a cascade event is distinct from topologies created by long range muons. The short development length of cascade leads to more spherical topologies. The cascade analysis (Ackermann et al., 2004a) delivers excellent energy resolution of 0.1-0.2 in $\log(\Delta E/E)$ and improved sensitivity to all ν-flavors. AMANDA is sensitive to neutrino-induced cascades from any direction, but with relatively modest angular resolution of 30°-40°. The diffuse flux limit from the cascade analysis is

$$E^2\Phi_{\nu_{all}}(E) \leq 8.6 \times 10^{-7} \text{ GeV} \cdot \text{cm}^{-2} \cdot \text{s}^{-1} \cdot \text{sr}^{-1} \qquad (1)$$

derived for an E^{-2} spectrum at 90% confidence level (C.L.). As indicated in Fig.2, 90% of the signal lies within the energy interval between 50 TeV $< E_\nu$ $<$ 5 PeV. This limit is an improvement with respect to previous results (Ahrens et al., 2003b, Ackermann et al., 2004a).

(ii) Ultra High Energy (UHE) Neutrinos. Analysis strategies were developed to separate signal from background based on the large amount of Cherenkov light generated by interactions of neutrinos with $E_\nu > 10^{15}$ eV (Ackermann et al., 2004b). On the basis of 131 days of livetime of AMANDA-B10 data, 5 events were observed, while 4.6 events were expected. This corresponds to a flux limit (see Fig.2) for an E^{-2} spectrum of

$$E^2 \Phi_{\nu_{all}}(E) \leq 0.99 \times 10^{-6} \text{ GeV} \cdot \text{cm}^{-2} \cdot \text{s}^{-1} \cdot \text{sr}^{-1} \tag{2}$$

at 90% C.L., in the energy range of 1 PeV $< E_\nu <$ 3 EeV. An ongoing analysis with the AMANDA-II will yield a gain in sensitivity due to longer exposure time and improved cuts. Is is expected that the UHE search will benefit from new Transient Waveform Recorders electronics that registers the entire waveform of the PMT pulse for 10 μs. The TWR system was installed in AMANDA-II in January 2003.

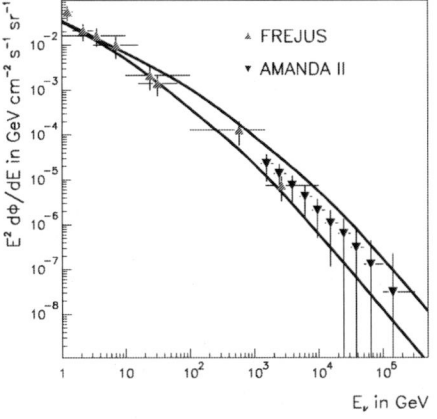

Figure 3. Atmospheric neutrino spectrum combined with measurements performed by the Frejus experiment.

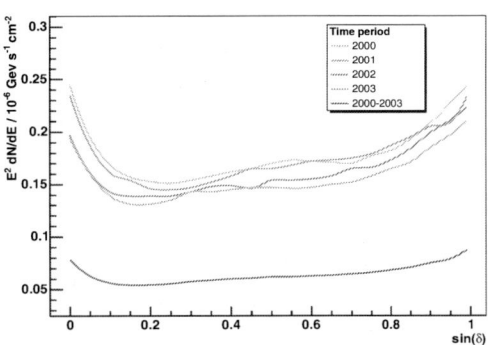

Figure 4. AMANDA-II sensitivities from a point source as a function of declination angle for $E^{-\bullet}$ spectra above $E_\nu = 10$ GeV at 90% C.L. Individual curves are presented for indicated years of operation. Bottom curve includes full data set.

(iii) Atmospheric Neutrinos. The energy spectrum of atmospheric neutrinos provides a third analysis tool to search for a diffuse source of high energy neutrinos. To measure the energy spectrum of the atmospheric ν_μ flux, a neural network for energy reconstruction was used (Geenen, 2002). Fig.3 shows preliminary results for data collected in 2000, combined with a previous measurement performed by the Frejus experiment (Daum et al., 1995). The measured atmospheric E_ν spectrum extends to 300 TeV, and agrees with theoretical parameterizations (Volkova, 1980). Since the spectrum of observed events is consistent with atmospheric neutrinos, it was used to limit a potential extraterrestrial ν_μ-flux with spectral dependence of E^{-2} to

$$E^2\Phi_{\nu_\mu}(E) \leq 2.6 \times 10^{-7} \text{ GeV} \cdot \text{cm}^{-2} \cdot \text{s}^{-1} \cdot \text{sr}^{-1} \qquad (3)$$

at 90% C.L., in the energy range of 100 TeV $< E_\nu <$ 300 TeV, which is approximately a factor 3 better than previously reported (Ahrens et al., 2003a).

Search for Point Sources. AMANDA-II has surveyed the entire northern sky since 2000 for non-statistical excesses in small regions of the sky (Ahrens et al., 2004d), none were found so far. The cut selection was optimized according to different declination bands and signal spectra from E^{-2} to E^{-3}. Fig. 4 summarizes the results of a 4-year combined analysis showing the expected sensitivity, based on the assumption of no signal as a function of declination for E^{-2} spectra above $E_\nu = 10$ GeV. Fig. 5 shows the statistical fluctuation for an unbinned sky search (color scale) and the 3369 observed events, which is in good agreement with the expected number of 3438 from atmospheric-ν's. No fluctuation exceeded 3.4σ, which is compatible with random fluctuation in the spatial distribution of atmospheric neutrinos.

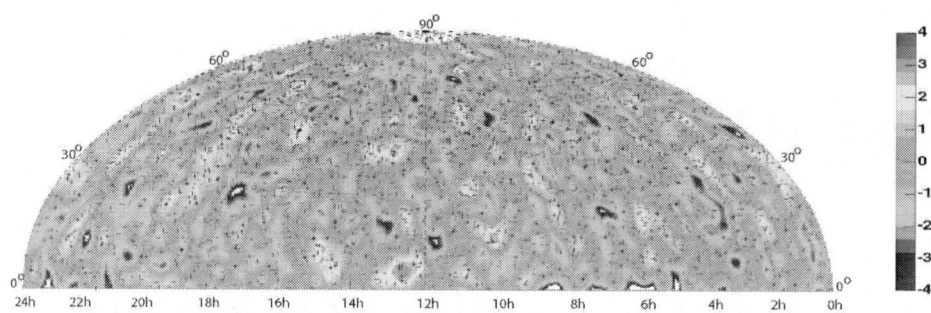

Figure 5.   Sky map (4 year data) plotted in coordinates of right ascension and declination overlaid with observed events (+ symbols). The scale on the right reflects excess or deficit in terms of standard deviation with respect to the mean background events.

Search for GRB Neutrinos.   The AMANDA GRB search for correlated ν emission relies on temporal (typically less than 100 s in duration) and angular information provided by BATSE and other satellites in the IPN network (Hurley et al., 1998). Initial search strategies for data collected between 1997-2000 (Hardtke et al., 2003) and for individual GRBs (Stamatikos et al., 2004), have assumed nearly concurrent emission within the duration of prompt gamma-ray emission (T_{90}). A new search (Kuehn et al., 2004) also scans for neutrino emission prior to the T_{90} start time. No excess was observed for either search method above the expected background from atmospheric ν and poorly reconstructed atmospheric muons. Therefore a flux limit can be computed for a given model spectrum. Fig. 6 shows the Green Function Fluence for ν_μ as a function of E_ν. Assuming an energy spectrum proportional to E^{-2}, Fig. 6 was used to obtain the preliminary experimental limit

$$E^2 \Phi_{\nu_\mu}(E) \leq 1 \times 10^{-8} \text{ GeV} \cdot \text{cm}^{-2} \cdot \text{s}^{-1} \cdot \text{sr}^{-1} \tag{4}$$

at 90% C.L. Assuming the broken power-law spectrum predicted by Waxman-Bahcall (Waxman, 2003), the preliminary flux limit is

$$E^2 \Phi_{\nu_\mu}(E) \leq 2 \times 10^{-8} \text{ GeV} \cdot \text{cm}^{-2} \cdot \text{s}^{-1} \cdot \text{sr}^{-1} \tag{5}$$

at 90% C.L. These limits were calculated for 119 GRB observed between 2000-2003, which includes GRB 030329. The corresponding sensitivities cal-

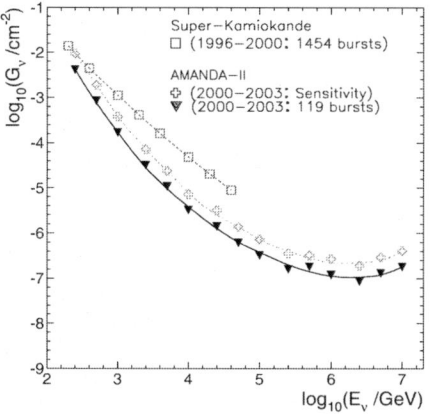

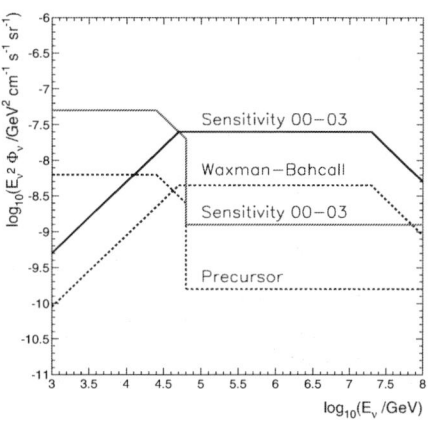

Figure 6. Green Function Fluence using 119 GRB as a function of E_{ν_μ}. Method developed by Fukuda et al., 2002.

Figure 7. Projected sensitivity using 149 GRB as a function of E_{ν_μ}. Detector sensitivity (solid curves) are compared to Waxman-Bahcall (Waxman, 2003) (rescaled for neutrino oscillation) and precursor (Razzaque and Mészáros, 2003, Razzaque et al., 2003) model spectra.

culated according to (Feldman and Cousins, 1998) are $E^2 \Phi_{\nu_\mu}(E) \leq 2 \times 10^{-8}$ GeV $\cdot$ cm$^{-2} \cdot$ s$^{-1} \cdot$ sr^{-1} and $E^2 \Phi_{\nu_\mu}(E) \leq 3 \times 10^{-8}$ GeV $\cdot$ cm$^{-2} \cdot$ s$^{-1} \cdot$ sr^{-1}, respectively. The quoted preliminary experimental limits are better than the sensitivities since fewer events were observed than expected from background. Work is continuing using 30 additional observable GRB with good localization and measurable duration which occurred between 2000 and 2003. Projected sensitivities, shown in Fig.7, were calculated according to the Waxman-Bahcall model, and to a precursor model which assumes a discontinuous spectrum (Razzaque and Mészáros, 2003, Razzaque et al., 2003).

Search for Dark Matter. The physics goals of AMANDA extend beyond the Standard Model. For example, AMANDA collaboration has reported results on searches for a class of dark matter candidates (Ahrens et al., 2002a) known as Weakly Interacting Massive Particles (WIMPs). One such candidate, the neutralino, arises from the Minimal Supersymmetric Extension of the Standard Model (MSSM). It is expected to be the lightest, and thus stable, supersymmetric particle, but massive enough to satisfy the requirements for dark matter (Lundberg and Edsjo, 2004). WIMPs populating the Galaxy occasionally are gravitationally trapped by the Sun or the planets. The neutralino WIMP density in the center of astronomical bodies grows until annihilation becomes comparable to the accumulation rate (Bergstrom et al., 1998). The annihilation process produces a variety of standard model particles that decay into neutrinos (Feng et al., 2001). A search for WIMPs from the Earth was performed on the

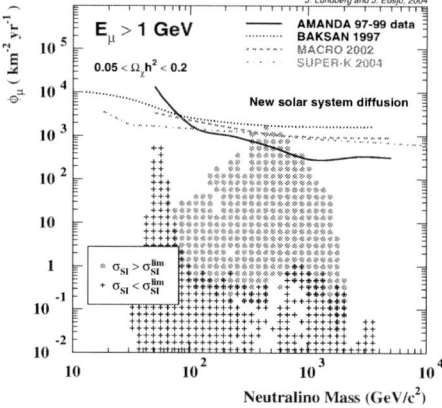

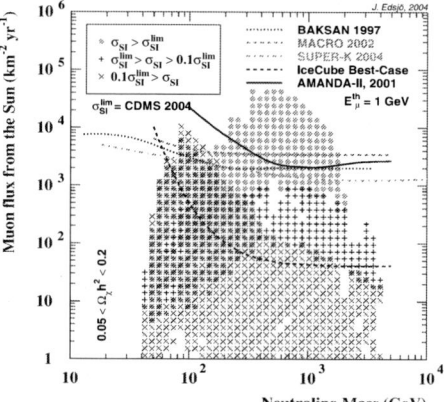

Figure 8. Muon flux from the Earth as a function of the neutralino mass. Dots and crosses represent model predictions from the MSSM.

Figure 9. Muon flux from the Sun as a function of the neutralino mass. Circles indicate predictions excluded by the CDMS experiment (Akerib et al., 2004).

basis of 422 days of livetime (collected between 1997-1999) with AMANDA-B10 data. No excess of events was observed and limits on the muon flux as a function of the neutralino mass were derived, (see Fig.8). A similar search for dark matter in the Sun was performed. The Sun, however, is visible only to $23°$ below horizon from AMANDA's location, which constrains the analysis to horizontal events. No excess above the expected atmospheric neutrino background was observed in 142 days of AMANDA-II livetime. Fig.9 shows that the AMANDA limits are competitive with other experiments.

Search for Supernovae. The copious flux of neutrinos originating from a SN produces low energy positrons that emit Cherenkov light throughout the detector. Coincidently with a SN explosion, the collective count rates summed over the stable PMT in the array would exceed the collective average dark noise rate. On the basis of 220 days of livetime with AMANDA-B10 data collected in 1997-98, AMANDA-B10 would have been able to observe SN similar to SN1987A to a distance of 9.8 kpc (Silvestri, 2000, Ahrens et al., 2002b), which includes 70% of the stars in the Galaxy. In 2003, the SN DAQ was upgraded to include all operating OMs, and preliminary estimates show that AMANDA now monitors 92% of the Galaxy. An online algorithm has been developed that allows AMANDA to join the SuperNova Early Warning System (SNEWS) (Antonioli et al., 2004).

Conclusion

AMANDA has yet to observe an extraterrestrial neutrino source, but *"she"* has demonstrated the cost-effectiveness and robustness of the technique. The detector is very versatile; it addresses many different neutrino physics subjects and sets the most stringent upper limits on Galactic and extragalactic neutrino fluxes. The improved search for diffuse fluxes, which has ruled out several predictions, along with the extended four-year search for point sources has started to constrain the enormous parameter space that exist in many models of neutrino production. The reported experimental limits on the diffuse neutrino flux are less than an order of magnitude above the Waxman-Bahcall bound (Waxman and Bahcall, 1999). As more of the data on tape is analyzed, AMANDA sensitivities will continue to improve. This is a very exciting time in neutrino astronomy and we look forward to neutrino astrophysics with next generation of neutrino telescopes.

Acknowledgments

This research was supported by the following agencies: National Science Foundation – Office of Polar Programs, National Science Foundation – Physics Division, University of Wisconsin Alumni Research Foundation, Department of Energy and National Energy Research Scientific Computing Center (supported by the Office of Energy Research of the Department of Energy), UC-Irvine AENEAS Supercomputer Facility, USA; Swedish Research Council, Swedish Polar Research Secretariat and Knut and Alice Wallenberg Foundation, Sweden; German Ministry for Education and Research, Deutsche Forschungsgemeinschaft (DFG), Germany; Fund for Scientific Research (FNRS-FWO), Flanders Institute to encourage Scientific and Technological Research in Industry (IWT) and Belgian Federal Office for Scientific, Technical and Cultural affairs (OSTC), Belgium; Fundación Venezolana de Promoción al Investigador (FVPI), Venezuela; D.F.C. acknowledges the support of the NSF CAREER program; E.R. acknowledges the support of the Marie-Curie fellowship program of the European Union; M.R. acknowledges the support of the Swiss National Science Foundation – A.S. acknowledges the support of the NATO Advanced Study Institute for providing the Full Scholarship at Erice.

References

Ackermann, M. et al. (2004a). *Astropart. Phys.*, **22**:127–138.

Ackermann, M. et al. (2004b). in press. *Astropart. Phys.*

Aharonian, F.A. et al. (2004). *Nature*, **432**:75–77.

Ahrens, J. et al. (2002a). *Phys. Rev. D*, **66**:032006.

Ahrens, J. et al. (2002b). *Astropart. Phys.*, **16**:345–359.

Ahrens, J. et al. (2003a). *Phys. Rev. Lett.*, **90**:251101.

Ahrens, J. et al. (2003b). *Phys. Rev. D*, **67**:012003.

Ahrens, J. et al. (2004a). *Nucl. Instrum. Meth.*, **A524**:169–194.

Ahrens, J. et al. (2004b). *Astropart. Phys.*, **21**:565–581.

Ahrens, J. et al. (2004c). *New Astron. Rev.*, **48**:519–525.

Ahrens, J. et al. (2004d). *Phys. Rev. Lett.*, **92**:071102.

Akerib, D. S. et al. (2004). astro-ph/0405033.

Andres, E. et al. (2001). *Nature*, **410**:441–443.

Antonioli, P. et al. (2004). *New J. Phys.*, **6**, 114.

Barwick, S. W. et al. (2002). astro-ph/0211269.

Bergstrom, L., Edsjo, J., and Gondolo, P. (1998). *Phys. Rev. D*, **58**:103519.

Conrad, J., Botner, O., Hallgren, A., and Perez de los Heros, C. (2003). *Phys. Rev. D*, **67**:012002.

Daum, K. et al. (1995). *Z. Phys. C*, **66**:417–428.

Feldman, G. J. and Cousins, R. D. (1998). *Phys. Rev. D*, **57**:3873–3889.

Feng, J. L., Matchev, K. T., and Wilczek, F. (2001). *Phys. Rev. D*, **63**:045024.

Fukuda, S. et al. (2002). *Astrophys. J.*, **578**:317–324.

Gaisser, T. K. (2004). *These Proceedings*.

Geenen, H. (2002). Diploma thesis: University of Wuppertal, Wuppertal, Germany.

Hardtke, R., Kuehn, K., and Stamatikos, M. (2003). *Proceedings: 28th International Cosmic Ray Conferences*, 2717–2720.

Hurley, K. et al. (1998). *ATel*, **19**.

Kuehn, K. et al. (2004). *Proceedings: Division of Particles and Fields*. to be published in IJMPA.

Lundberg, J. and Edsjo, J. (2004). *Phys. Rev. D*, **69**(12):123505.

Mannheim, K. (1995). *Astropart. Phys.*, **3**:295–302.

Nellen, L., Mannheim K. and Biermann, P. (1993). *Phys. Rev. D*, **47**:5270–5274.

Protheroe, R. J. (1996a). astro-ph/9612213.

Protheroe, R. J. (1996b). astro-ph/9607165.

Razzaque, S. and Mészáros, P. (2003). *Phys. Rev. D*, **68**:083001.

Razzaque, S., Mészáros, P., and Waxman, E. (2003). *Phys. Rev. Lett.*, **90**:241103.

Silvestri, A. (2000). DESY-THESIS-2000-028, Zeuthen. ISSN 1435-8085.

Stamatikos, M. et al. (2004). *AIP Conference Proceedings*, **727**(1):146–149, Sep. 28, 2004.

Stanev, T. (2002). Neutrinos from SNR. *NeSS 2002 International Workshop*.

Stecker, F. W. et al. (1991). *Phys. Rev. Lett.*, **66**:2697–2700. and Erratum (1992), ibid., **69**:2738.

Stecker, F. W. and Salamon, M. H. (1996). *Space Sci. Rev.*, **75**:341–355.

Szabo, A.P. and Protheroe, R.J. (1992). *Proceedings: High Energy Neutrino Astrophysics*.

Volkova, L. V. (1980). *Sov. J. Nucl. Phys.*, **31**:784–790.

Waxman, E. (2003). *Nucl. Phys. B Proc. Suppl.*, **118**, 353–362.
Waxman, E. and Bahcall, J. (1999). *Phys. Rev. D*, **59**:023002.
Woschnagg, K. et al. (2004). *in preparation*.

OBSERVATION OF VERY HIGH ENERGY GAMMA RAYS FROM THE REMNANT OF SN 1006 WITH HEGRA CT1

Vincenzo Vitale, for HEGRA CT1-Munich group
Max Planck Institute for Physics, Foehringer Ring 6, 80805 Munich
vitale@mppmu.mpg.de, vitale_tmp_2003@yahoo.it

Abstract   The Galactic Cosmic Rays (GCRs, those Cosmic Rays with intermediate energy between 10s of GeV and 100s of TeV per unit of charge) are studied from more then nine decades. The Supernova Remnants (SNRs) are the most likely GCRs candidate accelerators. But only very recently the Imaging Atmospheric Cherenkov Telescopes (IACTs) started to test directly this hypothesis. The HEGRA CT1 group observed the SN 1006 shell-type remnant, in an up to now unexplored energy region above 15-20 TeV. Deep exposures at large zenith angle (ZA) are possible from the telescope site. An excess of multi-TeV gamma rays is observed from the North East remnant cap. The observations, data analysis and results will be briefly discussed.

Keywords: gamma rays, cosmic rays, supernova remnant

Introduction

Strong shock waves are associated with the shell type SuperNova remnants. These shocks may accelerate the GCRs, via the Diffusive Shock Acceleration. For example, the SN 1006 remnant (G327.6+14.6, [1]) expands at 2890+-100 Km/sec [2], into a low density ISM. Electrons are accelerated up to 10-100 TeV by this source. Evidences of the electrons acceleration are found in the radio and the non-thermal X ray emissions. Such non-thermal radiations are observed mainly from the two shell caps (NE, SW [3]), and they are explained as synchrotron emission of the multi-TeV electrons. New results on this source have been recently obtained with the new X ray satellites (for example [4] and [5]). Gamma rays above few TeV are reported [6], but from only one of the caps (NE). These gammas may be produced: 1) via Inverse Compton scattering of TeV electrons on low energy photons; 2) by the freshly accelerated nuclei ($p \, p \to p \, p\pi_0$, $\pi_0 \to \gamma \, \gamma$).

M. M. Shapiro et al. (eds.), Neutrinos and Explosive Events in the Universe, 287–291.

Observation and Analysis

The telescope CT1 ([7], [8]) was part of the HEGRA complex on La Palma (28.75 N, 17.9 W, 2225m asl). The SN 1006 remnant (RA 15:02:48.8, Dec -41:54:42) has elevation $< 20°$, from La Palma. Therefore deep exposures at large zenith angle (ZA) are possible. The data, used for the present study, are: a) ON-source events, $\approx$346000, i.e. more than 220h of observations, from 1999 to 2001, with ZA between 71 and 73°; b) OFF-source events, a fraction of which between 71 and 73° of ZA; c) A large sample of muons passing through the camera (camera-muons), which leave narrow images and might mimic gamma events; d) A sample of simulated (Korsica) gamma events, among which $\approx$52000 survive the trigger conditions and $\approx$11000 are at 72° ZA. The data analysis proceeded in accordance with the following scheme:

I) Application of *Filter cuts* rejecting accidental noise triggers or events detected with non-optimal atmospheric condition. The trigger rate after filter is (0.5±0.1)Hz. Then the data pre-processing and image reconstruction are performed [10].

II) Gamma events selection, by means of rejection cuts against the dominant hadron background. At large ZA: a) the images of the air showers shrink ($\propto cos^{-1}(ZA)$); b) hadronic events may produce images small as low ZA gamma events. Then the selection cuts for low ZA are not effective for large ZA. Furthermore, small images, compared to the diameter of camera pixels, are less precisely measured. New selection cuts have been developed, with the study of the experimental hadronic events, spanning the full ZA dynamical range (from 0 to 73 °). The images transformation as a function of the zenith angle has been studied [10]. Such ZA dependence has been included into the selection cuts. It has been assumed that the images transformations are produced by geometrical factors, which are the same for gamma or hadron showers. An error in the new cuts extrapolation will result in a less than optimal gamma selection, and will not fake a gamma signal. A study of the camera-muons (rate=$5.1*10^{-3}$Hz, large ZA) has been performed, and specific rejection cuts have been obtained. The camera-muon rate after the rejection is $9.2*10^{-6}$ Hz.

III) Search for off-axis signal, within the telescope field of view (FOV). The so-called False Source Method (FSM) has been implemented. A grid of 0.1° step size was defined in the FOV, and at each knot the analysis for point-like source has been applied. The method works also for slightly extended sources underestimating somewhat the γ flux.

IV) Residual background suppression. The residual background has been evaluated with two different methods: the Alpha fit and the Ring method. 1) Alpha method: the distribution of the Alpha parameter has been fitted with a second order polynomial, without the linear term, between 20 and 80 ° (where only background is present). Then the fit extrapolation to the 0-12.5° region has been assumed as residual background. 2) Ring method: the data have been divided into two independent samples. A first sample is used to study the region of the putative source, while the second is used to study a Ring-shaped region, from which the background level is obtained (see also [10]). Both methods are in agreement. The Ring method increases the background statistics by a factor ≈ 3.

V) Telescope calibration with Monte Carlo simulations. The effective collection areas and an events energy estimate have been obtained.

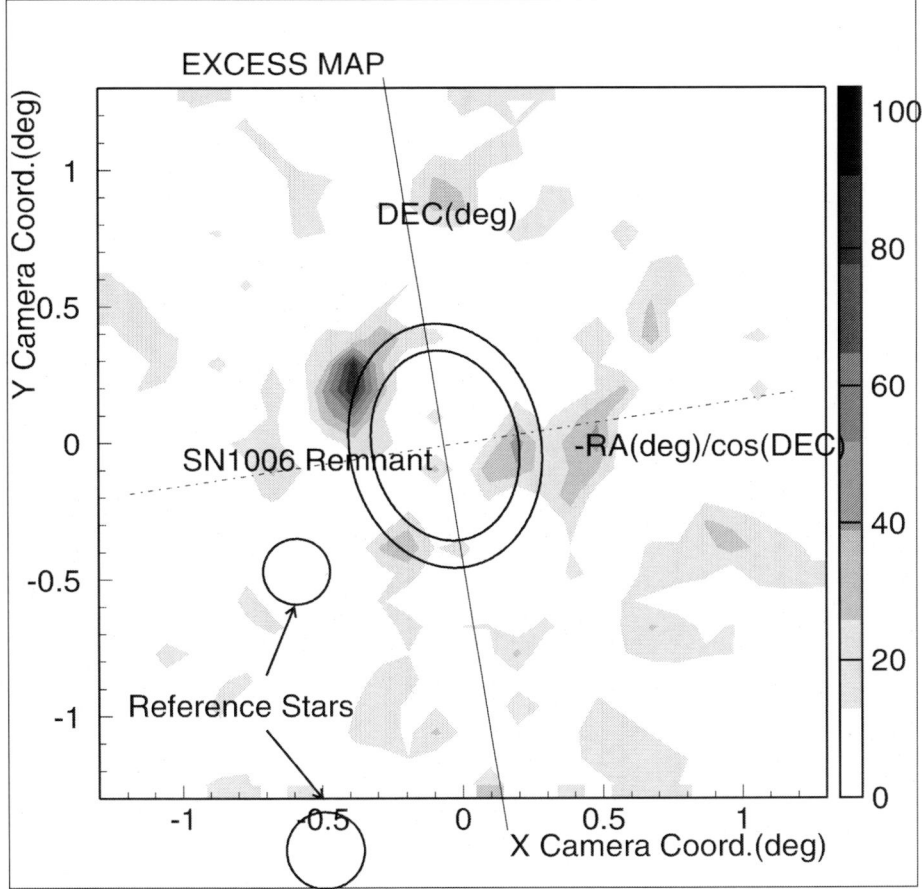

Figure 1. The sky map of the events excess. The Declination and Right Ascension axes have been reconstructed (the two axes in the map), by means of the reference stars. In the sky map it is evident a maximum, located at the NE region of the SN 1006 remnant shell (The SN1006 shell is represented with the black double ellipse).

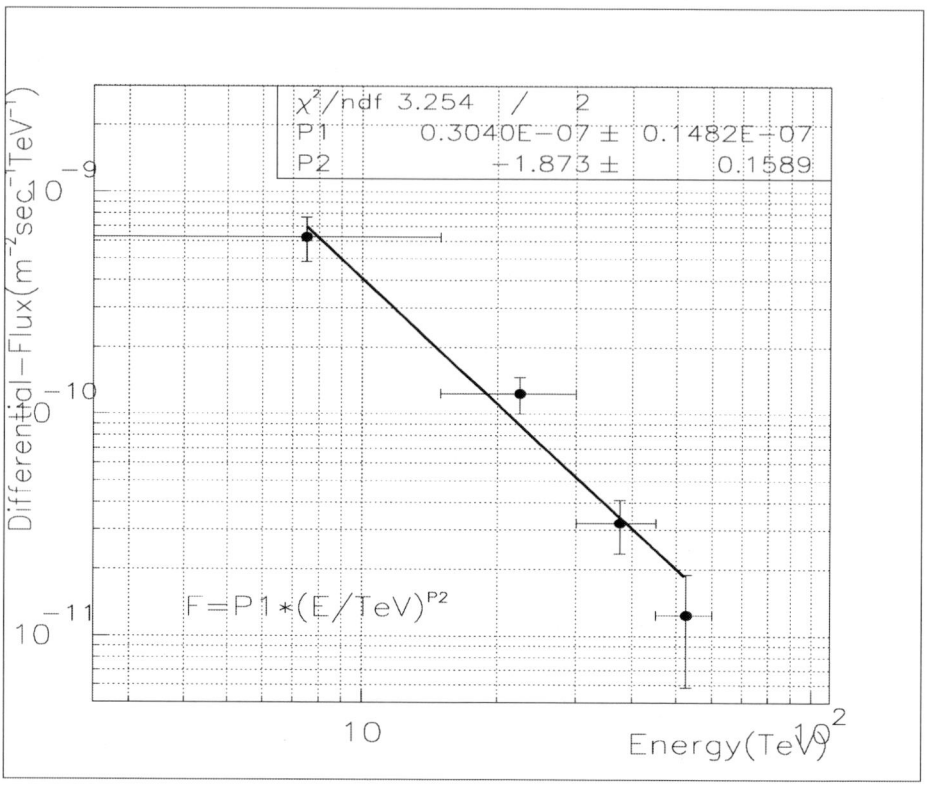

Figure 2.   Differential Flux versus Energy. The differential flux from the NE cap of SN1006 is shown. Data are grouped in four bins of energy. The differential flux is fitted with a pure power-law function. The statistical errors and those on the energy reconstruction are included. A 30% systematic error is not included. Further spectral studies are limited by the small statistics.

Results

With the above mentioned analysis, the following results have been achieved:

- The detection of TeV gamma rays from the North-East cap of the SN 1006 remnant. The TeV gamma rays are detected as an excess of events. The excess consists of 103 ± 17 events. The residual background consists of 225 ± 9 events.

- The excess significance is 5.1 standard deviations (σ), in accordance with the Li and Ma theory [9], which is usually applied within the gamma ray astronomy field. The initial significance was 5.6 σ, it has been reduced in order to account for the position indetermination of the source.

- The integral flux is : $\Phi(E>18\pm2 \text{ TeV}) = (2.35\pm0.4_{stat})*10^{-13}\text{cm}^{-2}\text{s}^{-1}$. This flux is obtained under the assumption that: the source emits a differential energy spectrum in the form of a pure power law, with spectral index Γ=-2.0. The Energy threshold for the same photon index is 18 ± 2 TeV. A systematic uncertainty of 30% has been considered for the fluxes calculations. With the inclusion of the systematic errors, the integral flux, for Γ=-2.0, becomes: $\Phi(E>18\pm2 \text{ TeV}) = (2.35\pm 0.4_{stat}\pm0.7_{syst})*10^{-13}\text{cm}^{-2}\text{s}^{-1}$

- The differential flux, in the form of a pure power law $A_{scale}*E^{-\Gamma}$, is: $\frac{d}{dE} = (0.30\pm0.15)*10^{-11} *(E/\text{TeV})^{-1.9\pm0.2}$ $\text{TeV}^{-1}\text{cm}^{-2}\text{s}^{-1}$. It should be noted that the differential flux at 1 TeV is an extrapolation, the result is valid for energies above the experimental threshold. The errors include the statistical ones, and those for the energy reconstruction. A 30% systematic error is not included.

- No signal has been found at the South-Western and centeral regions of the remnant. It is therefore possible to set the integral flux upper-limits for both: for the South-West cap $\Phi(E>18\pm2 \text{ TeV})_{SW} < (0.83)*10^{-13}\text{cm}^{-2}\text{s}^{-1}$; for the SNR central region $\Phi(E>18\pm2\text{TeV})_{CEN} < (0.66)*10^{-13}\text{cm}^{-2}\text{s}^{-1}$. Both upper-limits are at a confidence level of 95% and with an assumed photon index Γ=-2.0.

References

[1] Long, et al. 2003, Ap.J. 586
[2] Ghavamian, et al. 2002, Ap.J. 572
[3] Koyama, et al. 1995, Nature 378
[4] Rothenflug et al. A & A 425, 121
[5] Bamba,A. et al. 2004, Ad. Space Res. 33, issue 4, 376
[6] Tanimori, et al. 1998, Ap.J 497, L6
[7] Aharonian, F. et al 1999, A & A 349, 29
[8] Aharonian, F. et al 2003, A & A 410, 813
[9] Li,T. and Ma,Y. 1983, Ap.J. 272, 317
[10] Vitale,V. Ph.D. Thesis, 2004 Technische Universitaet Muenchen
    http://tumb1.biblio.tu-muenchen.de/publ/diss/ph/2004/vitale.html

OPTICAL ASTRONOMY WITH CHERENKOV TELESCOPES

E. de Oña-Wilhelmi[1,2], J. Cortina[3], O.C. de Jager[1] and V. Fonseca[2]

[1] *Unit for Space Physics, North West University, Potchefstroom 2520, South Africa*

[2] *Dept. de Física Atómica, Molecular y Nuclear, UCM, Ciudad Universitaria s/n, Madrid, Spain*

[3] *Institut de Fisica d'Altes Energies, UAB, Barcelona, Spain*

Abstract   The poor angular resolution of imaging γ-ray telescopes is offset by the large reflector areas of next generation telescopes such as MAGIC (17 m diameter), which makes the study of optical emission associated with some γ-ray sources feasible. Furthermore, the extremely fast time response of photomultipliers (PMs) makes them ideal detectors for fast (subsecond) optical transients and periodic sources like pulsars. The optical pulse of the Crab pulsar was detected with the HEGRA CT1 central pixel using a modified PM, similar to the future MAGIC camera PMs.

Introduction

Imaging Atmospheric Cherenkov Telescopes (IACTs) can be used to detect the optical emission of an astronomical object through the increased current if the camera pixels. Each of these pixels has a field-of-view (FOV) in the range of 0.1-0.3°. Hence, when an IACT follows a star, its optical emission is fully contained in the telescope central pixel.

The 3rd generation of IACTs are characterised by very large reflectors and fast response photo detectors (PMs), which make IACT fine instruments to detect transients, such as bright Active Galactic Nuclei (AGN) and Pulsars. For these type of telescopes, single photo counting techniques are expected to be applied. Nevertheless, using smaller reflectors, such as the HEGRA CT1 (Mirzoyan et al.) telescope, the optical emission can also be detected through the digitization of the central PM current.

Since the Crab pulsar shows pulsed emission in the optical wavelengths with the same frequency as in radio and most probably of VHE γ-rays, here we use the central pixel of HEGRA CT1 to monitor the optical Crab pulsation.

M. M. Shapiro et al. (eds.), Neutrinos and Explosive Events in the Universe, 293–296.
© 2005 *Springer. Printed in the Netherlands.*

The optical pulsed emission of the Crab pulsar has already been detected using IACTs (Srinivisan et at.) and other Cherenkov detectors (De Naurois et al.). The main purpose of such studies are to determine the timing parameters of the pulsar, with which the γ-ray arrival time can be folded, thus reducing the search for pulsed emission to a single statistical trial. It also allows us to test the accuracy of the used timing system and the solar system barycentric correction software, when comparing the timing parameters against contemporary radio data. This paper reports the optical timing signature of the Crab pulsar obtained, using the central pixel of CT1. This detection was used to derive the LONS around the Crab Nebula.

Experimental Procedure

Measurements were done with the HEGRA CT1 telescope during November 1-8th 2002. The standalone HEGRA CT1 telescope has a reflector area of 10 m^2 and a camera of 127 PMs, each one with 0.25^o FOV. Datasets were obtained with the Crab in the central pixel, and OFF source runs were also taken to check for any systematics. For optical observations the central pixel PM was modified. The DC branch, designed to monitor the DC current of the pixel, was adjusted to detect pulses of $\sim$3 ms (timescale of the pulse width), and the AC branch, which is designed to transmit the ns fast signals generated by the Cherenkov showers was removed (De Oña-Wilhelmi et al.).

Analysis and Results of the Crab Pulsed Optical Signal Measurements

The events' arrival times were recorded and transform to an inertial frame, the solar barycentre system. For this transformations we used the TEMPO software (Manchester et al.). The ephemeris (frequency, first frequency derivative and reference time) applicable to these observations were provided by Jodrell Bank ephemeris data base.

The arrival times t_i were folded to obtain the phases Φ_i.

$$\Phi_i = \Phi_0 + \nu(t_i - t_0) + \frac{1}{2}\dot{\nu}(t_i - t_0)^2, \tag{1}$$

where $\nu, \dot{\nu}$ are the Crab rotational frequency and frequency derivative, and t_0 is the reference epoch.

A phaseogram was produced for each independent frequency confined to the wide frequency range between 26.0 and 32.0 Hz. The independent Fourier spacing (IFS) between this range of independent frequencies is known to be $1/T$ (periodogram bin size), where T hours is the observation time (2 hours).

The folded intensities for each test frequency were tested against a uniform distribution by performing a χ^2 fit to a constant intensity. The reduced χ^2

was calculated (Figure 1). A maximum value of the reduced χ^2 of ~ 154 was found and the periodogram signal is contained within one IFS of the expected Jodrell Bank ephemeris Figure 1 shows our measured Crab pulse profile, with two sharp peaks, with a ~ 0.42 phase difference.

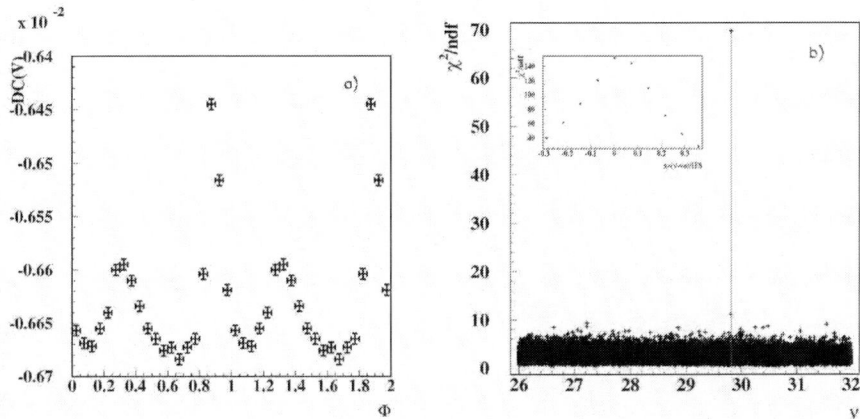

Figure 1. *Observations of the optical Crab pulsar with CT1. On the left, (a) the two peaks are clearly visible in the light curve with a separation in phase of ~ 0.4. On the right (b) the excess in the reduced $\chi^{\cdot}$ searching in a wide range of frequencies, using the 15th October ephemeris. The difference between the frequency found with 15th October ephemeris and 15th November is consistent with the drift in the digitization card clock. The inset shows the value of the reduced $\chi^{\cdot}$ vs. $x = (\nu - \nu_o)/IFS$, where ν_o corresponds to 15th October.*

Determination of the Galactic Anticenter LONS

Once we have detected the pulsed fraction of the signal, we will determine the galactic night sky background around the Crab pulsar. The point spread function (PSF) of the telescope and misspointing will have an influence on the among of light collected by the central pixel. The bright star ζ-Tauri, which appears in the camera field of view when CT1 points to the Crab pulsar, is used to measure these effects. The PSF of the telescope was determined from a raster scan with a resolution finer than the 0.25^o pixel size, with zero position corresponding to the true position of the star). Since the telescope is focused to ~ 6 km (maximum Cherenkov shower), the image of a point-like source will be smeared out around 0.14^o. The signal from ζ Tauri is slightly shifted to lower right ascension values, due to the telescope misspointing. This was accounted for by calculating the maximum value of the signal, and measuring the difference between the shifted measurement and the true position. Therefore to calculate the true signal collected by the central pixel, the signal intensity was fitted to a bidimensional Gaussian the fraction of the signal within the central pixel was measured to be:

$$\text{II} = \frac{\int_0^{r_{pixel}} \int_0^{2\pi} G(r,\theta)drd\theta}{\int_0^{\infty} \int_0^{2\pi} G(r,\theta)drd\theta} = 69\%. \tag{2}$$

From the misspointing correction factor II, it was possible to derive the LONS rate from:

$$p = \frac{R_{pulsed}}{R_{pulsed} + R_{nebula} + R_{LONS}}, \tag{3}$$

where p is the empirical pulsed fraction, R_{pulsed} is the expected rate of Crab pulsed events, R_{nebula} the expected rate from the Crab nebula and R_{LONS} is the galactic background rate around the Crab nebula. Using the pulsed fraction of the total Crab signal, the frequency integrated LONS (weighted with the frequency dependency of the PM quantum efficiency) is:

$$I_{LONS} = \int F_\nu^{LONS} d\nu = (4.6 \pm 0.1) \times 10^{12} \ ph \cdot m^{-2} \cdot s^{-1} \cdot sr^{-1} \tag{4}$$

Conclusions

The Crab pulsar was detected in optical wavelengths using the 10 m^2 reflector HEGRA-CT1 Cherenkov telescope, as a prototype to install a optical detector in the MAGIC camera central pixel. The pulsed signal was detected within 200 sec at 5σ level. The result was used to calculate the galactic LONS background on the surroundings of the Crab region. This value was found to be on average 2.7 times higher than the LONS value for a dark region. The advantage of this new method is that it relies on the stability of the clock, and is independent of other sky-techniques in the DC mode, provided that the PSF is known. The determination of the LONS for the Crab region is important, since the threshold energy determination and the spectrum depend on the level of the LONS.

References

De Naurois M. et al., CELESTE Coll. (P. Espigat, F. Mnz, A. Volte). ApJ 566 (2002) 343-357

De Oña-Wilhelmi E., J. Cortina and O.C. de Jager. MAGIC-TDAS 02-09 (2002)

Manchester R.N. et al. MNRAS, 328 (2001) 17-35

Mirzoyan R. et al, Nucl. Instr. and Meth. A, 351 (1994) 513.

Srinivisan R. et al., in Proc. "Towards a Major Atmospheric Cerenkov Detector V", p. 51, ed. O.C. de Jager (1997).

STATUS OF ANITA AND ANITA-LITE

Andrea Silvestri for the ANITA Collaboration
Department of Physics and Astronomy, University of California, Irvine, CA 92697, USA
silvestri@HEP.ps.uci.edu

The ANITA Collaboration:
S. W. Barwick[*], J. J. Beatty[*], D. Z. Besson[*], W. R. Binns[*], B. Cai[**], J. M. Clem[*], A. Connolly[*], D. F. Cowen[*], P. F. Dowkontt[*], M. A. DuVernois[**], P. A. Evenson[*], D. Goldstein[*], P. W. Gorham[*], C. L. Hebert[*], M. H. Israel[*], H. Krawzczynski[*], J. G. Learned[*], K. M. Liewer[*], J. T. Link[*], S. Matsuno[*], P. Miocinovic[*], J. Nam[*], C. J. Naudet[*], R. Nichol[*], M. Rosen[*], D. Saltzberg[*], D. Seckel[*], A. Silvestri[*], G. S. Varner[*], F. Wu[*].

[*] *Bartol Research Institute, University of Delaware, Newark, DE 19716, USA,* [*] *Dept. of Astronomy and Astrophysics, Penn. State University, University Park, PA 16802, USA,* [*] *Dept. of Physics and Astronomy, University of California, Irvine CA 92697, USA,* [*] *Dept. of Physics and Astronomy, University of California, Los Angeles, CA 90095, USA,* [*] *Dept. of Physics and Astronomy, University of Hawaii, Manoa, HI 96822, USA,* [*] *Dept. of Physics and Astronomy, University of Kansas, Lawrence, KS 66045, USA,* [*] *Dept. of Physics, Ohio State University, Columbus, OH 43210, USA,* [*] *Dept. of Physics, Washington University in St. Louis, MO 63130, USA,* [*] *Jet Propulsion Laboratory, Pasadena, CA 91109, USA,* [**] *School of Physics and Astronomy, University of Minnesota, Minneapolis, MN 55455, USA*

Abstract   We describe a new experiment to search for neutrinos with energies above $3 \times 10^{**}$ eV based on the observation of short duration radio pulses that are emitted from neutrino-initiated cascades. The primary objective of the ANtarctic Impulse Transient Antenna (ANITA) mission is to measure the flux of Greisen-Zatsepin-Kuzmin (GZK) neutrinos and search for neutrinos from Active Galactic Nuclei (AGN). We present first results obtained from the successful launch of a 2-antenna prototype instrument (called ANITA-lite) that circled Antarctica for 18 days during the 03/04 Antarctic campaign and show preliminary results from attenuation length studies of electromagnetic waves at radio frequencies in Antarctic ice. The ANITA detector is funded by NASA, and the first flight is scheduled for December 2006.

Keywords: High Energy Neutrinos, Neutrino Telescopes, Neutrino Astronomy, Antarctic Ice Attenuation, GZK Neutrinos, Radio Detection, ANITA

Introduction

ANITA is designed to search for particles that are created in the most powerful environments in the universe. Its primary goal is to discover neutrinos gen-

M. M. Shapiro et al. (eds.), Neutrinos and Explosive Events in the Universe, 297–306.
© 2005 *Springer. Printed in the Netherlands.*

erated by interactions between the highest energy cosmic rays and the cosmic microwave background radiation, as well as neutrinos created at the inner edge of supermassive black holes (SMBH). Flying at an altitude between 35-40 km above the Antarctic continent, the ANITA balloon-borne telescope will be sensitive to a target volume of $\sim 10^6$ km^3 of radio-transparent ice. The aperture of ANITA exceeds ~ 100 km^3·sr at $E_\nu = 3 \times 10^{18}$ eV, averaged over neutrino flavor and assuming equal fluxes of all flavors. The aperture continues to grow rapidly as E_ν increases, reaching the order of 10^5 km^3·sr at $E_\nu = 10^{21}$ eV. At EeV energies, the Earth is opaque to neutrinos so only horizontal or slightly downgoing neutrinos are detectable (Fig. 1). Neutrinos interact within the ice and generate compact particle showers, which emit coherent radio signals that are detectable in the radio-quiet environment of the Antarctic continent.

Science Goals

Standard Sources. High Energy neutrinos carry unique information from objects in the universe, and complement the information provided by UHE γ-ray astronomy. For example, neutrinos, unlike photons, can propagate throughout the universe unattenuated, but photon astronomy between 10-100 TeV is limited to distances less than a few hundred Mpc due to interactions with infrared background photons, and to even shorter distances at PeV energies due to interactions with the cosmic microwave background. Neutrinos may be the only method to shed light on acceleration processes associated with sources of the highest energy cosmic rays, which extend more than six orders of magnitude above energy of 100 TeV. Thus, the direct detection of high energy neutrinos (Stecker, 1968, Berezinsky and Zatsepin, 1969, Stanev *et al.* 1993) will complement the investigation of the GZK cutoff (Greisen, 1966, Zatsepin and Kuzmin, 1966, Kuzmin, 2004), one of the most controversial issues in cosmic ray physics.

Several theoretical models predict neutrino emission related to accreting SMBH, which arise from particle acceleration near the black hole at the center of AGN (Mannheim et al., 2001). In this environment ultra-high energy protons may escape from the source, and a fraction of them eventually interact to produce high energy neutrinos. Therefore, high-energy cosmic-ray observations can be used to set a model-independent upper bound on the high-energy neutrino flux (Waxmann and Bahcall, 1999).

Physics beyond the Standard Model. The ANITA science goals extend beyond the Standard Model if GZK neutrinos produce highly unstable micro black holes (BH) when they interact with ice (Feng and Shapere, 2002, Alvarez-Muniz et al., 2002). The decay of these highly unstable micro BH via Hawking radiation would generate energetic hadronic showers that ANITA would detect. The signature of such events is the observation of an enhanced detection rate that is strongly energy dependent. This observation would provide evidence of new phenomena, such as the existence of extra dimensions.

ANITA

The Concept. The concept of detecting high energy particles through coherent radio emission was first postulated by Askaryan (Askaryan, 1962) and has recently been confirmed in accelerator experiments (Saltzberg et al., 2001). Particle cascades induced by neutrinos in Antarctic ice are very compact, no more than a few centimeters in lateral extent (Zas et al., 1992). The resulting radio emission is coherent Cherenkov radiation which is characterized by a conical emission geometry, broadband frequency content, and linear polarization. ANITA (Barwick et al., 2003a) will observe the Antarctic ice sheet out to a horizon approaching 700 km, monitoring a neutrino detection volume of order 10^6 km^3. The direction to the event is measured by time differences between antennas in the upper and lower clusters (Fig. 2). The statistical distribution of

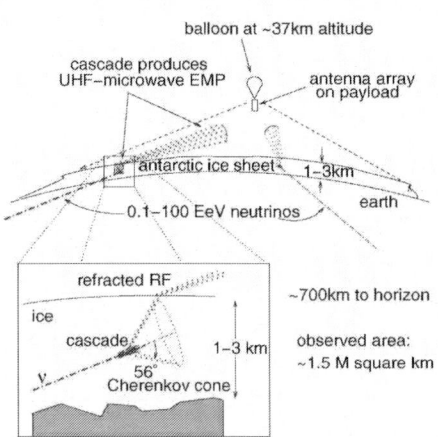

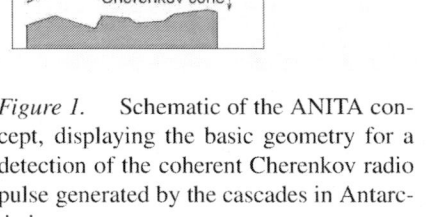

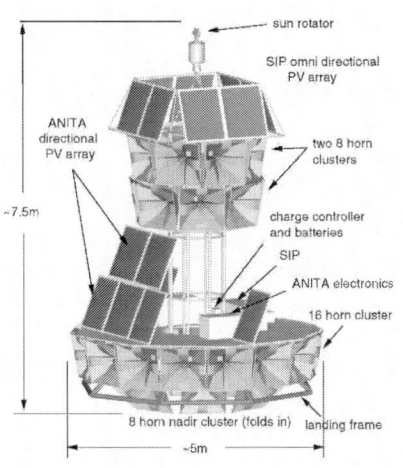

Figure 1. Schematic of the ANITA concept, displaying the basic geometry for a detection of the coherent Cherenkov radio pulse generated by the cascades in Antarctic ice.

Figure 2. Layout of the ANITA payload showing the geometry of dual-polarization, quad-ridged antennas.

events should correlate with ice thickness, averaged over observable volume. As illustrated in Fig. 1, ANITA will search for radio pulses that arise from electromagnetic and hadronic cascades within the ice. The signals propagate through 1-3 km of ice with little attenuation. At the energies of relevance to ANITA, the Earth strongly attenuates the neutrino flux, so ANITA is primarily sensitive to horizontal neutrinos. A radio pulse with zenith angle $< 34°$ will refract through the air-ice interface and may be observed by ANITA. Refraction and reflection effects due to small variations of the index of refraction in the layered structure of Antarctic ice are modest for the relevant incident angles. The RF emission pattern is peaked in a forward conical geometry, but considerable power remains within 5° of the Cherenkov cone. At ANITA energies, cascades initiated by electrons (e.g. in ν_e charged current events) are altered by LPM (Landau and Pomeranchuk, 1953, Migdal, 1957) effects which narrows the width of the Cerenkov Cone pattern to considerably less than 5°. Cascades initiated by recoil hadrons are not affected, and provide the bulk of observable events for neutrino energies above 1 EeV (Alvarez-Muniz and Zas, 1998).

Effect of Non-uniform Snow Surface.   The ANITA team is investigating the effect of surface features on signal characteristics. Surface structures that are somewhat larger than the wavelength band of the Cherenkov pulse, such as sastrugi, occupy only a small fraction of the surface. Random small-scale features are not expected to produce significant variations from average behavior. While the visual impact of surface roughness is quite apparent, the density contrast (and gradient of index of refraction) is only 10-30%. The broadband characteristics of the Cherenkov pulse also tend to mitigate amplitude variation due to interference. An event located at a typical depth of 1 km produces a coherent patch on the surface of $\sim$ 30 m in radius for the frequencies of interest. We have performed an analytic treatment of the air interface, assuming Gaussian fluctuations for the depth of snow, and find that the loss of phase coherence at the antenna reduces the detected power by less than 5%. We are pursuing the use of optics analysis software to numerically study surface effects.

The Detector.   The ANITA instrument (Barwick et al., 2003b), shown in Fig. 2, is a cluster of broadband quad-ridged horn antennas with a frequency range from 0.2 to 1.2 GHz. The beam width of the antenna is about 60°, with a gain of approximately 9 dBi at 300MHz; gain is roughly constant across the entire band. The detector geometry is defined by a cylindrically symmetric array of 2 levels of 8 antennas each on the upper portion, with a downward cant of about 10°, to achieve complete coverage of the horizon down to within 40° of the nadir. The antenna beams overlap, giving redundant coverage in the horizontal plane. A second array of 16 antennas on the lower portion of the payload provides a vertical baseline for time delay measurements which will determine pulse direction in elevation angle. The absolute azimuthal orientation is established by Sun sensors, and payload tilt is measured by differential GPS.

ANITA-lite

A two-antenna prototype of ANITA, called ANITA-lite, was flown for 18 days on a Long-Duration Balloon (LDB). It was launched on December 17, 2003 as a piggyback instrument onboard the Trans-Iron Galactic Element Recorder (TIGER) (Link et al., 2003). Both polarizations of each quad-ridged horn antenna were digitized. For the purpose of triggering, the four linear polarization channels were converted to right-handed and left-handed circular polarization for a total of four channels. The trigger criteria required that 3 of 4 waveforms exceed a specified voltage threshold, which varied throughout the flight. The ANITA-lite mission tested nearly every subsystem of ANITA, and monitored the Antarctic continent for impulsive Radio Frequency Interference (RFI) and ambient thermal noise levels as well as triggered events. We discuss a few results that impact the upcoming ANITA experiment.

Timing Analysis Results. While aloft, ANITA-lite received signals from a surface transmitter to measure the timing resolution. Signals were observed out to distances of 200 km. The waveforms were processed by digital filtering of the frequency components (Silvestri, 2004), and the results of a band-pass filter applied to the raw waveform can be seen in Fig. 3. The time ref-

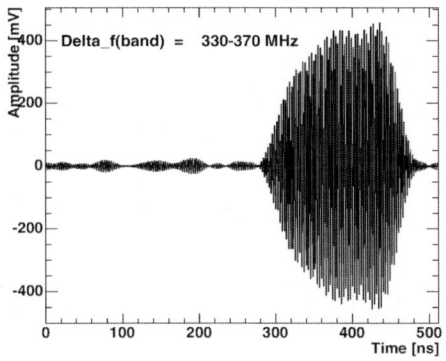
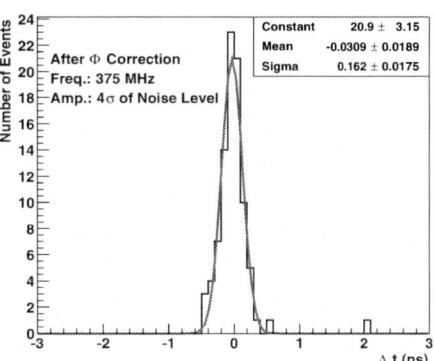

Figure 3. Calibration event at 350 MHz. Waveform after Band-Pass filter.

Figure 4. Distribution of time difference Δt between copolarized receiver channels.

erence for each antenna was extracted by interpolating the zero-crossing of a signal that exceeds $4V_{rms}$, where V_{rms} is the rms fluctuation of the noise voltage (Nam, 2004). The uncertainty in the time difference between the co-polarized channels of both antennas flown on ANITA-lite was $\sigma_{\Delta t} = 0.16$ ns (see Fig. 4) after correction for azimuthal variation. This value is based on the full set of calibration events that were acquired over a range of $40°$ in zenith angle θ, indicating that the systematic dependence of timing on θ is rather weak.

For ANITA, we expect the time resolution between antenna clusters to improve to $\sigma_{\Delta t} = 0.1$ ns due to the increase in the number of measurements by the full array. Using 3.3 m vertical separation between the upper and lower antenna arrays, d, the expected intrinsic zenith angle resolution σ_θ is

$$\sigma_\theta = (\sigma_{\Delta t} \frac{c}{d} 57.3°) = 0.5° \tag{1}$$

for events near the horizon. Similarly, azimuthal angular resolution σ_ϕ is estimated to be $1.5°$ using the ~ 1 m baseline separation between antenna elements in a ring.

Thermal Noise Analysis Results.   Another goal of Anita-lite was to investigate the level of ambient broadband noise in Antarctica at the balloon altitude and relate it to the absolute thermal noise background. Cosmic background radiation combined with the Galactic radio noise contributes noise of the order of 10-50K, and the average temperature of the Antarctic ice surface is in the range of 220-250K. The contribution of the Sun at 0.2-1.2 GHz is between 10^5 to 10^6 K depending on solar activity and radio frequency, however, its angular size in the radio band is of order $1°$ so its contribution is comparable to the Galaxy when averaged over the angular acceptance of the antenna. Both the Sun and the brightest region of the Galaxy were within the same field of view during the ANITA-lite flight. The results of this analysis are shown in Fig. 5,

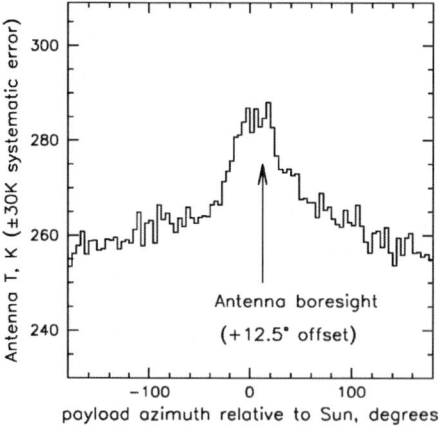
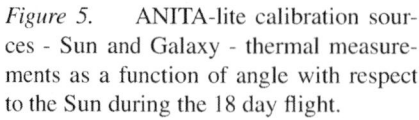

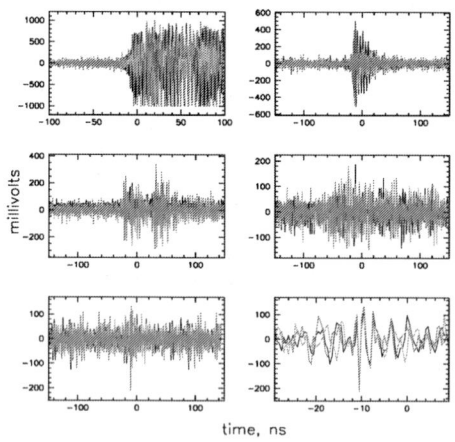

Figure 5.   ANITA-lite calibration sources - Sun and Galaxy - thermal measurements as a function of angle with respect to the Sun during the 18 day flight.

Figure 6.   Examples of ANITA-lite interference events (top 4 panes) and two views of an injected signal-like impulse (bottom panes).

where a clear excess of power in the solar direction is seen, consistent with the expectation (Miocinovic et al., 2004). This conclusion is encouraging for ANITA since it shows that the measured RF noise is consistent with expected noise levels.

Impulse Noise Analysis Results. ANITA-lite measured about 130,000 distinct events, with an approximate rate of 5 events per minute. Fig. 6 shows examples of the major classes of impulsive noise encountered. The vast majority of triggered events are due to local payload interference such as switching noise from the Support Instrument Package. The duration of this class of events exceeds several hundred nanoseconds and a few events exhibit circular polarization for a portion of the pulse. The bottom two panes show an example of a synthetic pulse injected into the data stream to simulate a true signal event. The pulse is coherent and aligned across all channels. Preliminary studies of event selection procedures were conducted. They yield no passing events, while 97% of simulated Askaryan-induced impulses with amplitude 5σ above noise survive (Miocinovic et al., 2004).

Attenuation Studies of Antarctic Ice

The attenuation length of the ice beneath the South Pole Station was measured in 2004 (Barwick et al., 2004). A pair of broadband Ultra-High Frequency (UHF) horns, with range from 200 to 700 MHz, were used to make echograms by reflecting radar pulses off the bottom of the ice cap (depth 2810 m), and measuring the return amplitude in a separate receiver. Antarctic ice exhibits a horizontally layered structure, which creates small variations in the index of refraction. The near vertical penetration of the radio signal through the ice strata minimizes the impact of reflections due to variation of the index refraction. By making the assumption that the reflection coefficient off the bottom is unity, one can derive a lower limit on the attenuation length. Because of the logarithmic dependence of the derived attenuation length on most of the parameters, the uncertainty in the attenuation length is relatively small. A summary of the results is shown in Fig. 7, where the error bars are an estimate of the 2σ systematic errors such as uncertainty in temperature profile of the ice, reflection coefficient, and antenna responses. The radio pulse propagates through ice which varies from $-50°C$ at the surface to near $0°C$ at the bottom. We correct the results to $-45°C$, a typical temperature for Antarctic ice. At this temperature the field attenuation length is ≥ 1 km between 200-700 MHz. These encouraging results define one of the most fundamental physical properties necessary for the success of ANITA.

ANITA Sensitivity

Fig. 8 shows various neutrino models, experimental limits, and a recent (late 2004) estimate of the sensitivity of ANITA for the baseline 3 LDB flights.

ANITA will achieve ~ 2 orders of magnitude improvement over existing limits on neutrino fluxes in the relevant energy regime. Assuming a 67% exposure over deep ice, typical of an IDB flight, we expect between 5-15 events from standard model GZK fluxes (Engel et al., 2001, Protheroe and Stanev, 1996), with the uncertainty primarily arising from assumptions about cosmic ray source evolution. Conversely, an observation of no events would reject *all* standard GZK models at the $\geq$ 99% C.L We stress that the GZK neutrino fluxes shown are

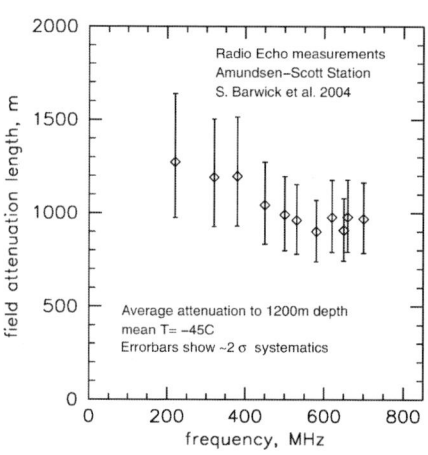

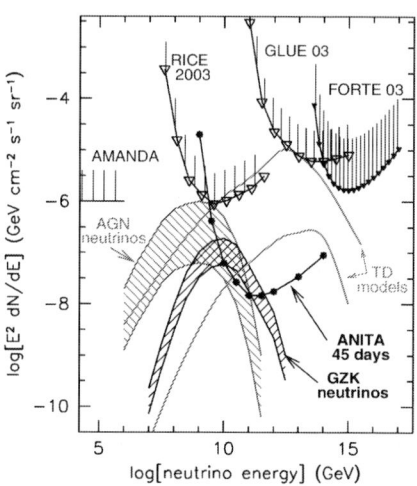

Figure 7. Field attenuation length versus radio frequency for 2004 measurements at the South Pole.

Figure 8. Neutrino models and experimental limits (90% C.L.) along with a recent estimate single-event sensitivity per energy decade of ANITA for the baseline mission of 3 flights that achieve 45 days of total exposure, assuming that ice is observable from ANITA payload 67% of the time.

predictions of great importance to both high energy physics and our understanding of cosmic-ray production and propagation. A non-detection of these fluxes by ANITA would suggest non-standard phenomena in either particle physics or the astrophysics of cosmic ray propagation.

Conclusion

ANITA will use radio emission from the cascade induced by a neutrino interaction within the Antarctic ice cap to detect UHE neutrino interactions that occur within a million square km area. The remarkable transparency of Antarctic ice to radio waves makes this experiment possible, and the enormous volume of ice that can be simultaneously monitored leads to an unparalleled sensitivity to neutrinos in the energy range of 0.1 to 100 EeV. Three separate flights provide 45 days of livetime, which can be increased by multiple orbits around the South Pole. It is possible, under optimal conditions and the use of multiple orbits around Antarctica, to obtain an integrated livetime approaching 100 days. In December 2001 the TIGER instrument obtained a record 31.8 day flight by circling twice around the Antarctic continent. Even the first ANITA flight will constrain model predictions of GZK neutrino fluxes and provide insight to particle physics at the energy frontier.

Acknowledgments This research was supported by the following agencies: NASA Research Opportunities in Space Science (ROSS) – The UH grant number is NASA NAG5-5387, Research Opportunities in Space Science – NSF and Raytheon Polar Services for Antarctic Support – NSBF for Balloon Launching and Operations – TIGER Collaboration for allowing ANITA-lite to fly as a piggyback – NATO Advanced Study Institute for providing the Full Scholarship at Erice.

References

Alvarez-Muniz, J., Feng, J. L., Halzen, F., Han, T., and Hooper, D. (2002). *Phys. Rev. D*,**65**:124015.

Alvarez-Muniz, J. and Zas, E. (1998). *Phys. Lett. B*, **434**:396–406.

Askaryan, G.A. (1962). *Sov. Phys. JETP*, **14**:441.

Barwick, S.W., Besson, D., Gorham, P., and Saltzberg, D. (2004). Submitted to *J. Glaciol.*

Barwick, S.W. et al. (2003a). Overview of the ANITA project. *Proceedings of SPIE*, vol. **4858** *Particle Astrophysics Instrumentation*. edited. by Peter W. Gorham (SPIE, Bellingham, WA, 2003), 265-276.

Barwick, S.W. et al. (2003b). Antarctic Impulsive Transient Antenna (ANITA) Instrumentation. ibid., 277-283.

Berezinsky, V.S. and Zatsepin, G.T. (1969). *Phys. Lett.*, **28B**:453.

Engel, R., Seckel, D., and Stanev, T. (2001). *Phys. Rev. D*, **64**:093010.

Feng, J.L. and Shapere, A.D. (2002). *Phys. Rev. Lett.*, **88**:021303.

Greisen, K. (1966). *Phys. Rev. Lett.*, **16**:748.

Kuzmin, V.A. (2004). *These Proceedings.*

Landau, L. D. and Pomeranchuk, I. (1953). *Dokl. Akad. Nauk Ser. Fiz.*, **92**:735–738.

Link, J. T. et al. (2003). *Proceedings of the 28th International Cosmic Ray Conferences*, **4**:1781.

Mannheim, K., Protheroe, R.J., and Rachen, J.P. (2001). *Phys. Rev. D*, **63**:023003.

Migdal, A. B. (1957). *Sov. Phys. JETP*, **5**:527.

Miocinovic, P. et al. (2004). Results of ANITA-lite prototype antenna array. *Neutrino 2004 Proceedings.*

Nam, J. (2004). Time calibration study for ANITA-lite
    http://www.ps.uci.edu/~jwnam/anita/timecal/tcal.html.

Protheroe, R.J. and Stanev, T. (1996). *Phys. Rev. Lett.*, **77**:3708.

Saltzberg, D., Gorham, P., Walz, D., et al. (2001). *Phys. Rev. Lett.*, **86**:2802.

Silvestri, A. (2004). Timing analysis of ANITA-lite
    http://www.ps.uci.edu/~silvestri/ANITA.html.

Stanev, T. (2004). *These Proceedings.*

Stecker, F.W. (1968). *Phys. Rev. Lett.*, **21**:1016.

Waxmann, E. and Bahcall, J. (1999). *Phys. Rev. D*, **59**:023002.

Zas, E., Halzen, F., and Stanev, T. (1992). *Phys. Rev. D*, **45**:362–376.
Zatsepin, G.T. and Kuzmin, V.A. (1966). *JETP Letters*, **4**:78.

OPTICAL SETI WITH MAGIC

Alexandre Armada, Juan Cortina, Manel Martinez
Institut de Fisica d'Altes Energies (IFAE), Universitat Autonoma de Barcelona,
08193 Bellaterra, Spain.
armada@ifae.es, cortina@ifae.es, martinez@ifae.es

Abstract   For almost half a century (since the 1960's) we have been looking for extraterrestial signals originated by extraterrestrial intelligent life in the radiofrequency (RF) range. However the development of the laser has made us aware that pulsed narrow beams may be ideal carriers for interstellar communications. Those features of the optical transmissions and the amazing improvement of the laser technology and power in the last years (it has practically doubled capability every two years) suggests that if some intelligent extraterrestrial life is trying to communicate with us, they probably do it in the optical range instead of the RF. In this sense, the MAGIC Telescope may be the best tool at the moment for the detection of these pulsed signals in the optical range, due to its large reflective surface ($234m^*$) and sensitivity to pulses as short as $1ns$.

Introduction

Philosophically, the known as *Principle of Mediocrity* tell us about the possible fact that Earth is not a special place in the universe, and life on it is just the coincidence of some environmental conditions that could be found elsewhere in the universe [1]. Furthermore, from a scientific point of view, the recent development in Astrometry and Planetary Search has led to the discovery of Jupiter-like planets so far, but Earth-like planets are expected to be also detected. In the same way, recent studies have pointed to the fact that life can also evolve in different environments taking profit of some different energy sources, enhancing the probability of planets evolving life [2].

Optical SETI

The optical range presents some relevant advantages with respect to the RF/MicroWave range, as it is a higher bandwith, that let us transfer information faster, and the fact that the interstellar medium produces a smaller perturbation on it. These features of the optical communication were known in 1959 when the SETI projects in the RF began, but only with the development of LASER

M. M. Shapiro et al. (eds.), Neutrinos and Explosive Events in the Universe, 307–310.
© 2005 *Springer. Printed in the Netherlands.*

devices in the last years it is possible to assert that the increasing power of LASERs makes them more suitable than MASERs for interstellar communication [3]. Furthermore now we know more about the natural phenomena in the universe, and it seems to be only one that produces optical pulses lasting a few nanoseconds: the Cherenkov flashes coming from the cascades generated by cosmic rays arriving at our atmosphere [4]. The expected background is thus lower than in the RF range if the communication is based on nanosecond optical pulses.

OSETI with MAGIC

The MAGIC (*Major Atmospheric Gamma-ray Imaging Cherenkov*) Telescope presents some special features: MAGIC has the largest optical reflector ever built (*i.e.*, $234 m^2$ mirror surface), with very high reflectivity and a special system to preserve the parabolic shape, giving a high concentration of light in the camera plane. Furthermore, the pixels (PMTs) in the camera are specially designed to detect pulses of a few photons. We can conclude that the photon sensitivity of MAGIC is extremely high to short pulses. Finally, the time resolution of the telescope is about 1 ns, enough to resolve the expected signals.

Detection Method

If the energy of the pulse, assumed to be of 1 ns of duration and also collected entirely in one pixel, is able to produce a signal above the single pixel trigger (*SPT*) threshold (currently 40 photons), then we get a single pixel trigger. In this way, the analog pulse is transformed to a logic one. A counter measures the rate of these logic pulses.

Minimum Pulse Energy to get a *SPT*

Assuming a mean value of the quantum efficiency of the pixel of 20%, a transmission of the signal in the telescope (mirrors and Winston Cones) of 80%, and the wavelength of the optical radiation about $500 nm$, we obtain that the minimum energy per unit area at the Earth to get a SPT is $8 \times 10^{-20} J/m^2$ per pulse. It is important to note that this extremely small amount of energy only depends on the characteristics of the telescope. But to get a complete estimation about the power of the LASER device the extraterrestial civilization should use, we must consider the divergence of the beam while traveling through interstellar space, given by

$$A_{di} = 4 \cdot \pi \cdot D^2 \cdot \left(\frac{\lambda}{d_t}\right)^2 \tag{1}$$

where A_{di} is the area of the diverged beam in a distance D from the LASER, λ is the wavelength of the transmitted signal, and d_t is the transmission aperture of the device. For example, taking a transmission aperture of $1m$, and a distance of $10pc$, the minimum energy emitted per pulse by their device should be about $24kJ$ to generate a SPT in MAGIC. This is not a huge amount of energy: a inertial fusion pulsed LASER project called *Helios* already exists, and is expected to have a $3ns$ pulses each containing $3.6MJ$, although with only $10Hz$ rate.

Backgrounds. As we will see, the main background signal for this kind of measurements with the MAGIC Telescope stems from the accidental SPT due to the stars present in the field of view. Let us suppose an F type star (like our Sun) with an apparent magnitude 5.5^m. Then the flux of photons arriving to Earth from the star is given by

$$F_{Earth} = 2.52 \cdot 10^{-8} \cdot 10^{-0.4\,m}\,J \cdot s^{-1} \cdot m^{-2} \tag{2}$$

and reconstructing the signal generated by these photons through the telescope we get 13.5 photo-electrons per ns, clearly above the possible contribution of the *Night Sky Background*, measured to be about 0.13 photo-electrons per ns. Assuming that this is the mean value for the assumed gaussian distribution of photo-electrons arriving from the star, the probability of a fluctuation of more than 8 photo-electrons above this mean level is $1.5 \cdot 10^{-2}$. We get an accidental SPT rate of $15MHz$. This rate is five orders of magnitude larger than the SPT rate that would be caused by the Cherenkov flashes if they were intense enough (around few hundred Hz).

For MAGIC the maximum readable SPT rate is expected to be approximately $200MHz$. Then, a significant detection of a signal must be a rate above the possible accidental rate due to presence of a star in the same pixel and also below the saturation value for the counter.

Conclusions

From the energetics of the SPT and the rates discussed before, we can represent the reach D of MAGIC detections from the power used by the emitting LASER devices assuming a frequency of $100MHz$ (figure 2). It must be noted that the main power limitation of this method applied to MAGIC is the quantum efficiency of the photomultiplier-tubes.

The MAGIC Telescope is still in the commissioning phase, but data for the SPT rates are already taken. We intend to analyze these data and come up with upper limits on pulsed emission of candidate planets on the field of view of observed gamma ray sources in the next months.

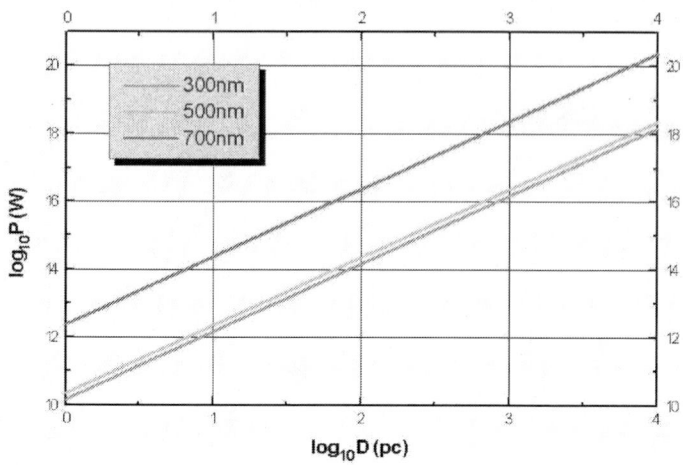

Figure 1.   Detectability of the signal by its power and distance.

References

[1] Webb, S. (2001). . Where is everybody? 50 solutions to the *Fermi's Paradox. Copernicus Books & Praxis Publishing Ltd.*.

[2] Sozzetti, A., Casertano, S., Lattanzi, M. G. and Spagna, A.. The GAIA astrometric survey of the solar neighborhood and its contribution to the target database for Darwin/TPF. *Astro-ph*, 0305111 v1.

[3] Tarter, J.. The Search for Extraterrestrial Intelligence (SETI). *Annu. rev. Astron. Astrophys.* 2001 39:511-48.

[4] Eichler, D. and Beskin, G.. Optical SETI with Air Cherenkov Telescopes.*Astro-ph*, 0111081 v1.

IV

ENERGETIC PARTICLES: FROM THE HELIOSPHERE TO THE GALAXY AND BEYOND

ELEMENTAL COMPOSITION AND PROPAGATION OF COSMIC RAYS AT HIGH ENERGIES

Dietrich Muller
Enrico Fermi Institute, University of Chicago
dmuller@uchicago.edu

Abstract   Detailed observations of the individual energy spectra and relative intensities of the cosmic-ray nuclei are required in order to unfold the changes in the cosmic ray population during propagation and hence, to understand how and at which rate the primary cosmic rays are generated at their sources. For the limited energy region where accurate data exist, composition and energy spectra at the sources and at the observer are characteristically different. We summarize the significance of currently available information and the challenge of new measurements.

INTRODUCTION

Ever since cosmic rays were discovered, the question arose how the medium through which the particles had traveled might have affected their observed composition and energy spectra. Hence, the problem of cosmic-ray propagation is an old one. It was a dominant question in early research in the first half of the 20th century, when observations made on ground or at relatively low elevations in the atmosphere detected particles (muons, electrons, photons) that had little resemblance to the parent radiation entering the top of the atmosphere. Studies of the cosmic-ray interactions during propagation through the atmosphere were an important field of research at that time, and led to pioneering discoveries in high-energy particle physics.

When we now use the term *"primary cosmic rays"*, we usually mean those particle species that are accelerated at some cosmic ray source. *Major* primary cosmic rays include protons, α-particles, the even-Z nuclei carbon, oxygen, neon, magnesium, silicon, and iron, as well as electrons. In order to deduce the nature of the source and acceleration mechanism, one needs to determine the relative abundances and the individual energy spectra of the primary components *at the sources*, i.e., before the particles start their voyage through the

M. M. Shapiro et al. (eds.), Neutrinos and Explosive Events in the Universe, 313–326.
© 2005 *Springer. Printed in the Netherlands.*

galaxy. Up to energies of at least 10^{15} eV, we believe that the sources are located in our galaxy.

The primary cosmic rays propagate through the interstellar medium (ISM) until they either escape into extragalactic space, or are removed by interaction or energy losses in the ISM. Their interstellar equilibrium intensity may be recorded with a detector which is usually carried above the earth's atmosphere on spacecraft or balloon. *Secondary cosmic rays* are those that are generated as products from interactions of the primaries in the ISM: positrons and antiprotons mostly come from interactions of primary protons, while the secondary nuclei such as Li, Be, B, and the elements just below iron, which cannot be produced by primary nucleosynthesis, are the products of spallation reactions of heavier primaries in the ISM. The overall arriving cosmic-ray intensity represents a mix of primary and secondary particles.

At low energies, below a few GeV/nucleon, the cosmic ray flux becomes strongly influenced by the interaction of the particles with the solar magnetic field and the solar wind. These processes of "solar modulation" will not be discussed in this article.

At very high energy, between 10^{15} eV and 10^{20} eV, the picture changes again: here, we cannot exclude that particles originating in extragalactic space contribute to the cosmic-ray intensity, and their propagation history could be quite different from that of galactic particles at lower energy. At these energies, one has to rely on observations of air showers on the ground, and unfortunately, an event-by- event identification of the arriving cosmic-ray particle is presently not possible.

A FEW HISTORIC COMMENTS

The development of cosmic-ray astrophysics has been guided by a few far-sighted ideas. For instance, Hess [1], [2] noted in his historical discovery papers that the radiation must originate in space beyond our nearest star, the Sun. Already in 1934, the generation of cosmic rays in supernovae was suggested by Baade and Zwicky [3]. However, the mechanism of acceleration remained mysterious until Fermi [4] proposed a process that involved collisions of particles with large-scale magnetic fields in the galaxy. There was an obvious conflict: the Baade/Zwicky suggestion required localized cosmic ray accelerators in the galaxy, while Fermi's mechanism seemed to involve the galaxy as a whole.

In the meanwhile, significant progress in the phenomenological understanding of the arriving cosmic-ray intensity, and in the identification of the individual components was made. Nevertheless, it took until the late 1970's before the currently accepted model of cosmic-ray origin was developed. This model reconciles the Baade/Zwicky and Fermi proposals by describing a first-order

Fermi acceleration mechanism that operates in shockfronts which are powered by supernova explosions and propagate from the supernova remnant (SNR) into the ISM ([6], [7], [8], [9]and others). This shock-acceleration model also predicts that cosmic rays are produced with a power-law energy spectrum, albeit with a power-law slope that is considerably harder than that actually measured for the arriving cosmic ray intensity. This problem could be reconciled with data from earlier observations [5], which clearly indicated that during propagation through the ISM the more energetic cosmic rays produce fewer secondaries, and hence, seem to be removed from the galaxy more readily than the low-energy ones. Consequently, the energy spectrum of primary cosmic rays *at their sources* must indeed be much harder than that observed near Earth.

THE L/M RATIO

Figure 1 shows the well-known diagram of nuclear cosmic ray abundances (around ~ 1 GeV/nucleon) in comparison with the abundances of the elements in the solar system. The most striking feature is the near absence of the light elements Li, Be, and B in the solar system. The cosmic-ray abundance of these elements is a good measure for the rate of spallation reactions of their parent nuclei (mostly carbon and oxygen) during propagation through the ISM. The abundance ratio Li+Be+B/C+O (the *'L/M ratio"*), or simply, the ratio B/C, is then used to determine the *propagation pathlength* Λ, with an approximate value of $\Lambda \approx 7g/cm^2$ at low energy (around 1 GeV/nucleon).

A compilation of measurements of the B/C ratio over a large range of energies is shown in Figure 2 [11]. Obviously, the ratio changes with energy: it seems to reach a maximum around 1 GeV/nucleon, but decreases both towards lower and towards higher energies. At low energies, solar modulation and resonances in the spallation cross section complicate the interpretation of the data. At high energies, which are the subject of this paper, we can assume that the cross sections reach energy independent geometrical values, and therefore, that the decrease of the B/C ratio indicates a decrease in the propagation pathlength. In fact, the energy dependence $\Lambda = \Lambda(E)$, is a key to the understanding of high-energy cosmic rays.

The mathematical description of the propagation of cosmic rays through the galaxy usually assumes dynamic equilibrium between *production* of a cosmic ray species either by acceleration in a source (primary cosmic rays), or by spallation of a parent species in the ISM (secondary cosmic rays), and *loss* by diffusion or convection, energy loss, radioactive decay, or spallation (e.g. [10]). Many of the parameters in the balance equation, in particular diffusion coefficients and boundary conditions, are poorly known. For our discussion here, we grossly simplify the situation: Instead of dealing with a diffusion co-

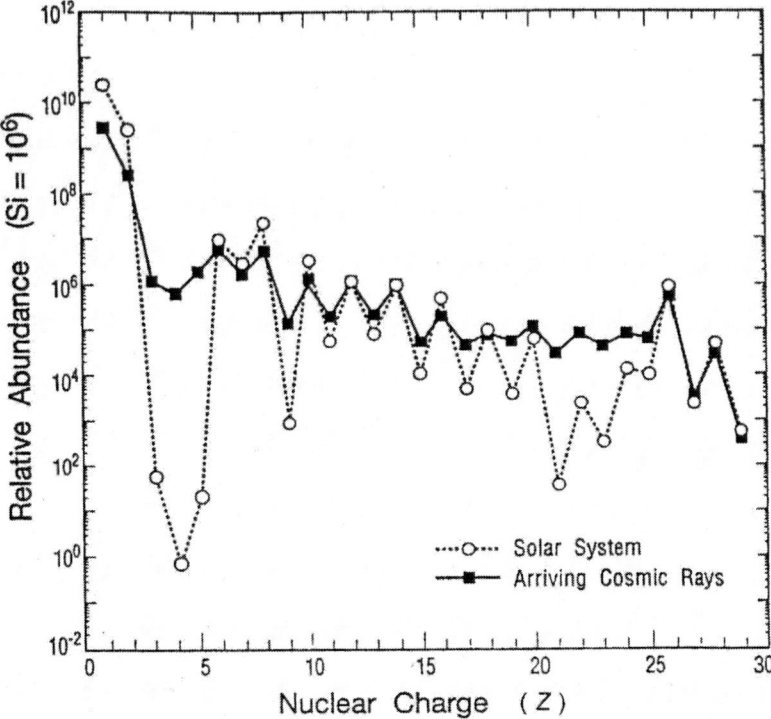

Figure 1.   Relative elemental abundances, normalized to silicon, in the cosmic ray nuclei, and in the material of the solar system.

efficient, we use the containment time τ, and we consider a system of just two cosmic ray components, a primary parent nucleus with number density N_p, and a purely secondary daughter nucleus with number density N_s. For example, N_p and N_s may represent the pair carbon and boron.

Ignoring energy loss, convection, or radioactive decay, we then have the equilibrium conditions:

$$\frac{N_p}{\tau} + n\beta c\sigma_p N_p = Q_p \tag{1}$$

$$\frac{N_s}{\tau} + n\beta c\sigma_s N_s = n\sigma_{ps}\beta c N_p \tag{2}$$

with n = number density of the ISM, βc = speed of particle, σ_p and σ_s = spallation cross sections of primary and secondary species; σ_{ps} = spallation cross section for the primary species N_p, to produce N_s, and Q_p = source production rate of the primary species.

With the escape pathlength $\Lambda = \rho\beta c\tau (\rho = nm = mass\ density\ of\ the$ *ISM)*, and the interaction pathlength $\Lambda_i = m/\sigma$, and with a differential energy spectrum at the source in form of a power law $Q_p = KE^{-\Gamma}$, equation (1) becomes

$$N_p \left(\frac{1}{\Lambda} + \frac{1}{\Lambda_i} \right) = \frac{1}{\rho\beta c} KE^{-\Gamma} \qquad (3)$$

The observed cosmic ray intensity $N_p(E)$ is determined by the competition between escape and spallation. It is important to realize that the numerical values of Λ_i and Λ at 1 GeV/nucleon are of comparable magnitude, for instance $\Lambda_i \approx 6.5 g/cm^2$ for nitrogen, and $2.3 g/cm^2$ for iron, and, as mentioned, $\Lambda \approx 7 g/cm^2$. While we may take Λ_i to be energy independent, Λ apparently is not. Hence, the *measured* energy spectrum $N_p(E)$ does not necessarily follow the same power law as the source spectrum.

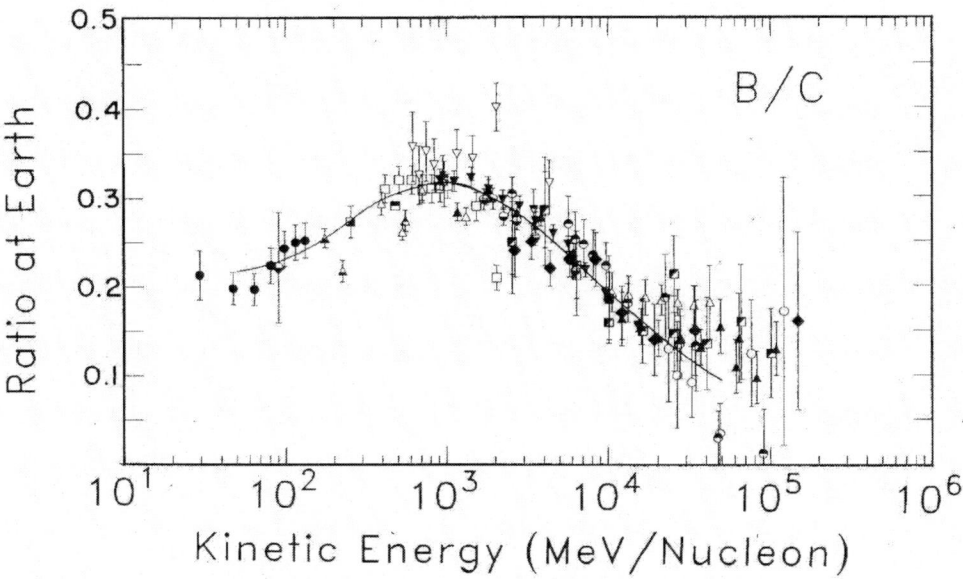

Figure 2. Compilation of the measured abundance ratio of boron to carbon nuclei as a function of energy. Figure from Garcia Muñoz et al., 1987.

Obviously, it is important to determine the exact form of the energy dependence $\Lambda(E)$. The data in Figure 2 have large error bars and support a wide range of interpretations or "fits" that can be found in the literature. Figure 3 shows a set of results on the B/C ratio [12] which come from measurements in space on HEAO-3, and which have the smallest systematic and statistical uncertainties of all data in the high-energy region. On the basis of these results, one concludes an energy dependence $\Lambda(E) \propto E^{-\alpha}$, with $\alpha \approx 0.6$ for the range 1-100 GeV/nucleon. The highest-energy results from Spacelab-2 [13], also shown in Figure 3, are consistent with this parameterization, but have large uncertainties and certainly do not strongly constrain the behavior of $\Lambda(E)$ at those energies.

With $\Lambda(E) \propto E^{-\alpha}$, equation (3) predicts that the observed spectrum reaches an asymptotic slope with spectral index $(\Gamma + \alpha)$, at sufficiently high energy (where $\Lambda(E) \ll \Lambda_i$). With a measured value $\Gamma + \alpha \approx 2.7$, and $\alpha \approx 0.6$, the source index Γ should be about $\Gamma \approx 2.1$. Such a value is favored in shock acceleration models for fairly strong shocks. If this source index is the same for all primary species, their observed energy spectra would be expected to slowly steepen until they all reach the same asymptotic slope $(\Gamma + \alpha)$ at high energy. The heavier cosmic rays, like iron, with relatively smaller values of Λ_i, would reach the asymptotic slope at higher energies than the lighter nuclei. The purely secondary nuclei would approach slopes with power law index $(\Lambda + 2\alpha) \approx 3.3$ (see eq. 2).

The model just described is oversimplified. For instance, cosmic rays propagating through the galaxy may not be just subject to diffusion and convection, and to the various loss processes mentioned above, but they may also experience a certain amount of reacceleration when encountering turbulent regions in the ISM. In fact, reacceleration has been discussed quite frequently in recent literature. While the decreasing L/M ratio suggests that reacceleration cannot be significant in the 1-100 GeV/nucleon region, Bereshko et al. [14] propose that the situation at higher energies could become different. They predict that reacceleration of ambient cosmic ray particles encountering shock fronts in the ISM, and spallation reactions inside the acceleration region would mainly affect the observed energy spectrum of secondary cosmic rays above about 100 GeV/nucleon. As a consequence, the L/M ratio would not be expected to continue falling with energy above 100 GeV/nucleon as steeply as observed at lower energy. Unfortunately, there are no measurements available at present to test such predictions.

With this discussion, we wish to emphasize two important conclusions:

(1) Over the limited region of energies for which we have detailed information on the cosmic ray composition, the character, i.e., the energy spectrum at the source, is significantly different from that observed in the arriving cosmic-ray intensity. This difference is of key importance for interpretations of cosmic ray acceleration in SNR shocks. However, there is a vast region of higher energies where the arriving overall energy spectrum has been observed, but not the particle composition. We therefore do not know how the observed spectrum may vary from that generated at the source in this energy region. This simple fact is often overlooked.

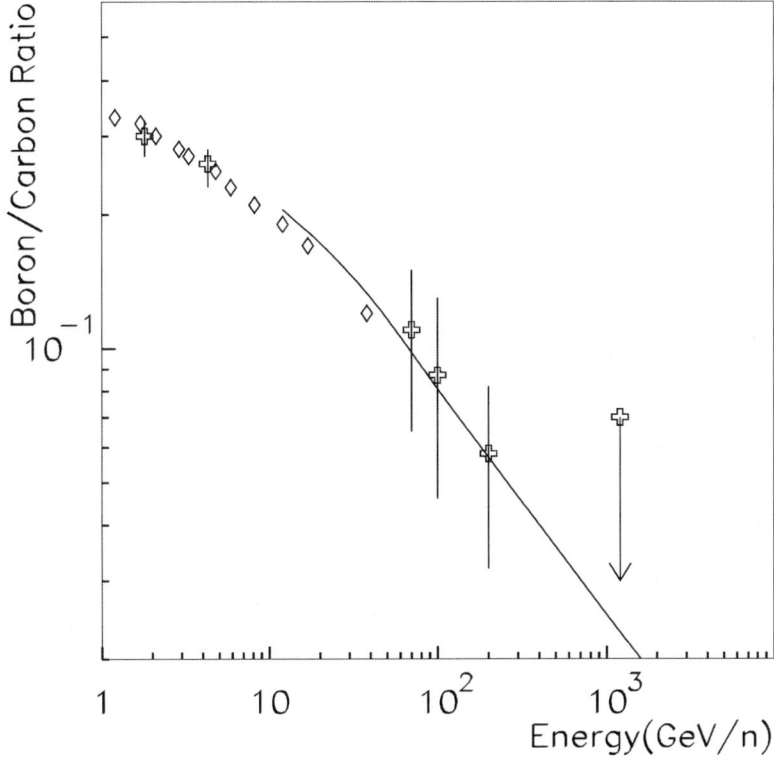

Figure 3. Abundance ratio boron to carbon at high energies. Shown are data from measurements in space, on HEAO-3 (diamonds, Engelmann et al., 1990) and with CRN/Spacelab 2 (crosses,Swordy et al., 1989).

(2) It is only a relatively small energy interval below ~ 100 GeV/nucleon, where the spectra and relative elemental abundances at the cosmic-ray source have been inferred from measurements of the intensities of primary and secondary nuclei. Extending this energy range is one of the most important, but unfortunately, also most difficult tasks in cosmic-ray astrophysics.

THE HIGHEST ENERGIES

The highest energy cosmic rays observed through air showers have energies in excess of 10^{20} eV. There is no understood or generally accepted mechanism of acceleration of these particles, and their mass composition is highly uncertain (see, e.g., [26]). It is quite possible, if not likely, that their sources are not constrained to objects in our galaxy but are to be found in extragalactic space. Entirely new phenomena will affect the propagation of these particles. If they are protons with energies above $\sim 5 \cdot 10^{19}$ eV, they will lose energy rapidly through photo-pion production with the cosmic microwave background [15],

[16]. If they are heavy nuclei, they will undergo photodisintegration in the far infrared and microwave backgrounds. In either case, a particle with primary energy in excess of 10^{20} eV will have its energy significantly degraded after traversing ~ 100 Mpc of intergalactic space [17]. Hence, the energy spectrum produced by hypothetical sources in extragalactic space could be drastically different from that observed near Earth, and the deformation of the spectrum would depend on the spatial distribution of the sources. For isotropically distributed sources, one would observe a pronounced cutoff of the observed intensity below 10^{20} eV (the "Greisen/Zatsepin/Kuzmin cutoff" or "GKZ cutoff"), even if the primary spectrum extended to much higher energies. Whether such a cutoff is consistent with the available measurements, is currently a much debated question.

CURRENT MEASUREMENTS

Measurements attempting to extend the range over which individual energy spectra and relative abundances of the cosmic ray nuclei are precisely known represent a considerable challenge to the experimenter. Observations of air showers on the ground may reveal the total energy of the particle (the *"energy per nucleus"*), but encounter fundamental difficulties in obtaining individual charge- or mass-resolution. Direct measurements above the atmosphere (which typically measure the *"energy per nucleon"* or the Lorentz factor $\gamma = E/mc^2$) can, in principle, identify each individual particle traversing the detector, but are limited to sensitive areas of, at most, a few square meters, and to exposure times of up to a few weeks on balloons, or a few years on spacecraft. With these limitations, direct measurements may reach energies in the region of the "knee" in the overall energy spectrum (just above 10^{15} eV per nucleus) for the primary cosmic ray nuclei, and perhaps 10^{14} eV per nucleus for the secondaries if their spectrum continues with the same steep slope that is measured below 10^{12} eV per nucleus. Higher energies must remain the domain of air shower observations.

For the direct measurements, the greatest experimental challenge is the measurement of the particle energy. The straightforward approach of using a total-absorption nuclear calorimeter is problematic at relativistic energies because of the enormous mass requirement: just one nuclear interaction length corresponds to absorber materials of the order of 1 ton per square meter. Therefore, techniques are more attractive that are based on electromagnetic interactions of the particle and non-destructive to the particle. A technique of this kind that is practical up to energies corresponding to Lorentz factors $\gamma \approx 10^5$ is based on the detection of transition radiation (see [19], for a recent discussion). Figure 4 shows a cosmic-ray detector (TRACER - "Transition Radiation Array for Cosmic Energetic Radiation") that uses this principle. This instrument has been successfully flown on a one-day balloon flight in 1999, and on a two-week long-duration balloon in Antarctica in December 2003. A description of the instrument is given by Gabhauer [18].

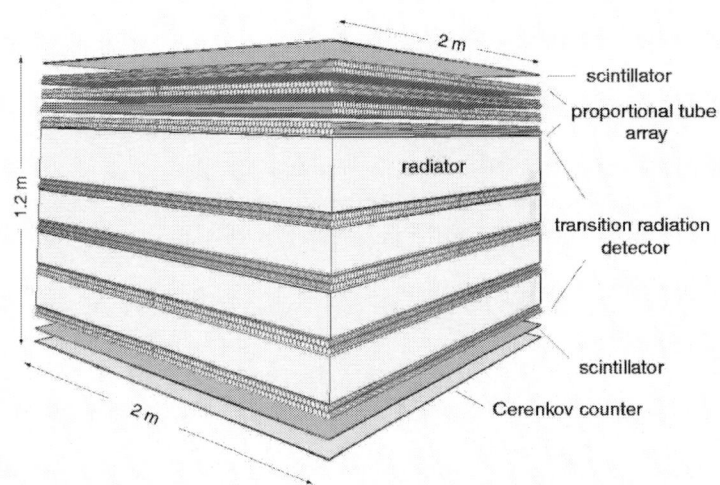

Figure 4. Schematic arrangement of the TRACER balloon instrument. The nuclear charge Z of traversing cosmic-ray nuclei is determined with plastic scintillators. An acrylic Cherenkov counter determines their Lorentz factor γ at low energies, and arrays of single-wire proportional tubes measure γ from the relativistic rise in specific ionization, and from transition radiation signals.

Transition radiation detectors are suitable for the heavier cosmic ray nuclei (with charge number $Z \geq 3$), but suffer from relatively large statistical fluctuations if observations of the abundant nuclei with low charge, H and He, are required. For these, a practical alternative to nuclear calorimeters does not exist currently. Hence, a comprehensive cosmic-ray detector for observations at very high energies might combine a small (but weighty) calorimeter with a TRD of much larger area. Such an instrument, under the acronym ACCESS, has been studied [20] as a candidate mission for space flight on the Space Station, or on free-flying satellite.

PROSPECTS OF FUTURE OBSERVATIONS

Let us assume that a space mission will be successful in the not-too-distant future to indeed measure the energy spectra of the individual cosmic ray nuclei with good statistical and systematic accuracy to energies corresponding to the cosmic-ray "knee" above 10^{15} eV per nucleus. What will we learn?

First, we can investigate if the current paradigm of supernova shock acceleration which predicts this mechanism to be efficient only up to some maximum rigidity is indeed true, and perhaps related to the spectral "knee" above 10^{15} eV/nucleus. According to this paradigm, one would expect a cutoff in the individual energy spectra at energies around $Z \cdot 10^{14}$ eV per nucleus. This cutoff would lead to a change in the cosmic-ray charge composition over the $10^{14} - 10^{16}$ eV per nucleus region: above 10^{14} eV, the proton intensity should drop more rapidly than at lower energies, and similar cutoffs would be expected for the heavier elements at higher energies, for instance for iron the cutoff would occur at $\sim 5 \cdot 10^{15}$ eV per nucleus.

Important detail can be derived from the knowledge of the spectral shapes for the elements below these cutoff energies. For instance, it is plausible that the energy dependence of the propagation pathlength $\Lambda(E) \propto E^{-\alpha}$ does not persist up to the very highest energies where Λ would become extremely small. Rather, it has already been suggested long ago (e.g., [21]) that the pathlength Λ may approach a finite high-energy value Λ_0.

We might consider a model in which the cosmic ray source produces primary cosmic rays with an energy spectrum $\propto E^{-\Gamma}$, but that the propagation pathlength is parameterized as $\Lambda(E) = CE^{-\alpha} + \Lambda_0$. Equation (3) then predicts that the *observed* spectrum would have a spectral index Γ at low energies where $\Lambda \gg \Lambda_i$, and steepen to a spectral index $\Gamma + \alpha$ at energies where $\Lambda(E) \ll \Lambda_i$, but again approach the source spectral index Γ at still higher energy, when $\Lambda(E) \approx \Lambda_0 \ll \Lambda_i$. This is illustrated in Figure 5, where we compare measured intensities for iron (multiplied with $E^{2.5}$) with the expected spectral shapes for $\Lambda_0 = 0, 0.15 g/cm^2$, and $0.5 g/cm^2$, respectively. The data points at the highest energies come from the CRN measurement on the Space Shuttle ([22]), and from the first balloon flight of he TRACER detector ([18]). It is obvious that the data in Figure 5 are not of sufficient statistical quality to discriminate between the three models. However, the results of the long-duration balloon flight of TRACER which are presently being analyzed are expected to drastically improve this situation.

Let us assume that indeed, a hardening of the energy spectrum of iron and of that of other primary nuclei will be observed at high energy. In fact, measurements made by the JACEE ([28]) and the RUNJOB ([27]) collaborations suggest such a behavior. In order to make sure that this is the consequence of a finite propagation pathlength Λ_0, one will need to measure the L/M ratio directly, and one would expect that the ratio becomes constant at high energy, when $\Lambda \approx \Lambda_0$. While extremely important, this is a difficult measurement because of the low particle intensities, and because of systematic background in balloon observations due to spallation reactions in the residual atmosphere. Preferably, such a measurement should be conducted with a space-borne detector.

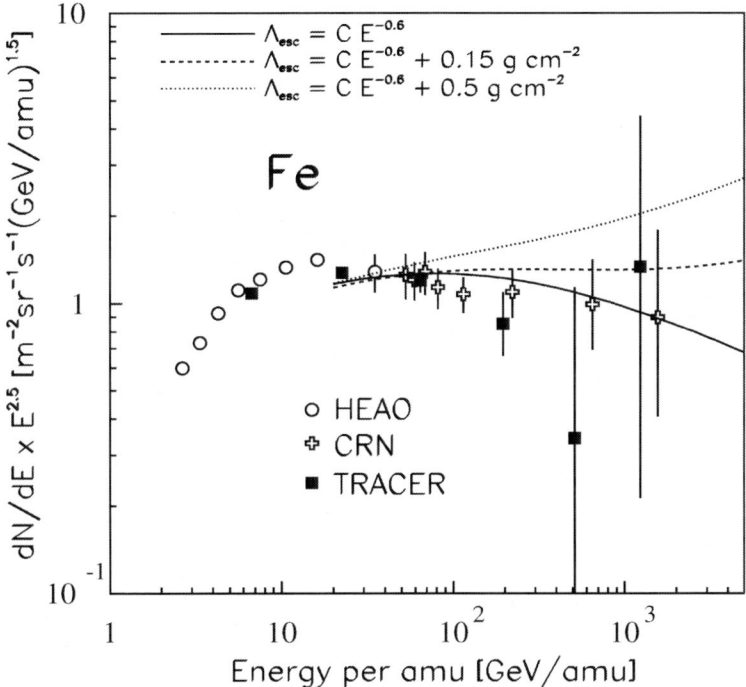

Figure 5. The differential intensity of iron nuclei at very high energy. The data are from HEAO-3, CRN/Spacelab 2 and TRACER. The lines indicate predicted spectral shapes for different assumptions about the energy-dependence of the propagation pathlength .

We may consider another scenario which predicts a flattening of the L/M ratio at high energy due to reacceleration of the secondary cosmic rays. This would not appreciably affect the observable spectral slope of the primary species ([14]). Hence, for this case, and if we assume $\Lambda_0 = 0$, we would expect no upturn in the energy spectra of the primary cosmic rays in spite of the flattening of the L/M ratio.

While more complicated scenarios may be possible, these examples illustrate the kind of conclusions that one may reach when more accurate measurements become available. The urgent need for such measurements is undeniable.

Currently, there exist several other long-duration balloon detectors that will address this question with new observational results, including ATIC [23] and CREAM [24]. However, it is clear that the full potential of direct measurements must be exploited with a large-area detector in space flight of several years duration.

Even then, we must be content with the fact that direct measurements of the cosmic-ray composition may be possible up to the "knee", but will not cover the vast energy range up to 10^{20} eV. However, the direct measurements around the knee would be invaluable in calibrating the response of air shower installations in this region of overlap. The notorious difficulty in air shower measurements caused by the uncertainty in the proper nuclear interaction model at high energy (e.g., [25]) may then be removed, and the interpretation of air shower data, even at higher energies, may become more straightforward. Whether the accuracy of the analysis will suffice to unravel the propagation history of these particles remains to be seen.

CONCLUSION

In this lecture, we have tried to summarize the importance of accurate measurements of the cosmic-ray composition in order to understand the nature of the cosmic-ray source. We recall that the discovery of the decrease of the relative abundance of secondary cosmic rays at high energies, i.e., the energy dependence of the propagation pathlength $\Lambda(E)$, came as a complete surprise in the early 1970's. However, this discovery provided major observational support to supernova shock acceleration models which, a few years later, predicted that the cosmic-ray sources should be characterized by relatively hard energy spectra.

The overall cosmic-ray energy spectrum is known over an enormous range of energies (see Figure 6). It is sobering to realize that for only a limited range of energies, below a few 10^{12} eV, relatively precise composition measurements exist that permit an extrapolation of the data to determine the spectra and particle intensities created in the cosmic ray sources. And it is quite revealing that neither the particle composition nor the shape of the individual energy spectra at the source are the same as those observed in ambient space. There is a large range of higher energies for which the overall energy spectrum is measured, but for which only rudimentary and contradictory information on the particle composition exists. The information available is insufficient to understand the propagation of these particles from their sources to the observer on Earth and to detect the changes in composition and energy spectra in this region that may occur during propagation.

In the energy region around the knee, the most extensive studies to understand the particle composition have been made with the KASKADE air shower array ([29]). As mentioned, definite conclusions can still not be made because the hadronic interaction modes governing the generation of air showers are not well enough known (Hörandel , 2003). Direct measurements from balloons or in space in this energy region would provide the necessary cross-calibration

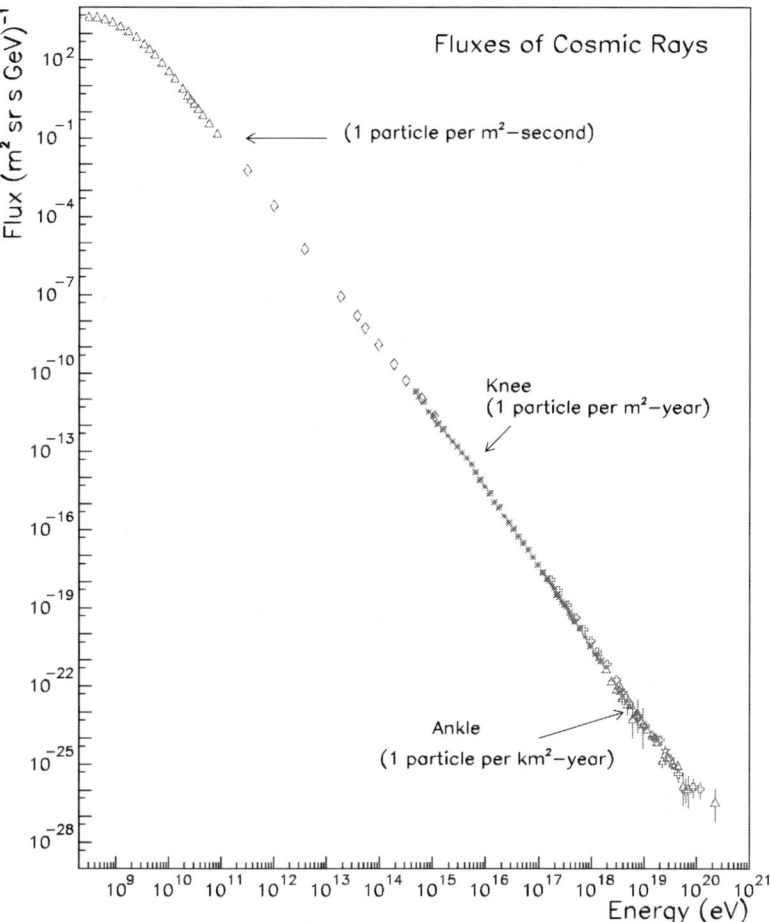

Figure 6. Compilation of the differential all-particle energy spectrum of cosmic rays over a very wide range of energies.

and would help in the interpretation of air shower measurements even at much higher energy.

Much work remains to be done in order to understand the cosmic-ray phenomenon at high energies. To make progress, difficult but possible new observations are required: the ball is solidly in the experimentalist's court.

References

[1] Hess, V. *Phys. Z.*, 13, 1084, 1912.

[2] Hess, V. *Phys. Z*, 14, 610, 1013.

[3] Baade, W. and F. Zwicky *Proc. Nat. Acad. Sci.* 20,259,1934.

[4] Fermi, E. *Phys. Rev.* 75, 1169, 1949.

[5] Juliusson, E., P. Meyer, D. Müller *Phys. Rev. Lett.* 29, 445, 1972.

[6] Axford, W.I., E. Leer, G. Skadron *Proc. 15th Int. Cosmic Ray Conf.* 11, 132, 1977.

[7] Krymsky, G.F.*Dok. Acad. Nauk USSR* 234, 1306, 1977.

[8] Bell, A.R. *Mon. Not. R. Astr. Soc.* 182, 147, 1978.

[9] Blandford, R.D., and J.P. Ostriker *Ap.J.* 221, L29, 1978.

[10] Ginzburg, V.L. and V.S. Ptuskin (eds.) *Astrophysics of Cosmic Rays*, p. 43ff. (North Holland) 1990.

[11] Garcia Muñoz, M., J.A. Simpson, T.G. Guzik, J.P. Wefel, & S.H. Margolis *Ap. J. Suppl.* 64, 269, 1987.

[12] Engelmann, J.J., Ferrando, P., Soutoul, A., Goret, P., & Juliusson, E. *Astronomy & Astrophys.* 233, 96, 1990.

[13] Swordy, S.P., D. Müller, P. Meuer, J. L'Heureux, & J.M. Grunsfeld *Ap.J.* 349, 625, 1990.

[14] Bereshko, E.G., L.T. Ksenofontov, V.S. Ptuskin, V.N. Zirakashvili, & H.J. Völk *Astronomy & Astrophys., 2003* 410, 189, 2003.

[15] Greisen, K. (1966) *Phys. Rev. Lett.* 16, 748, 1966.

[16] Zatsepin, G.T. & V.A. Kuzmin *JETP Lett.* 4, 78, 1966.

[17] Cronin, J.W. *The Pierre Auger Project Design Report*, Fermilab, 1995.

[18] Gahbauer, F., G. Hermann, J.R. Hörandel, D. Müller, & A. Radu *Ap.J.* 607, 333, 2004.

[19] Wakely, S.P., S. Plewnia, D. Müller, J.R. Hörandel, & F. Gahbauer *Nucl. Inst. & Meth.* A 531, 435, 2004.

[20] Israel, M. et al. *ACCESS: A Cosmic Journey, Report of NASA Working Group*, 2000.

[21] Cowsik, R. and L.W. Wilson *Proc. 13th Int. Cosmic Ray Conf.*, 1, 500, 1973.

[22] Müller, D., S.P. Swordy, P. Meyer, J. L'Heureux, & J.M. Grunsfeld *Ap.J.* 374, 356, 1991.

[23] Guzik, T.G. et al. *Proc. 26th Int. Cosmic Ray Conf.*, 5, 9, 1999.

[24] Seo, E.S. et al. *Adv. Space Res.*, 30, 1263, 2002.

[25] Hörandel, J.R. *J. Phys. G: Nucl. Part. Phys.* 29, 2439, 2003.

[26] Watson, A.A. *astro-ph/0410514 v. 1, talk at XIII ISVHECRI, Pylos*, 2004.

[27] Apanasenko, A.V et al. *Astropart. Phys.* 16, 13, 2001.

[28] Takahashi, Y. et al. *Nucl. Phys. B. Proc. Suppl.* 60(3), 83, 1998.

[29] Antoni, T. et al. *Nucl. Inst. & Meth.* A 513, 490, 2003.

TEV-SCALE GRAVITY: DETECTING BLACK HOLES WITH COSMIC RAY AIR SHOWERS

Eun-Joo Ahn[1,2], Marco Cavaglia[3]

[1]*Dept of Astronomy and Astrophysics, University of Chicago*
5640 S. Ellis Ave, Chicago, IL, 60637, USA

[2]*Kavli Institute for Cosmological Physics, University of Chiacgo*
5640 S. Ellis Ave, Chicago, IL, 60637, USA

[3]*Dept of Physics and Astronomy, The University of Mississippi*
University, MS 38677-1848, USA

Abstract   In models of large extra dimensions, gravitational effects may become relevant to particle processes at the TeV scale. In this scenario, extra-dimensional graviton production and graviton exchange events occur at sub-TeV scales, whereas black holes and branes are formed at super-TeV scales. Black holes could soon be detected in particle colliders and in Earth's atmosphere by cosmic ray telescopes on the ground and in space. The phenomenology of gravitational events is significantly different from that of standard model processes. We review the differences between extensive air showers generated by black holes and standard model interactions.

Keywords: High-energy Physics, Black Holes, Cosmic Rays.

Introduction

The fundamental scale of gravity may be much lower than the observed gravitational scale, even of the order of a few TeV. (For general references see *Cavaglià 2003a*). In braneworld scenarios with large extra dimensions, the observed weakness of the gravitational field is due to the "leakage" of gravity in the extra dimensions: Standard Model (SM) fields are constrained to four-dimensions (brane), whereas gravitons are allowed to freely propagate in the higher-dimensional spacetime (bulk). Strong gravitational effects become manifest at this scale. Super-Planckian collisions would then take place in next-generation particle colliders, such as the Large Hadron Collider[1] (LHC) (Dimopoulos, Landsberg 2001; Giddings, Thomas 2002), and in the at-

M. M. Shapiro et al. (eds.), Neutrinos and Explosive Events in the Universe, 327–334.

mosphere, induced by ultrahigh energy cosmic rays (UHECRs) (Feng, Shapere 2002). Black holes (Banks, Fischler 1999) and branes (Ahn, Cavaglià, Olinto 2003, Ahn, Cavaglià 2002) could be created from these collisions.

UHECRs are attractive because of their high centre-of mass energy. The proton-nucleon cross section for BH formation is very small compared to other hadronic processes. The neutrino-nucleon cross section for BH formation may be higher than the SM process, thereby giving interest to neutrino interaction. However, rate counting is not sufficient to prove black hole formation in the atmosphere. Discovery of BHs in UHECRs requires discrimination of BH and SM air showers. The extensive air shower (EAS) characteristics of these processes will differ, and with new detector methods and enough statistics expected from the new generation cosmic ray observatories, it may be possible to detect BH-induced EASs.

Large extra dimensions

According to Grand Unification Theories (GUT), EW and strong interactions unify at around $E_{GUT} \sim 10^{13}$ TeV. Unification of GUT interactions with gravity is expected few orders of magnitude above this scale, at around $E_G \sim 1.22 \times 10^{16}$ TeV $\equiv M_{Pl}$, where M_{Pl} is the Planck mass. Alternatively, gravity can be brought down to the EW scale by the existence of extra dimensions. Unification of all interactions occurs at this scale. In this scenario, the universe is d-dimensional, with the length scale of at least some of the extra dimensions larger than the fundamental Planck length. Gravity propagates in all dimensions whereas SM fields are confined to the $3 + 1$ dimensional brane. The gravitational interaction becomes strong above the fundamental Planck mass $M_\star$. In models with flat extra dimensions (which we consider henceforth) $M_\star$ is related to the observed Planck mass M_{Pl} by the relation

$$M_{Pl}^2 = M_\star^{d-2} V_{ed} \,, \tag{1}$$

where V_{ed} is the volume of the extra dimensions.[2] In particle processes with energy just below $M_\star$, perturbative events such as Kaluza-Klein modes and virtual graviton exchange are expected to happen. Events such as BH and brane formation take place at energies above $M_\star$, where physics is non-perturbative.

The size of extra dimensions, or alternatively the fundamental Planck mass, is constrained by experiments and observations. Direct tests of gravity such as the Cavendish experiment (Hoyle 2004) set the maximum length of equal-size extra dimensions to $160 \mu m$. Non-observation of perturbative events in particle colliders gives constraints of $M_\star \geq 0.62$ TeV for $n = 2$ and $M_\star \geq 0.31$ TeV for $n = 5$ (Giudice, Strumia 2003). Constraints from astrophysical and cosmological observations are much higher than laboratory constraints when the number of extra dimensions is $n \leq 3$ (see *Cavaglià 2003a* for a brief summary of astrophysical and cosmological constraints). However, uncertainties in the modelling of these processes makes these predictions less reliable than those obtained in the laboratory.

Black hole formation and evaporation

The cross section of a $(n + 4)$-dimensional Schwarzschild BH formation is given by

$$\sigma_{\nu i \to BH} = F \frac{1}{\pi M_\star^2} \left[\frac{8 \Gamma \left(\frac{n+3}{2} \right)}{(n+2)} \right]^{\frac{\cdot}{n\cdot\cdot}} \left(\frac{M_{BH}}{M_\star} \right)^{\frac{\cdot}{n\cdot\cdot}} , \tag{2}$$

where M_{BH} is the mass of the BH and F is a form factor. The Schwarzschild radius r_s is of the order $M_\star \sim$ TeV, thus Eqn. (2) must be interpreted at parton level. The total cross section for a neutrino-nucleon cross section ignoring inelasticity effects is given by

$$\sigma_{\nu p \to BH} = \sum_i \int_{x_m}^1 dx \, q_i(x, -Q^2) \, \sigma_{\nu i \to BH} , \tag{3}$$

where $-Q^2$ is the four-momentum transfer squared, $\sqrt{x}$ is the fraction of the nucleon momentum carried by the parton, $\sqrt{s x_m} = M_{BH,min}$ is the minimal mass for BH formation. When inelasticity effects are considered (Yoshino, Nambu 2003), the lower cutoff in the integral (3) depends on the impact parameter. A proposal for the inelastic cross section is (Anchordoqui et al. 2003):

$$\sigma'_{\nu p \to BH} = \sum_i \int_0^1 2z dz \int_{x'_m}^1 dx \, q_i(x, -Q^2) \, \sigma_{\nu i \to BH} , \tag{4}$$

where z is the impact parameter normalized to its maximum value and $x'_m = x_m / y^2(z)$, $y(z)$ being the fraction of centre of mass energy that is trapped into the BH. Eqn. (4) gives the total cross section as an impact parameter-weighted average over parton cross sections. Eqn. (4) lowers the total BH cross section of Eqn. (3).

The BH will evaporate via Hawking radiation into both brane and bulk. The particle emission rate per species i for a BH with temperature T is described by (Ahn et al. 2003)

$$\frac{dN_i}{dt} = c_i \Gamma_{s_i} f_i \frac{\zeta(3) T}{16\pi^3} \frac{(n+1)(n+3)^{\frac{n\cdot\cdot}{n\cdot\cdot}}}{2^{\frac{\cdot}{n\cdot\cdot}}} , \tag{5}$$

where c_i, s_i, Γ_{s_i} are the degree of freedom, spin, greybody factor of the species, and $f_i = 1 \, (3/4)$ for bosons (fermions).[3] The SM particles will generate cascades on the brane. It is widely accepted that the BH evaporates mainly on the brane because of the predominance of brane degrees of freedom over bulk degrees of freedom (Emparan, Horowitz, Myers 2000). However, we do not know the n-dimensional graviton greybody factor, and whether there exists other non-SM quanta that can propagate in the bulk at the TeV scale. Depending on these two factors, BHs may or may not evaporate mainly on the brane (Frolov, Stojkovic 2000a; 2002b; Cavaglià 2003b). This point has been overlooked in the literature.

Thus Eqns. (2)-(5) provide only a rough estimation of BH formation and evaporation. Uncertainties can be listed as below.

- Ignorance of fundamental parameters, e.g. n, $M_\star$, topology of the extra dimensions.

- Ignorance of gravitational physics at high energies, e.g. minimum BH mass $M_{BH,min}$, cross section form factor, higher dimensional graviton greybody factor.

- Ignorance of particle physics at high-energies, e.g. parton distribution functions, additional field degrees of freedom above Planckian energies.

- Ignorance of phenomenological parameters, e.g. choice of momentum transfer between inverse Schwarzschild radius and black hole mass.

The effect of some of these uncertainties on the black hole cross section are plotted in Fig. 1. Even if these uncertainties are removed, observation of BHs with neutrino telescopes presents additional uncertainties related to the experiment acceptance and the flux of neutrinos at high energies. Cosmogenic neutrinos which are created from photopion production of UHECRs interacting with the cosmic microwave background are the best candidate for these high energy neutrinos. Various models give a wide range of flux (Protheroe, Johnson 1996; Engel, Seckel, Stanev 2001; Ahn, Cavaglià, Olinto 2003), differing between one to three orders or magnitude as shown in Fig. 2. As cosmogenic neutrinos have not been detected, the flux may even be zero.

Setting limits on $M_\star$ from non-observation of UHECR BH-induced EAS

The event rate of BH induced air showers can be used to set a limit on $M_\star$ The number of neutrino-nucleon BH events detected by a cosmic ray detector in time T is (Anchordoqui et al. 2002; 2003):

$$N = N_A T \int dE_\nu \, \sigma_{\nu p \to BH} \frac{d\Phi}{dE_\nu} \, A(yE_\nu) \,, \qquad (6)$$

where $A(yE_\nu)$ is the experiment acceptance for an air shower energy yE_ν, N_A is Avogadro's number, and $d\Phi/dE_\nu$ is the source flux of neutrinos.

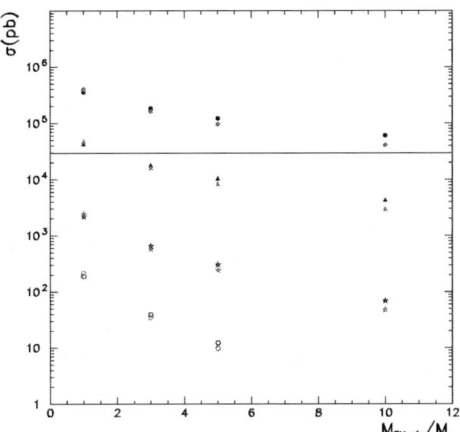

Figure 1. BH cross section with no inelasticity at $E_\nu = 10^\cdot$ TeV for $n = 3$ (black) and $n = 6$ (red), with $M_\star = 1, 2, 5, 10$ TeV (different symbols from top to bottom), plotted for various values of $M_{BH,min}/M_\star$. The horizontal line is the SM charged current cross section at the same energy. (Adapted from *Ahn et al. 2003.*)

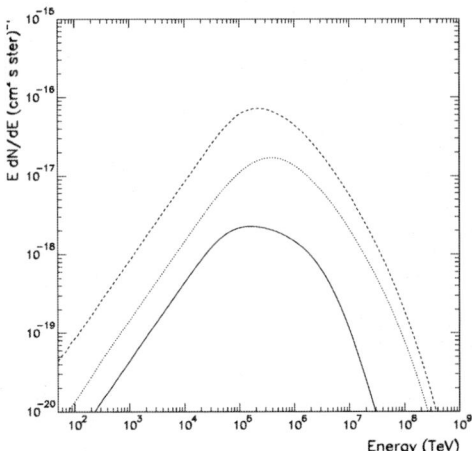

Figure 2. Cosmogenic neutrino flux of the ν_μ family for various models; Protheroe-Johnson (dashed line), Engel-Seckel-Stanev (dotted line), and the minimal (solid line) flux. (From *Ahn, Cavaglià, Olinto 2003.*)

Despite the enhancement, the cross section for BH formation is too low to have a definite interaction point in the atmosphere, thus only deeply penetrating horizontal air showers are considered. If the BH cross section is higher than the SM cross section for neutrinos, BH formation will dominate over SM-charge current (CC) interaction and the event rate for horizontal air showers will increase. The Pierre Auger Observatory[4] and other existing experiments such as AGASA[5] and RICE[6] should be able to detect this increase. Current non-observation of deeply penetrating horizontal air showers sets a limit on $M_\star$.

However, we have seen in the above sections that the cross section of BH formation cannot be precisely determined by current models. Even worse, it can be lower by a few orders of magnitude than the SM cross section. If the source flux is lower than predicted, the rate will be reduced. The Protheroe-Johnson flux leads to a minimum bound of $M_\star \gtrsim 0.35 - 0.6$ TeV for a 10-dimensional spacetime. This falls to $M_\star \gtrsim 0.1$ TeV when the minimal flux is used. This limit is less stringent than collider limits (Giudice, Strumia 2003).

EAS simulations

Both BH and SM induced air showers are simulated to compare their detectable characteristics. The most relevant SM process for the comparison is the ν_e-CC interaction producing an electron air shower that is modelled with AIRES (Sciutto 1999). For the BH production process, the secondary particles from BH evaporation are hadronised with PYTHIA (Sjöstrand et al. 2001) before producing an air shower with AIRES. Fig. 3 shows air showers generated

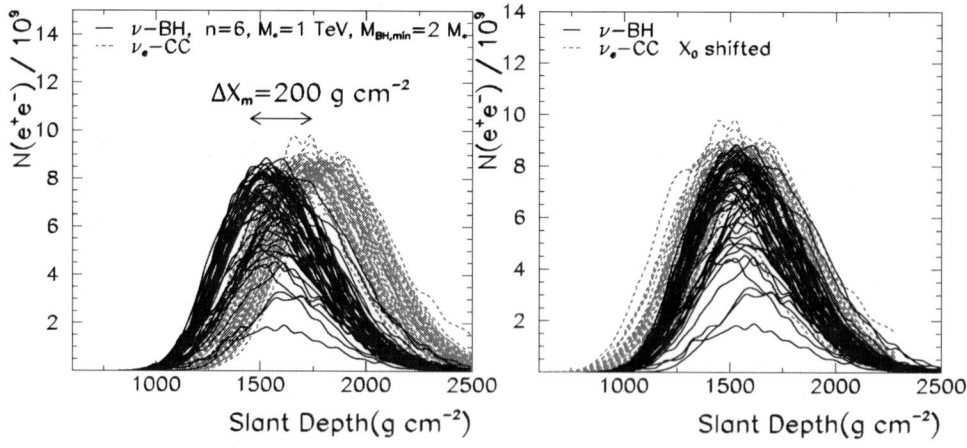

Figure 3. $e^+ e^-$ pairs as a function of slant depth for BH and ν_e-CC air showers at $E_\nu = 10^\cdot$ TeV. On the left, $X_\cdot$ of BH and CC are equal, while on the right, $X_\cdot$'s are shifted such that X_m of BH and CC are equal. (From *Ahn et al. 2003.*)

by neutrino primaries with $E_\nu = 10^7$ TeV for $n = 6$, $M_{BH,min} = 2\,M_\star = 2$ TeV in the case of no inelasticity (Eqn. (3)) and particles evaporating mainly on the brane. On the left panel, the first interaction point (X_0) is fixed to be 10 km above sea level with shower zenith angle of $70°$, corresponding to a slant depth of 780 g cm^{-2}, for both BH and SM air showers. The shower maxima (X_m) of the BH and ν_e-CC air showers are clearly different by ~ 200 g cm^{-2}. However, the interaction length for neutrinos in the atmosphere is large, thus X_0 is not fixed for either CC or BH. By shifting X_0 such that the X_m for both cases match, the difference in air shower development is much harder to distinguish as seen in the right panel of Fig. 3. Given a large number of horizontal neutrino air showers, one can distinguish the SM and the BH cases by studying the rise, $X_m - X_{0.1}$, where $X_{0.1}$ is the slant depth containing 10 % of particles of X_m. In addition, the muon content of BH air showers is larger than the SM case, since BHs produce hadrons while CC air showers do not. A deep horizontal air shower accompanied by many muon secondaries is a sign of BH formation. Due to its hybrid detectors, the Pierre Auger Observatory might be able to detect BHs.

Conclusions

BH formation may soon be observed in particle collisions if the fundamental Planck scale is $\sim$ TeV. UHECRs are good candidates to observe these events. Given the uncertainties in the BH cross section and the flux of high-energy cosmic neutrinos, distinguishing TeV extra dimensional models from the SM via neutrino induced EASs rate counting is likely to be unattainable. Detection of BH events is possible, albeit difficult, by looking at the characteristics of BH- vs. SM-induced EASs. BH air showers develop faster that the SM ones, leading to a significant difference in the shower profile. However, the variation in the first interaction point of neutrino EASs make the distinction between BH and SM air showers quite difficult. The detection of a large number of neutrino EASs with measured muon content and rise parameter is necessary for a clear identification of atmospheric BH events.

Acknowledgments

The content of this work is partially based on previous results with Maximo Ave and Angela Olinto and we are grateful to them for interesting discussions and suggestions. We would like to thank the organisers of the Erice school for giving financial support and providing a wonderful atmosphere to work. This research was carried out at the University of Chicago, Kavli Institute for Cosmological Physics and was supported in part by grant NSF PHY-0114422. KICP is a NSF Physics Frontier Center.

Notes

1. http://lhc-new-homepage.web.cern.ch/lhc-new-homepage/

2. Note that different notations are used in the literature. Conversion factors between different definitions of the fundamental Planck scale can be found in *Cavaglià 2003a*.

3. In *Ahn et al. 2003*, right below Eqn. (12), Γ_{s_i} should be changed to s_i. Also, the degree of freedom of gluon is 16 in Table II. We thank L.A. Anchordoqui for pointing out this typographical error to us.

4. http://www.auger.org

5. http://www-akeno.icrr.u-tokyo.ac.jp/AGASA

6. http://kuhep4.phsx.ukans.edu/ iceman/

References

E. J. Ahn and M. Cavaglià, Gen. Rel. Grav. **34**, 2037 (2002).

E. J. Ahn, M. Ave, M. Cavaglià and A. V. Olinto, Phys. Rev. D **68**, 043004 (2003).

E. J. Ahn, M. Cavaglià and A. V. Olinto, accepted to Astropart. Phys. arXiv:hep-ph/0312249.

L. A. Anchordoqui, J. L. Feng, H. Goldberg and A. D. Shapere, Phys. Rev. D **65**, 124027 (2002).

L. A. Anchordoqui, J. L. Feng, H. Goldberg and A. D. Shapere, Phys. Rev. D **68**, 104025 (2003).

T. Banks and W. Fischler, arXiv:hep-th/9906038.

M. Cavaglià, Int. J. Mod. Phys. A **18**, 1843 (2003) 2003a.

M. Cavaglià, Phys. Lett. B **569**, 7 (2003) 2003b.

S. Dimopoulos and G. Landsberg, Phys. Rev. Lett. **87**, 161602 (2001).

R. Emparan, G. T. Horowitz and R. C. Myers, Phys. Rev. Lett. **85**, 499 (2000).

R. Engel, D. Seckel and T. Stanev, Phys. Rev. D **64**, 093010 (2001).

J. L. Feng and A. D. Shapere, Phys. Rev. Lett. **88**, 021303 (2002).

V. P. Frolov and D. Stojkovic, Phys. Rev. D **66**, 084002 (2002).

V. P. Frolov and D. Stojkovic, Phys. Rev. Lett. **89**, 151302 (2002).

S. B. Giddings and S. Thomas, Phys. Rev. D **65**, 056010 (2002).

G. F. Giudice and A. Strumia, Nucl. Phys. B **663**, 377 (2003).

C. D. Hoyle *et al.* Phys. Rev. D **70**, 042004 (2004).

R. J. Protheroe and P. A. Johnson, Astropart. Phys. **4**, 253 (1996).

S. J. Sciutto, arXiv:astro-ph/9905185.

T. Sjöstrand *et al.* Computer Physics Commun. **135** (2001) 238

H. Yoshino and Y. Nambu, Phys. Rev. D **67**, 024009 (2003).

ULTRA HIGH ENERGY COSMIC RAYS: ANALYSIS OF DIRECTIONAL CORRELATIONS WITH ASTROPHYSICAL SOURCES

Ioana C. Maris[1,2] and Peter L. Biermann [2]

[1] *Universität Karlsruhe*

[2] *Max Planck Institut für Radioastronomie, Bonn*

Abstract   We searched for directional correlations between Ultra High Energy Cosmic Rays (UHECRs) and different types of sources like Compact Steep Spectrum and Gigahertz Peaked Spectrum sources, clusters of galaxies, Gamma Ray Bursts, BL Lac objects, infrared (IRAS) bright galaxies, flat spectrum radio sources, and others. We show that there is no plausible association between the cosmic ray events and the sample of sources used because of the poor statistics. However various source candidates can be used to develop toy models. One possible toy model for accelerating particles to such high energies could be the activity of the relativistic jets in Active Galactic Nuclei. We computed the maximum energy the particles can be accelerated to and the cosmic ray flux contribution for the compact core sources that are in directional correlation with UHECRs.

Introduction

A key region of the spectrum of cosmic rays is its highest energies [1]. The flux here is so low that the small event statistics in our observations to date leave us uncertain of the spectral structure above the key energy of 4×10^{19} eV, where a spectral downturn has been predicted [18; 19], due to the interaction with 2.7K photons from the cosmic microwave background.

If the spectrum of cosmic rays does extend well beyond 10^{20} eV, determining the origin of these particles could have important implications for astroparticle physics.

In the AGASA experiment a non-uniform distribution of arrival directions was found, which may lead us to search for point sources. Finley and Westerhoff [14] showed that the autocorrelation in the AGASA data is is weaker than previous claimed.

Searches for point sources were made before. Tinyakov and Tchakev have found a strong correlation between UHECRs an BL Lac objects [17]. Farrar and Biermann searched for correlations with compact sources [16].

We hypothesize that the particles responsible for the observed events are stable and neutral. So we do not take into account the particle trajectory deviation

M. M. Shapiro et al. (eds.), Neutrinos and Explosive Events in the Universe, 335–341.

Table 1.   Probability using first only AGASA data and then all available data in a angular diameter of 5 degrees. Col. 1 the number of the catalogue, col. 2 the number of sources in the catalogue, col. 3 the number of sources in the defined sample, col. 4 the angular diameter, col. 5 (col. 8) the number of hits taking into account the double counting due to doublets, col. 6 (col. 9) the probability (%) to get N_{Hit}, col. 7 (col. 10) number of hits without double counting, col. 8 (col. 11) the probability (%) to get $N^{\cdot}_{Hit}$

Cat	N_{cat}	N_s	N_{Hit}	P	$N^{\cdot}_{Hit}$	$P^{\cdot}$	N^{all}_{Hit}	P^{all}	$N^{all_1}_{Hit}$	P^{all_1}
0	517	38	3	24.8160	4	9.2529	3	26.2849	4	10.0989
1	608	19	3	5.2570	5	0.1440	3	5.7360	5	0.1640
2	354	39	6	0.7550	9	0.0070	7	0.1700	10	0.0030
3	67	47	2	63.6039	2	63.6039	4	17.9089	4	17.9089
4	272	145	4	90.0610	5	79.3550	7	52.5399	8	37.2750
5	94	62	2	78.0319	4	31.3129	3	56.2049	5	16.7160
6	1699	89	4	58.2109	4	58.2109	6	24.0810	6	24.0810
7	631	113	4	76.0160	6	41.0910	8	16.3519	10	4.0670
8	5250	157	8	42.4710	9	28.9440	10	20.4689	11	12.0260

due to the magnetic field [20; 21]. After finding the sources that are in directional correlations with UHECRs we computed the maximum contribution that they can give to the cosmic ray flux.

Directional correlations

To obtain the samples of sources we perform a scan over the flux of sources at different wavelength, magnitude, redshift, depending on the catalogue in such a way to maximize the directional correlation signal.

The samples obtained have the minimum number of sources (N_s) obeying the conditions and the maximum number of hits (N_{Hit}) in an error cone diameter D, from the UHECRs events.

For each sample we generate a large number of simulated data sets (10^5) and we calculate $P_{MC}(N_s, D, N_{Hit})$, the fraction of sets of N_s data in which we obtained N_{Hit} in a diameter D.

We count the number of hits for each (N_{cat}, N_s, D) and the probability P for observing N_{Hit} or more is calculated as:

$$P(N_{cat}, N_s, D) = \sum_{n=N_{Hit}}^{\infty} P_{MC}(N_s, D, n) = 1 - \sum_{n=0}^{N_{Hit}-1} P_{MC}(N_s, D, n)$$

(1)

We make two calculations, first using the AGASA data, and then using all available and reliable data (marked in the variables with $_{all}$).

The next computations we made were to see the influence that we obtain from the claim that there may be doublets and the triplet in the AGASA data.

If a source is in directional correlation with one of the UHECR event from a doublet it is almost sure in directional correlation with its companion so we get a double counting due to the doublets. If we take into account the doublets, the number of hits from the sample $N^1_{Hit} = N_{Hit} - n_k$, where n_k is the number of doublets and triplets in positional coincidence with any source. In this case the probability $P^1(N_{cat}, N_s, D) > P(N_{cat}, N_s, D)$ so the correlation is weaker.

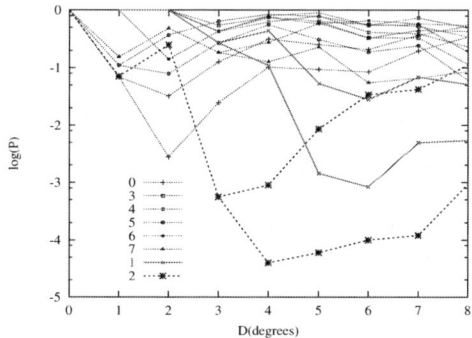

Figure 1. Probability distribution as a function of the angular diameter

In the Table 1 and Figure 1 the catalogues and the samples are referred as:
0- The catalogue of extragalactic sources having flux densities greater than 1Jy at 5GHz [2]. Conditions: magnitude > 18.5, spectral index < 0.5 and redshift> 0.15
1- The catalogue of quasars and active galactic nuclei: 10th edition [3]. We used the BL Lac objects. Conditions: magnitude < 18, the flux at 6cm > 0.17Jy, redshift> 0.1
2- Radio Identifications of extragalactic IRAS sources [4]. Conditions: flux at $12\mu m > 1.2Jy$, at $25\mu m > 2.5Jy$, at $60\mu m > 25Jy$ and at $100\mu m > 40Jy$
3- The Compact Steep-Spectrum and Gigahertz Peaked-Spectrum Radio Sources [5]. Condition: redshift> 0.4
4- The third EGRET catalog of high-energy gamma-ray sources[6], with no conditions.
5- A quantitative evaluation of potential radio identifications for 3EG EGRET sources [7]. Condition: the sources to be quasars.
6- RBSC-NVSS Sample. radio and optical identifications of a complete sample of 1556 X-ray sources [8]. Condition: Xray flux $> 100mJy$
7- A 1.45 GHz Atlas of the IRAS bright galaxy sample [9]. Conditions: Xray flux > 100 mJy, redshift< 0.3, and magnitude > 16
8- A catalog of rich clusters of galaxies [10]. Conditions: redshift < 0.1, magnitude < 16

The UHECR events used are 61 events from AGASA ([11] and AGASA web page), 12 events Yakutsk [12], 6 events from Volcano Ranch [13] and 1

Table 2.   Sources in directional correlation with UHECRs and with compact core flux.

Name	Distance	S	E	F
(-)	(Mpc)	(mJy)	$(10^{''} eV)$	(M87)
3C 120	140.47	3605.42	42.2	0.287
N 3690	44.52	392	9.36	0.141
3C 371	216.28	711.36	32.8	0.073
4C 39.25	2524.91	7097.06	362	0.065
M106	6.38	10.29	0.763	0.045
3C 147	2058.68	1188.35	175	0.022
3C 278	64.09	27.45	49.2	0.018
3C 449	72.94	18.45	4.7	0.013
0954+65	1452.04	119.72	64.4	0.006
3C 119	3424.74	129.67	117	0.0037
4C 62.22	1674.73	4.94	24.5	0.0006

event from Fly's Eye experiment. The energy cut is 4×10^{19} eV. The exception is the one event from AGASA with the energy of 3.8×10^{19} eV.

In Table 1 we show the probabilities obtained using AGASA data and then all available and reliable data. P_1 is the probability double counting the hits due the doublets from the data. We get very low probabilities for the radio identifications of IRAS sources (0.007%). We can see that the probability of correlation becomes weaker if we do not double count (0.75%). The first probability is 100 times smaller than the second, so it is a product of the second probability and the probability of having clusters in UHECRs (of order 10^{-2}). Another low probability to be the correlation by chance is with the Véron catalogue $P_1 = 0.14\%$, but $P = 5.25\%$.

The probability distribution, using the AGASA data, as a function of the angular diameter is shown in Figure 1. The lower probabilities are with double counting and the higher ones are without. We can see again that the difference between the two probabilities is of the same order as the probability of having doublets in the case of the lowest probabilities on the graph.

Toy model

This model contains the idea that particles get to such high energy in the relativistic jets from black holes. The theory [22; 23; 24] allows us to estimate the maximum particle energy and also the maximum contributed flux. It is used the well tested concept of shock acceleration in the relativistic jet and its shock waves, including the final hot spot, arising from the inner ring of the accretion disk, very close to a rapidly spinning black hole.

Irrespective of any acceleration mechanism, the maximum energy of charged particles is given by the spatial limit in the comoving frame. The gyromotion must fit the available space: $E_{\max} = K_g\, e\, c \cdot B\, r$. Here e is the charge of an electron, B is the magnetic field, r is the radius of the jet and the geometric factor $K_g \leq 1$. Introducing the magnetic field and the radio core flux the energy becomes:

$$E_{\max} = 9.68 \cdot 10^{19}\mathrm{eV}\ \left(\tfrac{\gamma_e}{100}\right)^{1/3} \left(\tfrac{S_\nu}{1.0\mathrm{Jy}}\right)^{1/3} \left(\tfrac{D}{8\mathrm{Mpc}}\right)^{2/3} \left(\tfrac{M_\bullet}{3\cdot10^\bullet\,M_\odot}\right) \quad (2)$$

where γ_e is the minimum Lorentz factor of the electrons in the jet, S_ν is the radio flux, D is the distance to the source, normalized to 8Mpc, and $M_\bullet$ is the black hole mass, assumed in the calculations to be $10^8 M_\odot$ (solar masses). The assumed mass enters in the expression for the maximum energy, but not in the expression for cosmic ray flux contribution.

In Table 2 the columns are: (1) Source name, (2) distance, (3) radio core flux; we use intercontinental radio interferometry or radio data from the Very Large Array ,(4) maximum energy, (5) cosmic rays flux relative to M87 and at some reasonably low particle energy such as 10^{18} eV.

The most important contributions to the cosmic ray flux could be from 3C 120, N3690, 3C371 and 4C 39.25. Another interesting source is 3C147, argued before as a possible origin of UHECRs [15]. It is a beam dump radiogalaxy, so the production of the exotic particles that we need may be caused by the collision of the very high energetic particles from the jet with the intergalactic cloud.

Conclusions

What is striking is that we find a probability that a coincidence is random of 0.007 % for the Condon survey of radio sources in positional coincidence with extragalactic infrared sources (IRAS). This probability becomes 0.755 % if we assume each doublet and the triplet as one event. The probability has lower significance because we didn't include any penalty factor for making the cuts in the catalogue.

The conclusion is that the low probabilities that we obtained with different samples of sources is strongly influenced by the probability that we have doublets and triplets. This is because of the poor statistics of the available data. The AUGER experiment [25] will certainly provide us with much more data on UHECRs, and this analysis shows how to search for any correlations in direction with an a priori method.

One possible toy model for accelerating particles to such high energies could be the activity of the relativistic jets in Active Galactic Nuclei. Such candidates for UHECRs might be 3C120, NGC 3690 and others which can have a good contribution to the cosmic ray flux. 3C147 and 4C39.25 are, for example, very well known extreme relativistic jet sources.

Acknowledgments:
Some of this work has been supported by an AUGER grant to prof. Peter L. Biermann.

References

[1] M. Nagano, A.A. Watson: Reviews of Modern Physics, **72** (2000)

[2] H. Kühr et al: Astron. Astrophys. Suppl. Ser. **45** 367-430 (1981)

[3] M-P. Véron-Cetty, P. Véron: Astron. Astrophys. **374** 92-94 (2001)

[4] J.J. Condon et al: Ap.J. **109** 6 (1995)

[5] Ch. O'Dea: P.A.S.P, **110** 439-532 (1998)

[6] R.C. Hartman et al: Ap. J. **123** 79-202 (1999)

[7] J.R. Mattox, R.C. Hartman, O. Reimer: Ap. J. Suppl. Ser. **135** 155-175 (2001)

[8] F.F. Bauer et al.: Ap.J. Suppl.Ser. **125** 547-562 (2000)

[9] J.J. Condon et al: Ap. J. Suppl. Ser. **73** 359-400 (1990)

[10] G.O. Abell, H.G. Corwin Jr., R.P. Olowin: Ap. J. Suppl. Ser. **70** 1-138 (1989)

[11] M. Takeda et al: Ap. J. **522** 225-237 (1999)

[12] B.N. Afanasiev et al.: Proc. Int. Symp. on Extremely High Energy Cosmic Rays : Astrophysics and Future Observatories (1996)

[13] Linsley J.:Catalog of Highest Energy Cosmic Rays, World Data Center of Cosmic Rays, Institute of Physical and Chemical Research N3 Tokyo (1980)

[14] Ch.B. Finley, St. Westerhoff: Astropart. Phys **21** 4 359-367 (2004)

[15] T. Stanev et al.: P.R.L. **75**, 17 (1995)

[16] G. Farrar, P.L. Biermann:Phys. Rev. Lett. **81** 17 (1998)

[17] Gorbunov D. S., Tinyakov P.G., Tkachev I.I., Troitsky S.V.:Ap. J., **577**:L93-L96 (2002)

[18] K. Greisen: Phys. Rev. Letters **16** 17 (1966)

[19] G.T. Zatsepin, V.A. Kuzmin: JETPh Lett., **4** 78 (1966)

[20] G.A. Medina-Tanco: A.J., **549** 711-715 (1999)

[21] D. Harari, S. Mollerach, E. Roulet, F.Sanchez: astro-ph/0202362 v2 (2002)

[22] H. Falcke, P.L.Biermann: Astron. Astrophys. **293** 665 (1995)

[23] H. Falcke, M.A.Malkan, P.L.Biermann: Astron. Astrophys. **298**,375 (1995)

[24] H. Falcke: The Galactic Black Hole, Lectures on General Relativity and Astrophysics (2002)

[25] J. Blümer for the Pierre Auger Collaboration: J.Phys. G: Nucl. Part. Phys. **29** 867-879 (2003)

ARE CLUSTERS INDICATORS OF THE COSMIC RAY ANISOTROPY ?

A.A. Mikhailov, N.N. Efremov

Yu.G. Shafer Institute of Cosmophysical Research and Aeronomy, 31 Leini ave., 677890 Yakutsk, Russia

Abstract: The clusters (doublets) in ultrahigh energy cosmic rays are considered based on Yakutsk and AGASA extensive air shower array data. It is discussed a problem of origin clusters. It is found that arrival directions of the clusters can point to a cosmic rays anisotropy. As a result of analysis for cluster arrival directions, the conclusion on the composition and origin of ultrahigh energy cosmic rays been made.

Key words: cosmic rays of ultrahigh energy, extensive air showers, anisotropy, clusters, doublets.

1. INTRODUCTION

In the ultrahigh energy region at $E > 4.10^{19}$ eV 7 clusters were found by AGASA extensive air shower (EAS) array data[1], 2 clusters by Yakutsk EAS array data[2]. We assume[2] that the clusters are formed as a result of fragmentation of super heavy nuclei into more light.

2. EXPERIMENTAL DATA

According the hypothesis[2], among the particles, forming as result of fragmentation, the most energetic particle must arrive to the Earth first. Here we check this supposition[4]. For this purpose, along with AGASA data[1] we have considered Yakutsk EAS array data beginning from $E > 5.10^{18}$ eV. Data the period of 1974-2002 are used and the arrival directions of $< 60°$ zenith angle showers (n=1863) have been analysis. Among them there are 5157 doublets (clusters having the great number of particles, triplet and etc, are not considered). The direction between the showers in doublets is $< 5°$. The total number of doublets is separated into 2 parts: doublets $N_1(E_1)$ whose particle having the energy E_1 greater than the energy E_2 of neighboring doublet particle ($E_1 > E_2$) arrives first to the Earth, $N_2(E_1)$ are remaining doublets. As a known, the shower energy is determined with some error. Sometimes there is no way telling what doublet particles is of them greater energy in fact. Therefore, doublets, whose particles are energies close to each other within $\lg(E_1/E_2) < 0.2$, are excluded.

A ratio R of the number of these doublets is presented in Fig.1 (circles –Yakutsk, triangles – AGASA: $R = N_1(E_1)/N_2(E_1)$, where $N_1(E_1)$ are doublets whose particles of E_1 greater than the neighboring particle of E_2 in doublet

M. M. Shapiro et al. (eds.), Neutrinos and Explosive Events in the Universe, 343–347.

$(E_1 > E_2)$ arrives to the Earth first, $N_2(E_1)$ are doublets whose particle of energy E_1 arrives second.

The doublet[1] A6 has been excluded from AGASA data because the doublet particles have close to each other energies. As seen from Fig.1, $E < 2.10^{19}$ eV doublets arrive uniformly on the average independent of the energy of leading particles. As $E > 2.10^{19}$ eV the number of doublets, whose particles of E_1 is leader, is ~70% fro Yakutsk array and ~80% for AGASA. So, one can conclude that doublets of $E > 2.10^{19}$ eV are most likely formed as a result of superheavy nuclei fragmentation. Apparently, at $E < 2.10^{19}$ eV doublets are formed by other means than doublets of the higher energies.

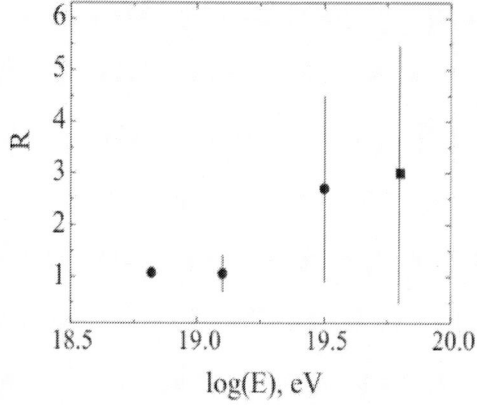

Figure 1. Ratio of the number of doublets $R = N_1(E_1)/N_2(E_1)$

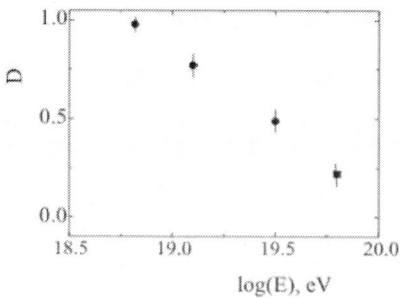

Figure 2. A portion of showers forming doublets N_1 to the total number of showers N_2.: $D = N_1/N_2$

Fig.2 presents a portion of showers forming doublets relative the total number of showers: D= N1/N$_2$, where N$_1$ is the number of showers forming doublets, N$_2$ is the total number of showers. As seen from Fig.2, a portion of showers forming doublets decreases with energy based on Yakutsk data.

Further, we considered the distribution of doublets in galactic latitude for three energy intervals (Fig.3): a) $10^{18.7}$–10^{19} eV – n=4822 doublets (1407 showers), b) 10^{19} – $10^{19.3}$eV – n=293 (337 showers), >$10^{19.3}$ eV– n=42 (59 showers). The histogram presents averaged latitudes for particles of doublets. In Fig.3a for E=$10^{18.7}$–10^{19} eV the galactic plane is seen. At latitudes $|b|$<3° the excess of the number of doublets observed relative to the number expected ones in the case of isotropy is 11σ=(474-288.9)/$\sqrt{288.9}$. The number of doublet expected is found by method taking into account the exposure of array on the celestial sphere (see in detail[6]). At latitudes of 24°>b>9° a maximumin the doublet distribution is also observed. More detail consideration shows that this maximum is limited by 150°>b>120° in galactic longitude. The maximum in doublet distribution repeats a maximum in the distribution od showers for the above coordinates at E~10^{19} eV (Fig.2 of paper[7]). At $|b|$<3° the number of doublets in the energy interval of 10^{19}-$10^{19.3}$ eV exceeds the number of doublets expected by 8.7σ=(54-17.5)/$\sqrt{17.5}$. Note that we found[8] the encreased flux of particles on the galactic plane side at E=$10^{18.9}$-$10^{19.6}$ eV, the number of showers observed exceeds one expected in the case of isotropy by 4.2σ.

The distribution of doublets at E>$10^{19.3}$ eV (Fig.3c) is more isotropic. It is possible that at these energies the formation of doublets starts at the expense of superheavy nuclei fragmentation. To appreciate the origin of doublets we considered the distribution of showers parallel with the distribution of doublets (Fig3b) in the same energy interval. In the distribution of showers the some maxima are observed: at $|b|$<3° (exceese of the number of showers relative to expected is 2.7σ=(29-17.5)/, at 21°>b>15° etc. However, the maximum in the distribution of doublets is seen at $|b|$<3° only. This means that the maximum number of doublets appears where there is the exceed particle flux. Other maxima in the particle distribution are most likely formed by accident and therefore maxima from these directions in the distribution of doublets do not observed. So, the arrival directions of doublets (clusters) can be an indicator of cosmic rays anisotropy. It is important in the case when the cosmic ray anisotropy cannot be detected because of the small statistic.

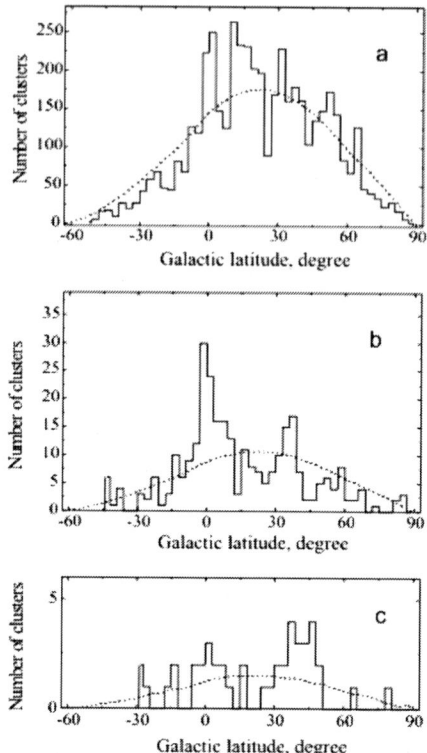

Figure 3. The distribution of doublets in galactic latitude the energy interval: a) $10^{18.7}$–10^{19} eV, b) 10^{19}– $10^{19.3}$ eV, c) E>$10^{19.3}$ eV.

Thus, the distribution of doublets in galactic latitude indicates that at 10^{19} eV cosmic rays are galactic. These result confirm conclusion obtained by us earlier using other methods[8,9] et al. As indicated above, at this energy ~ 80% of showers form doublets. From this it follows that the doublets are mainly tha same showers but are not formed by neutral particles etc.

In the galactic model for the cosmic ray origin, in the case of protons on the galactic plane side, a broad maximum of ~20° in galactic latitude is expected and for the iron nuclei it is the uniform distribution of particle fluxes in latitude[10]. The narrow maximum in the distribution of doublets on the galactic plane side (Fig.3b) can testify that the primaries are iron nuclei. In the absence of the regular magnetic field in the galactic plane (it can be absent because of permanent explosions of supernovae), the iron nuclei can propagated along it. The narrowness in the distribution of doublets is consistent with the conclusion that cosmic rays at ~ 10^{19} eV are iron nulei[10].

From the above analysis we conclude:

1)doublets (clusters) at $E<2.10^{19}$ eV are on whole the same ordinary showers, 2) clusters at $E>2.10^{19}$ eV are most likely formed by means of superheavy nuclei fragmentation, 3) a portion of showers forming clusters decreases with energy, 4) the distribution of doublets in the galactic latitude indicaters that cosmic rays at $E\sim10^{19}$ eV consist of iron nuclei most likely and are galactic, 5) arrival directions of doublets can be indicators of the cosmic ray anisotropy.

References

1. M. Takeda and AGASA Collaboration, AGASA Results (Anisotropy), *Proc. Int. Workshop on Extr. High Energy Cosm. Rays. Journ. of the Phys. Soc. Japan*, **70**, 15-21 (2001).
2. A.A. Mikhailov, Clusters in Very High Energy Cosmic Rays, Proc. Russian Conf. Cosm. Rays, Moscow, 306-308 (2004); astro-ph/0403231.
3. G. Sigl, D.N. Schramn, S. Lee, C.T. Hill. Proc. Natl. Acad. Sci., **94**, 1501 (1997).
4. G. Medina Tanco, On the Signicance of the Observed Clustering of Ultra-High Energy Cosmic Rays, *Astrophysical Journal*, **495**, L71-L74 (1998).
5. A.A. Mikhailov, N.N. Efremov, G.V. Nikolaeva, Proc. Russian Conf. Cosm. Rays, Moscow, 2004, 306; astro-ph/0403231.
6. N.N. Efimov, A.A. Mikhailov, M.I. Pravdin, Anisotropy of Cosmic Ray of Superhigh Energies, Proc. 18-th ICRC, Bangalore, **2**, 149-152 (1983).
7. A.A. Mikhailov, G.V. Nikolayeva, Proc. 28-th ICRC, Tsukuba, 2003, **1**, 417.
8. A.A. Mikhailov, On the Galactic Origin of Ultrahigh Energy Cosmic Rays, *Proc. 27-th ICRC, Hamburg*, **5**, 1772-1775 (2001).
9. A.D. Erlykin, A.A. Mikhailov, A.W. Wolfendale, Ultra High Energy Cosmic Rays and Pulsars, Phys. Nucl. Part. Phys., **28**, 2225-2233 (2002).
10. A.A. Mikhailov, Arrival Directions and Chemical Composition of Ultrahigh-Energy Cosmic Rays, *JETF Letters*, **72**, 160-162 (2000).

USING COSMIC RAY FOR MONITORING AND FORECASTING DANGEROUS SOLAR FLARE EVENTS

Lev I. Dorman

Israel Cosmic Ray & Space Weather Center and Emilio Segre' Observatory, affiliated to Tel Aviv University, Technion, and Israel Space Agency; Qazrin, ISRAEL and Cosmic Ray Department of IZMIRAN, Russian Academy of Sciences, Troitsk, Moscow region, RUSSIA; E-mail: lid@physics.technion.ac.il

Abstract: It is well known that in periods of great Solar Energetic Particles (SEP) fluxes of energetic particles can be so big that memory of computers and other electronics in space may be destroyed, satellites have serious anomalies: according to NOAA Space Weather Scales are dangerous Solar Radiation Storms S5-extreme with flux level of particles with energy > 10 MeV more than $10^5 \text{cm}^{-2}\text{s}^{-1}$, S4-severe (flux more than 10^4) and S3-strong (flux more than 10^3). In these periods is necessary to switch off some part of electronics for few hours to protect computer memories. These periods are also dangerous for astronauts on space-ships, and passengers and crew in commercial jets (especially during S5 storms). The problem is how to forecast exactly these dangerous phenomena. We show that exact forecast can be made by using high-energy particles (few GeV/nucleon and higher) which transportation from the Sun is characterized by much bigger diffusion coefficient than for small and middle energy particles. Therefore high energy particles came from the Sun much more early (8-20 minutes after acceleration and escaping into solar wind) than main part of smaller energy particles caused dangerous situation for electronics (about 30-60 minutes later). We describe here principles and experience of automatically working of program "SEP-Search". The positive result which shows the exact beginning of SEP event on the Emilio Segre' Observatory (2025 m above sea level, $R_c = 10.8$ GV), is determined now automatically by simultaneously increasing on 2.5 St. Dev. in two sections of neutron supermonitor (we determine also the probabilities of false and missed alerts). The next 1-min data the program "SEP-Search" uses for checking that the observed increase reflects the beginning of

M. M. Shapiro et al. (eds.), Neutrinos and Explosive Events in the Universe, 349–363.

real great SEP or not. If yes, automatically starts to work on line the programs "SEP-Research" for determining spectrum of SEP out of atmosphere, spectrum in source, diffusion coefficient. To continue the SEP spectrum determining in low energy range, we use also available from Internet satellite data. Obtained results allow making forecast of radiation hazard for spacecrafts and aircrafts for about two days ahead on the basis of the first 30-40 minutes cosmic ray (CR) data. The work of NM on Mt. Hermon is supported by Israel (Tel Aviv University) – Italian (Roma-Tre University and IFSI-CNR) collaboration. This research is in the frame of INTAS-810 and COST-724.

Key words: space weather, cosmic rays, radiation hazard, monitoring, forecasting

1.   SEP AND SATELLITE ANOMALIES; RADIATION HAZARD FOR SPACECRAFTS AND AIRCRAFTS AND GROUND BASED OBJECTS

In papers Belov et al. (2003), Iucci et al. (2004) was formatted a large database of anomalies (about 6000), registered by 220 satellites in different orbits over the period 1971-1994. For the first time, data of 49 Russian Kosmos satellites have been included in a statistical analysis. The database also contains a large set of daily and hourly space weather parameters. A series of statistical analyses made it possible to quantify, for different satellite orbits, space weather conditions in the days characterized by anomaly occurrences. In particular, very intense fluxes (>1000 pfu at energy >10 MeV) of solar protons are linked to anomalies registered by satellites in high-altitude (>15,000 km) near-polar (inclination >55°) orbits typical for navigation satellites such as those used in the GPS network, NAVSTAR, etc. (the rate of anomalies increases by a factor ~20), and to a much smaller extent to anomalies in geostationary orbits (they increase by a factor ~4). The efficiency in producing anomalies is found to be negligible for proton fluences <100 pfu at energies >10 MeV. Elevated fluxes of energetic (>2 MeV) electrons >10^8 $(cm^2 \cdot day \cdot sr)^{-1}$ are observed by GOES on days with satellite anomalies occurring at geostationary (GOES, SCATHA, METEOSAT, MARECS-A etc) and low-altitude (<1,500 km) near-polar (>55°) orbits (Kosmos, SAMPEX, etc.).

Fig. 1 shows the dependence of anomaly frequencies for satellites with different orbits in dependence of SEP flux. From Fig. 2 can be seen that the probability of anomaly for satellites with high altitude – high inclination orbits increased very much with increasing of proton fluxes (up to 100%).

Great SEP in some more rarely cases gave dangerous radiation also for aircraft's computers and navigation systems as well as for crew and passengers health (in dependence of aircraft position at moment of event), and even for ground based high level technology and people health.

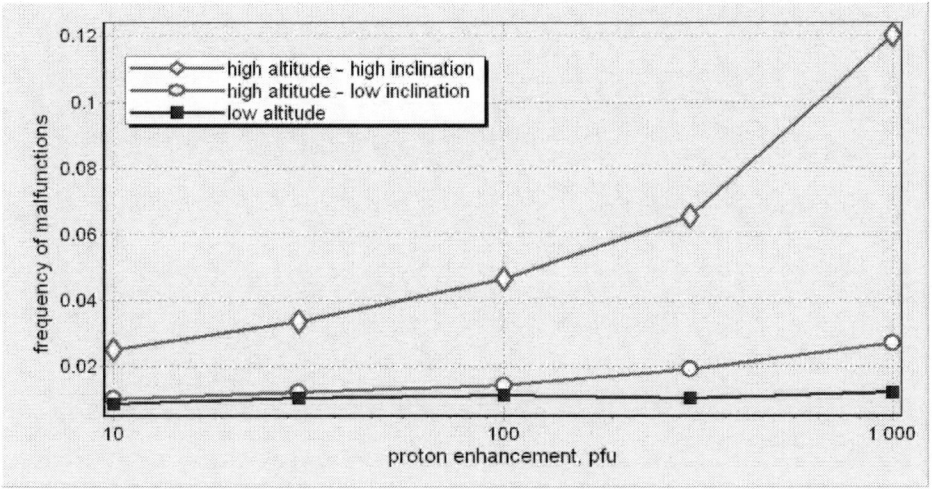

Figure 1 Mean satellite anomaly frequencies in 0- and 1-days of proton enhancements in dependence on the maximal > 10 MeV flux.

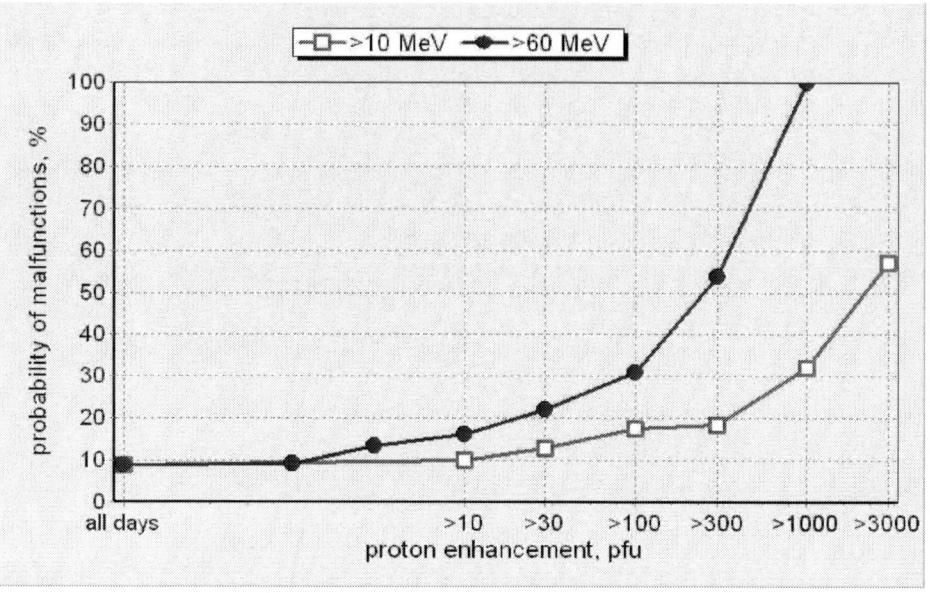

Figure 2. Probability of any anomaly (high altitude – high inclination group of satellites) in dependence on the maximal proton >10 and >60 MeV flux

2.   MAIN STEPS OF GREAT SEP FORECAST

The main steps of the great SEP forecast are as following:

1. Automatically determination of the SEP event starting by neutron monitor data.

2. Determination of energy spectrum out of magnetosphere by the method of coupling functions.

3. In the frame of the simplest model (not dependence of diffusion coefficient from the distance to the sun) determination of time of ejection, source function and parameters of propagation on the basis of found energy spectrums at different moments of time.

4. Checking the model by experimental data and extending it by taking into account possible dependence of diffusion coefficient from the distance to the Sun.

5. Additional extending of the model to the small energy range by using on-line one-minute data simultaneously from ground based neutron monitors and CR detectors on spacecrafts.

6. On the basis of found properties of source function and propagation parameters forecasting of radiation hazard for about two days ahead (on the

basis of 30-40 min CR data) for spacecrafts in the interplanetary space at different distances from the Sun, for satellites in the Earth's magnetosphere with different orbits, for aircrafts in dependence of altitude and cutoff rigidity, for ground objects in dependence of cutoff rigidity and air pressure.

3. AUTOMATICALLY DETERMINATION OF THE SEP EVENT STARTING BY NM DATA

3.1 Algorithms, real-time data, formation of Alerts

Let us describe the principles and on-line operation of program "SEP-Search", developed and checked in the Emilio Segre' Observatory of Israel Cosmic Ray & Space Weather Center. The determination of increasing flux is made by comparison with intensity averaged from 120 to 61 minutes before the present Z-th one-minute data. The program for each Z-th minute determines the values

$$D_{AZ} = \left[\ln\left(I_{AZ}\right) - \sum_{k=Z-120}^{k=Z-60} \ln\left(I_{Ak}\right)/60 \right] / \sigma, \tag{1}$$

$$D_{BZ} = \left[\ln\left(I_{BZ}\right) - \sum_{k=Z-120}^{k=Z-60} \ln\left(I_{Bk}\right)/60 \right] / \sigma, \tag{2}$$

where I_{Ak} and I_{Bk} are one-minute total intensities in the sections of neutron monitor A and B, and σ is the St. Dev. for one min data. If simultaneously

$$D_{AZ} \geq 2.5, \ D_{BZ} \geq 2.5, \tag{3}$$

the program "SEP-Search" repeat the calculation for the next Z+1-th minute and if Eq. 3 is satisfied again, the onset of great SEP is determined, in the website at his moment is automatically shown Alert and programs "SEP-Research" (described below), are started. In the website http://www.tau.ac.il/institutes/advanced/cosmic/icrc.htm of Israel Cosmic Ray & Space Weather Center are given one-min data (for the last 6 hours) and one-hour data (for the last 6 days) in real-time scale, upgraded each minute, and information on Alert of SEP starting (Yes or No).

3.2   The probability of false alarms

Because the probability function $\Phi(2.5) = 0.9876$, that the probability of an accidental increase with amplitude more than 2.5σ in one channel will be $(1 - \Phi(2.5))/2 = 0.0062\,\mathrm{min}^{-1}$, that means one in 161.3 minutes (in one day we expect 8.93 accidental increases in one channel). The probability of accidental increases simultaneously in both channels will be $((1 - \Phi(2.5))/2)^2 = 3.845 \times 10^{-5}\,\mathrm{min}^{-1}$ that means one in 26007 minutes ≈ 18 days. The probability that the increases of 2.5σ will be accidental in both channels in two successive minutes is equal to $((1 - \Phi(2.5))/2)^4 = 1.478 \times 10^{-9}\,\mathrm{min}^{-1}$ that   means   one   in 6.76×10^8 minutes ≈1286 years. If this false alarm (one in about 1300 years) is sent, it is not dangerous, because the first alarm is preliminary and can be cancelled if in the third successive minute is no increase in both channels bigger than 2.5σ (it is not excluded that in the third minute there will be also an accidental increase, but the probability of this false alarm is negligible: $((1 - \Phi(2.5))/2)^6 = 5.685 \times 10^{-14}\,\mathrm{min}^{-1}$ that means one in 3.34×10^7 years). Let us note that the false alarm can be sent also in the case of solar neutron event (which really is not dangerous for electronics in spacecrafts or for astronauts health), but this event usually is very short (only few minutes).

3.3   The probability of missed triggers

The probability of missed triggers depends very strong from the amplitude of the increase. Let us suppose for example that we have a real increase of 7σ (that for ESO corresponds to an increase of about 10%). The trigger will be missed if in any of both channels and in any of both successive minutes if as a result of statistical fluctuations the increase of intensity is less than 2.5σ. For this the statistical fluctuation must be negative with amplitude more than 4.5σ. The probability of this negative fluctuation in one channel in one minute is equal $(1 - \Phi(4.5))/2 = 3.39 \times 10^{-6}\,\mathrm{min}^{-1}$, and the probability of missed trigger for two successive minutes of observation simultaneously in two channels is 4 times larger: 1.36×10^{-5}. It means that missed trigger (delay in the starting of Alert on two first minutes) is expected only one per about 70000 events with amplitude 10%.

4. DETERMINATION OF ENERGY SPECTRUM OUT OF MAGNETOSPHERE BY THE METHOD OF COUPLING FUNCTIONS

Based on latitude survey data of Aleksanyan et al. (1985), Dorman et al. (2000) the polar normalized coupling functions for total counting rate and different multiplicities m can be approximated by special normalized function ($\int_{0}^{\infty} W_{om}(R)\,dR = 1$) introduced in Dorman (1969):

$$W_{om}(R) = a_m k_m R^{-(k_m+1)} \exp\left(-a_m R^{-k_m}\right), \qquad (4)$$

where m = tot, 1, 2, 3, 4, 5, 6, 7, ≥ 8. The details of coupling functions method and its broad applications are described in Dorman (2004). Here we will present shortly only final results. The coupling function for point with cut-off rigidity R_C, will be

$$W_m(R_C,R) = a_m k_m R^{-(k_m+1)}\left(1 - a_m R_C^{-k_m}\right)^{-1} \exp\left(-a_m R^{-k_m}\right), \quad (5)$$

if $R \geq R_C$, and $W_m(R_C,R) = 0$, if $R < R_C$. In the first approximation the spectrum of primary variation of SEP event can be described by function

$$\Delta D(R)/D_o(R) = bR^{-\gamma}, \qquad (6)$$

where $\Delta D(R) = D(R,t) - D_o(R)$, $D_o(R)$ is the differential spectrum of galactic CR before the SEP event (for this time the coupling function of Eq. 1 and 2 must be determined, see below), and $D(R,t)$ is the spectrum at a later time t. In Eq. 6 parameters b and γ depend on t.

The expected variation in total counting rate or in multiplicity m will be

$$\Delta I_m(R_C)/I_{mo}(R_C) = bF_m(R_C,\gamma), \qquad (7)$$

where m = tot, 1, 2, 3, 4, 5, 6, 7, ≥ 8, and

$$F_m(R_C,\gamma) = a_m k_m \left(1 - \exp\left(-a_m R_C^{-k_m}\right)\right)^{-1} \int_{R_C}^{\infty} R^{-(k_m+1+\gamma)} \exp\left(-a_m R^{-k_m}\right) dR \quad (8)$$

Let us compare data for multiplicities m and n. According to Eq. 7 we obtain

$$[\Delta I_m(R_c)/I_{mo}(R_c)]/[\Delta I_n(R_c)/I_{no}(R_c)] = \Psi_{mn}(R_c,\gamma), \qquad (9)$$

where special functions

$$\Psi_{mn}(R_c,\gamma) = F_m(R_c,\gamma)/F_n(R_c,\gamma) \qquad (10)$$

may be easy calculate by using Eq. 8. Comparison of experimental results (left side of Eq. 9) with the function $\Psi_{mn}(R_c,\gamma)$ gives the value of γ, and then, from Eq. 7, the value of parameter b. The observed SEP increase for different multiplicities allows the determination of parameters b and γ for the SEP event beyond the Earth's magnetosphere.

5. DETERMINATION OF TIME OF EJECTION, SOURCE FUNCTION AND PARAMETERS OF PROPAGATION (THE SIMPLEST MODEL)

Let us consider the simplest model of SEP generation and propagation: isotropic diffusion from the pointing instantaneous source:

$$Q(R,r',t') = N_o(R)\delta(r')\delta(t'). \qquad (11)$$

In this case the expected SEP rigidity spectrum on the distance r from the Sun in the time t after ejection will be

$$N(R,r,t) = N_o(R) \times \left[2\pi^{1/2}(K(R)t)^{3/2}\right]^{-1} \times \exp\left(-\frac{r^2}{4K(R)t}\right), \qquad (12)$$

where $K(R)$ is the diffusion coefficient in the interplanetary space in the period of SEP event (not dependent from the distance to the Sun). At $r = r_1 = 1\,\text{AU}$ and at some moment t_i the spectrum determined in Section 4 will be described by the function

$$N(R,r_1,t_i) = b(t_i)R^{-\gamma(t_i)}D_o(R), \qquad (13)$$

where $b(t_i)$ and $\gamma(t_i)$ are parameters determined the observed rigidity spectrum in the moment t_i, and $D_o(R)$ is the spectrum of galactic CR before event. Let us suppose that the UT time of ejection T_e, source function $N_o(R)$, and diffusion coefficient $K(R)$ are unknown. In this case for determining on-line simultaneously these three parameters we need information on SEP spectrum at least at three moments of time T_1, T_2 and T_3 (all times T are in UT scale). In this case we will have for times after SEP ejection into solar wind:

$$t_1 = T_1 - T_e = x,\ t_2 = T_2 - T_1 + x,\ t_3 = T_3 - T_1 + x, \tag{14}$$

where $T_2 - T_1$ and $T_3 - T_1$ are known values, and x is unknown value what we need to determine in the first. From Eq. 12 with taking into account Eq. 13 and Eq. 14 we obtain

$$\frac{T_2 - T_1}{x(T_2 - T_1 + x)} = -\frac{4K(R)}{r_1^2} \times \ln\left\{\frac{b(T_1)}{b(T_2)}(x/(T_2 - T_1 + x))^{3/2} R^{-[\gamma(T_1) - \gamma(T_2)]}\right\}, \tag{15}$$

$$\frac{T_3 - T_1}{x(T_3 - T_1 + x)} = -\frac{4K(R)}{r_1^2} \times \ln\left\{\frac{b(T_1)}{b(T_3)}(x/(T_3 - T_1 + x))^{3/2} R^{-[\gamma(T_1) - \gamma(T_3)]}\right\}. \tag{16}$$

After dividing Eq. 15 on Eq. 16 we obtain

$$x = [(T_2 - T_1)\Psi - (T_3 - T_1)]/(1 - \Psi), \tag{17}$$

where

$$\Psi = \frac{T_3 - T_1}{T_2 - T_1} \times \frac{\ln\left\{\frac{b(T_1)}{b(T_2)}(x/(T_2 - T_1 + x))^{3/2} R^{\gamma(T_2) - \gamma(T_1)}\right\}}{\ln\left\{\frac{b(T_1)}{b(T_3)}(x/(T_3 - T_1 + x))^{3/2} R^{\gamma(T_3) - \gamma(T_1)}\right\}}. \tag{18}$$

Eq. 17 can be solved by the iteration method: as the first approximation, we can use $x_1 = T_1 - T_e \approx 500$ sec what is a minimum time of relativistic particles propagation from the Sun to the Earth's orbit. Then by Eq. 18 we determine $\Psi(x_1)$ and by Eq. 17 determine the second approximation x_2, and so on. After solving Eq. 17 and determining the time of ejection, we compute very easy diffusion coefficient from Eq. 15 or Eq. 16:

$$K(R) = -\frac{r_1^2\left(T_3 - T_1\right)/4x\left(T_3 - T_1 + x\right)}{\ln\left\{\dfrac{b(T_1)}{b(T_3)}\left(x/\left(T_3 - T_1 + x\right)\right)^{3/2} R^{\gamma(T_3) - \gamma(T_1)}\right\}}. \quad (19)$$

After determining time of ejection and diffusion coefficient it is very easy to determine the SEP spectrum in source ($i = 1, 2, 3$):

$$N_o(R) = 2\sqrt{\pi}b(t_i) R^{-\gamma(t_i)} D_o(R) \times \left(K(R)t_i\right)^{3/2} \exp\left(r_1^2/\left(4K(R)t_i\right)\right). (20)$$

6.   CHECKING THE MODEL AND ACCOUNT NON-CONSTANT OF DIFFUSION COEFFICIENT

For the checking of used above model of SEP propagation in the interplanetary space, we determined in the first values of $K(R)$. These calculations we made according to the procedure described above at supposition that $K(R)$ does not depend from the distance to the Sun. Results for the event of 29 September 1989 are shown in Fig. 3.

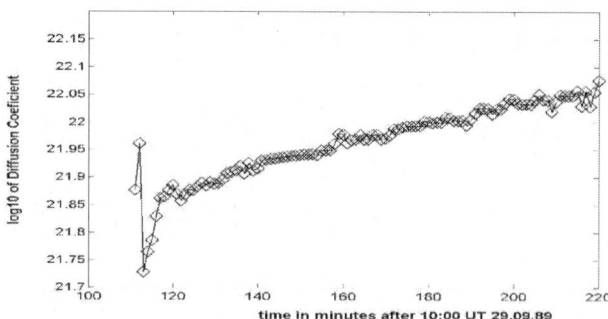

Figure 3. The behavior of $K(R)$ for $R \sim 10$ GV with time.

It can be seen that in the beginning of event obtained results are not stable what are caused by a big relative statistical errors. After few minutes amplitude of CR intensity increase became many times bigger than σ, and we see systematical increase of the diffusion coefficient with time: really its reflects the increasing of $K(R)$ with the distance to the Sun.

Let us suppose, according to Parker (1963), that the diffusion coefficient

$$K(R,r) = K_1(R) \times (r/r_1)^\beta . \tag{21}$$

In this case instead of Eq. 12 we obtain

$$N(R,r,t) = \frac{N_o(R) r_1^{\frac{3\beta}{2-\beta}} (K_1(R)t)^{\frac{-3}{2-\beta}}}{(2-\beta)^{\frac{4+\beta}{2-\beta}} \Gamma\left(\frac{3}{2-\beta}\right)} \exp\left(-\frac{r_1^\beta r^{2-\beta}}{(2-\beta)^2 K_1(R)t}\right) . \tag{22}$$

If we know N_1, N_2, N_3 at moments of time t_1, t_2, t_3, the final solutions for β, $K_1(R)$, and $N_o(R)$ will be

$$\beta = 2 - 3\left[\ln\left(\frac{t_2}{t_1}\right) - \frac{t_3(t_2 - t_1)}{t_2(t_3 - t_1)}\ln\left(\frac{t_3}{t_1}\right)\right]\left[\ln\left(\frac{N_1}{N_2}\right) - \frac{t_3(t_2 - t_1)}{t_2(t_3 - t_1)}\ln\left(\frac{N_1}{N_3}\right)\right]^{-1} , \tag{23}$$

$$K_1(R) = \frac{r_1^2\left(t_1^{-1} - t_k^{-1}\right)}{3(2-\beta)\ln(t_k/t_1) - (2-\beta)^2\ln(N_1/N_k)} , \tag{24}$$

$$N_o(R) = N_k(2-\beta)^{\frac{4+\beta}{2-\beta}} \Gamma\left(\frac{3}{2-\beta}\right) r_1^{\frac{-3\beta}{2-\beta}} (K_1(R)t_k)^{\frac{3}{2-\beta}} \times \exp\left(\frac{r_1^2(2-\beta)^{-2}}{K_1(R)t_k}\right) \tag{25}$$

In Eq. 24 index $k = 2$ or 3, and in Eq. 25 index $k = 1, 2$ or 3.

By using of the SEP event first few minutes NM data we can determine by Eq. 23 – 25 effective parameters β, $K_1(R)$, and $N_o(R)$, corresponded to the rigidity about 7 – 10 GV, and then by Eq. 22 we determine the forecasting curve of expected SEP flux behavior for total neutron intensity. This curve we compare with time variation of observed total neutron intensity. Really we use data for more than three moments of time by fitting obtained results in comparison with experimental data to reach the minimal residual (see Fig. 4, which contains 8 panels for time moments $t = 110$ min up to $t = 220$ min after 10.00 UT of 29 September, 1989).

From Fig. 4 can be seen that using only the first 5 minutes of NM data (t = 110 min) does not enough: the obtained curve forecasts too low intensity. For t = 115 min the forecast shows little bigger intensity, but also not enough. Only from t = 120 min (15 minutes after beginning) we obtain about stable forecast with good agreement with observed CR intensity.

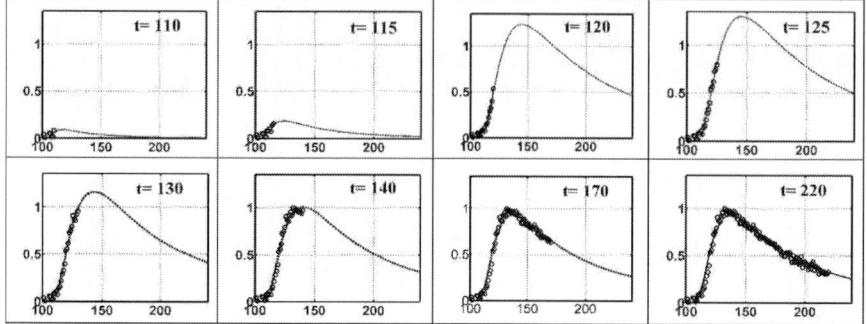

Figure 4. Forecasting of total neutron intensity : curves – forecasting, circles – observed total neutron intensity.

7.   EXTENDING TO SMALL ENERGY RANGE

The main idea of the model extending to the small energy range for using on-line available one-minute satellite data is as following: the source function relative to time remain to be as δ–function, but relative to energy is extended to the power function with an energy-dependent index γ ($\gamma = \gamma_o + a \ln(E_k / E_{k\,\max})$), i.e.:

$$N_o(R,T) = N_o \delta(T - T_e) \times R^{-(\gamma_o + a \ln(E_k / E_{k\,\max}))}. \qquad (26)$$

We suppose for the diffusion coefficient $K(R,r)$ in the interplanetary space power function from the distance to the Sun (as $\propto (r/r_1)^\beta$) and power function for the transport path, so that

$$K(R,r) = K_1(R) \times (r/r_1)^\beta \; ; \; K_1(R) = K_1 \times (v/c) \times (R/R_1)^\delta, \qquad (27)$$

and v is particle velocity, R_1 =1 GV. Parameter β was determined in Section 6 from NM data ($\beta \approx 0.6$). We suppose that for the low energy range parameter β is the same. Time of ejection T_e was also determined from NM data (see Section 5). Again we assume that it remains the same for the low energy range. The fitting of the parameters γ_o, a, $E_{k\,\max}$, δ was performed every 5 minutes, using more and more data: we start from $t = 105$ min (calculated from 10.00 UT September 29, 1989) and use the data of first five minutes; then our program performs a fit for $t = 110$ min using the first ten minute data, and so on. In Fig. 5 are shown the predicted SEP integral fluxes for $E_k \geq E_o = 0.1, 1, 3$ GeV, and comparison with the observed for $E_k \geq E_o = 0.1$ GeV by the GOES satellite.

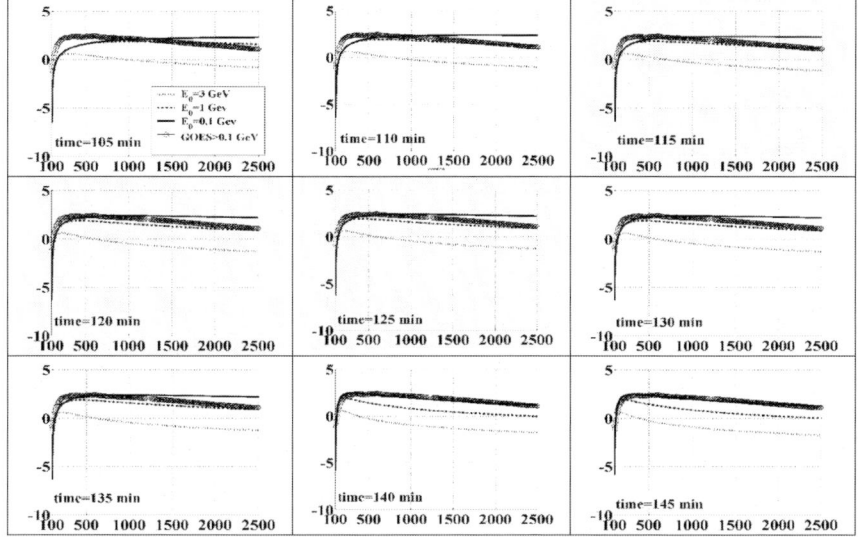

Figure 5. Predicted FEP integral fluxes for $E_k \geq E_o = 0.1, 1, 3$ GeV and comparison with observed on GOES. The ordinate is $\log_{10}$ of SEP integral flux (in $cm^{-2}sec^{-1}sr^{-1}$), and the abscissa is time in minutes from 10.00 UT of September 29, 1989.

From Fig. 5 it can be seen that the agreement between the predicted and observed FEP integral flux for $E_k \geq E_o = 0.1$ GeV is excellent after 30-40 minutes from the onset of the event. The agreement continues to more than 2500 minutes (about two days).

8. FORECASTING AND ALERTS

By using obtained results we can forecast of expected radiation hazard in space, in magnetosphere or in atmosphere. For example, for satellites at different orbits characterized by cut-off rigidities $R_c(T)$ inside the Earth's magnetosphere the expected SEP flux will be

$$I_S\left(T, R_c(T)\right) = \int_{R_c(T)}^{\infty} N_o(R) \frac{r_1^{\frac{3\beta}{2-\beta}} \left((T-T_e)K_1(R)\right)^{\frac{-3}{2-\beta}}}{(2-\beta)^{\frac{4+\beta}{2-\beta}} \Gamma\left(3/(2-\beta)\right)} \exp\left(-\frac{(2-\beta)^{-2} r_1^2}{(T-T_e)K_1(R)}\right) dR \quad (28)$$

The expected fluency (proportional to the radiation dose) will be

$$F_s\left(R_c(T)\right) = \int_{T_e}^{\infty} dT \int_{R_c(T)}^{\infty} N_o(R) \frac{r_1^{\frac{3\beta}{2-\beta}} \left((T-T_e)K_1(R)\right)^{\frac{-3}{2-\beta}}}{(2-\beta)^{\frac{4+\beta}{2-\beta}} \Gamma\left(3/(2-\beta)\right)} \exp\left(\frac{-(2-\beta)^{-2} r_1^2}{(T-T_e)K_1(R)}\right) dR \quad (29)$$

If the predicted fluxes and fluencies are expected to be dangerous for spacecraft and aircraft operations, for radiation situation on the ground, preliminary "FEP-Alert-1/Space", "FEP-Alert-1/Magnetosphere", "FEP-Alert-1/Atmosphere" and "FEP-Alert-1/Ground" will be sent in the first few minutes after the event beginning. As more data become available, better predictions of the expected fluxes will be made. On the basis of these results, more exact Alert-2 (on the basis of the first 10 min data), Alert-3 (on the basis of the first 15 min data) and so on will automatically be issued. These Alerts will give information on the expected level of dangerous situation for different objects in space, in magnetosphere, in atmosphere on different altitudes and at different cut-off rigidities; experts must decide what to do operationally in each concrete case.

9.   CONCLUSION

We show that by one-minute neutron monitor data and one-minute available from Internet CR satellite data for 20-30 min of great SEP event beginning is possible to determine the time of ejection, source function, and diffusion coefficient in dependence from particle energy and distance from the Sun. Then it is possible to forecast of expected SEP fluxes and fluencies in high and low energy ranges up to about two days for spacecrafts in space on different distances from the Sun, for satellites in magnetosphere with different orbits, for aircrafts in the atmosphere on different airlines, and on the ground in dependence of air pressure and cutoff rigidity. If the radiation situation is expected to be dangerous, each 5 minutes will be send corresponding Alerts, at the beginning of event very preliminary, but then in 20-30 min more and more exactly.

Acknowledgements. This research is partly supported by INTAS – 810 and COCT-724. The work of NM in ESO on Mt. Hermon is supported by the collaboration between Tel Aviv University (Israel) and Roma-Tre University and INFN (Italy). Our gratitude to Yuval Ne'eman and Abraham Sternlieb for constant support and attention. Many thanks to our colleagues N. Iucci, M. Murat, M. Parisi, L. Pustil'nik, M. Storini, G. Villoresi, and I. Zukerman for interesting discussions and collaboration. My great thanks also to M. Shapiro, J. Wefel, V. Ptuskin, T. Stanev, E. Berezhko, A. Mikhailov, V. Kuzmin for kind invitation and useful discussions.

References

Alexanyan T.M., L.I. Dorman, V.G. Yanke, and V.K. Korotkov, Proc. 19-th Intern. Cosmic Ray Conf., La Jolla, **5**, 300-303 (1985).

Belov A., L. Dorman, E. Eroshenko et al. "The relation between malfunctions of satellites at different orbits and cosmic ray variations", Proc. 28-th Intern. Cosmic Ray Conf., Tsukuba, 7, 4213-4216 (2003)

Dorman L.I., Cosmic Rays in the Earth's Atmosphere and Underground, Kluwer Ac. Publ., 2004.

Dorman L.I., Proc. of 11-th Intern. Cosmic Ray Conf., Budapest, Volume of Invited Papers and Rapporteur Talks, 381-444 (1969).

Dorman, L.I., G. Villoresi, N. Iucci, M. Parisi, N.G. Ptitsyna, *J. Geophys. Res.- Space Phys.*, **105**, 21047 (2000)

Iucci N., A.E. Levitin, A.V. Belov et al. "Space weather conditions and spacecraft anomalies in different orbits", Space Weather (2004) – accepted for publications.

Parker E.N. Dynamical Processes in the Interplanetary Medium, Intersci,, 1963.

THE KNEE IN THE ENERGY SPECTRUM
OF COSMIC RAYS

Jørg R. Hørandel
University of Karlsruhe, Institut für Experimentelle Kernphysik,
PO Box 3640, 76021 Karlsruhe, Germany
hoerandel@ik.fzk.de

Abstract   Results from direct and indirect measurements of cosmic rays are reviewed. Emphasis is given to the understanding of the knee in the energy spectrum. The data are compared to contemporary models for the knee. Implications on the present understanding of the origin of galactic cosmic rays are discussed.

Keywords: cosmic rays, energy spectrum, mass composition, knee

Introduction

The all-particle energy spectrum of cosmic rays (CRs) follows a power law $dN/dE \propto E^\gamma$ over many orders of magnitude. However, a close inspection reveals two structures, a change of the spectral index γ from about -2.7 to -3.1 at $\sim$ 4.5 PeV, commonly referred to as the knee, and a smaller structure at $\sim$ 400 PeV, the second knee. A compilation of the flux obtained by many experiments is presented in Fig. 1. The energy scale of the individual experiments has been slightly normalized ($\pm 10\%$) in order to match the flux obtained with direct measurements, see Hörandel 2003a. In this representation the good agreement between the experiments in the reconstructed shape of the spectrum is evident. An explanation for the origin of the structures in the energy spectrum is generally assumed to be a corner stone in the understanding of the origin of CR.

In the literature, many ideas have been sketched to explain the structures. Some authors propose the finite energy reached in the acceleration process, e.g. due to the limiting energy defined by the size and magnetic field strength of the acceleration region ($E_{max} < Z \times (B \times L)$), in supernova remnants (SNR) or γ-ray bursts to be responsible for the shape. Others discuss the diffusive propagation in the Galaxy, at high energies the particles cannot be magnetically bound efficiently by the Galaxy and consequently escape from it. A third group deals with the interactions of CRs with background particles during

M. M. Shapiro et al. (eds.), Neutrinos and Explosive Events in the Universe, 365–376.

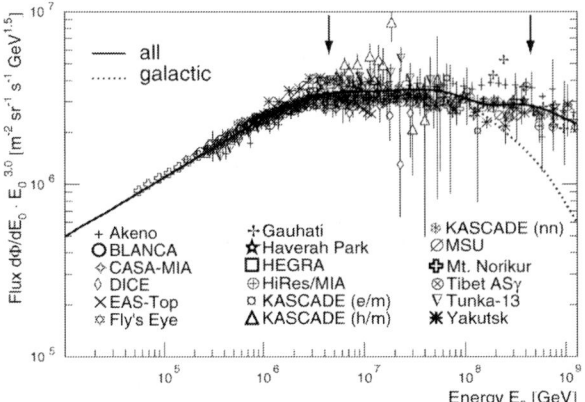

Figure 1 Normalized all-particle energy spectra from different experiments. The lines indicate the average all particle spectrum and the contribution of galactic cosmic rays. The knee at $E_k \sim 4.5$ PeV and the second knee at ~ 400 PeV$\approx 92 \cdot E_k$ are indicated. For details and references see Hörandel 2003a.

the propagation process. Finally, new types of interactions in the atmosphere are proposed to explain the observed structures. All these approaches result in sequential cut-offs for the energy spectra of individual elements, starting with the lightest elements hydrogen and helium.

In order to discriminate between the individual approaches one needs measurements of the energy spectra for individual elements, or at least elemental groups, the arrival directions of CRs, and their mass composition. In this lecture recent experimental results are reviewed and their implications on the contemporary understanding of the origin of the knee is discussed.

Ideally, one would like to measure the energy spectra for individual elements directly above the atmosphere. But the steeply falling spectrum requires large detection areas and long exposure times at high energies. For example, the integral flux amounts to about 0.5 particles/m² day above 1 PeV. At present, experiments exceeding several 10^4 m² can only be realized in ground based installations. They measure the secondary products generated by high-energy CR particles in the atmosphere – the extensive air showers. The challenge of these investigations is to reveal the properties of the shower inducing particle behind an absorber – the atmosphere – with a total thickness, corresponding to 11 hadronic interaction lengths or 30 radiation lengths. Two basic approaches can be distinguished: Measuring the debris of the particle cascade at ground level by registering the main shower components, the electromagnetic, muonic, and hadronic parts. Or measuring the longitudinal shower development in the atmosphere, exploring the Čerenkov or fluorescence light generated predominantly by the shower electrons. An astrophysical interpretation of air shower data requires detailed knowledge of the interaction processes in the atmosphere. The shower development is strongly influenced by the high-energy hadronic interactions and their particle production in the very forward kinematical region and by the low energy interaction models influencing mostly the lateral particle distributions. Unfortunately, both processes are poorly known, see

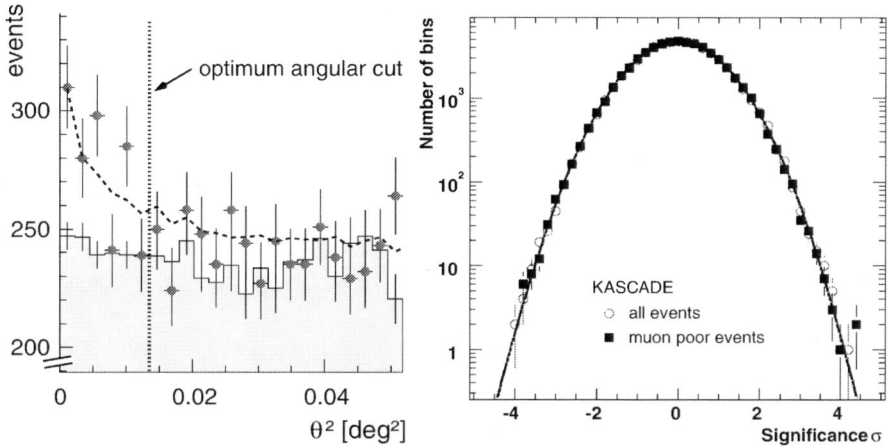

Figure 2. Left: Number of high-energy γ-rays as function of angular distance to Cassiopeia A as measured by the HEGRA experiment. The shaded area is a background estimate Aharonian *et al.* 2001. Right: Distribution of the significance values from a sky map of the arrival direction of CRs as measured by the KASCADE experiment for the complete data set (open circles) and a selection of muon poor showers (filled squares) Antoni *et al.* 2004b.

e.g. Drescher et al. 2004. This limits the interpretation of the measurements and stimulates the test and improvement of interaction models, e.g. Hörandel 2003c, Milke et al. 2000.

Experimental Results

A big step towards the understanding of CR sources would be their direct detection on the sky. Charged CRs are deflected in the galactic magnetic fields, the gyromagnetic radius of a proton with an energy of 1 PeV is about 0.4 pc. But neutral particles, like γ-rays, are good candidates for a point source search. Electromagnetic emission of SNR has been detected in a wide energy range from radio wave lengths to the x-ray regime. The observations can be interpreted as synchrotron emission from electrons, which are accelerated in these regions Berezhko *et al.* 2003. Recently, the HEGRA experiment has detected an excess of high-energy γ-rays from the SNR Cassiopeia A, see Fig. 2 (*left*) Aharonian *et al.* 2001. This is interpreted as evidence for hadron acceleration in the SNR. The hadrons interact with protons of the interstellar medium, producing π^0s, which decay into high-energy photons. The flux is compatible with a model of electron and hadron acceleration in shock fronts of the SNR Berezhko *et al.* 2003.

Despite of the above-mentioned deflection, it is of great interest to study the arrival direction of charged CRs as well. The result of such an analysis is depicted in Fig. 2 (*right*) Antoni *et al.* 2004b. Shown is the distribution of the significances from a sky map of the arrival direction of showers with ener-

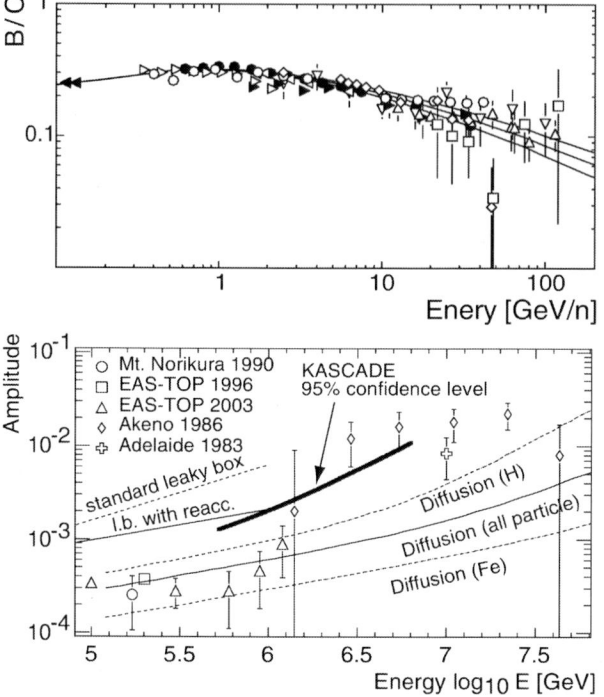

Figure 3 Top: Measured boron-to-carbon ratio as function of energy, the lines indicate model predictions, see Stephens & Streitmatter 1998. *Bottom:* Rayleigh amplitudes as function of energy for various experiments, for references see Antoni *et al.* 2004a. Additionally, model predictions for leaky box models Ptuskin *et al.* 1993 and a diffusion model are shown Candia *et al.* 2002. The lines indicate the expected anisotropy for primary protons, iron nuclei, and all particles.

gies above 0.3 PeV covering a region from $10°$ to $80°$ in declination. For an isotropic distribution the significances are expected to follow a Gaussian distribution as indicated by the solid line. Results for all events are presented, as well as for a selection of muon-poor showers. The latter are expected from potential primary γ-rays. No significant deviation of the data from the Gaussian distribution can be recognized. The analysis has been deepened by investigating a narrow band ($\pm1.5°$) around the galactic plane. Also circular regions around 52 SNRs and 10 TeV-γ-ray sources have been studied. None of the searches provided a hint for a point source. Upper limits for the fluxes from point-like sources are determined to be around 10^{-10} m^{-2}s^{-1}. In addition, no clustering of the arrival direction for showers with primary energies above 80 PeV is visible. Claims by the MAKET-ANI experiment for a point-source detection Chilingarian *et al.* 2003 have been withdrawn meanwhile Chilingarian 2004.

Information on the propagation pathlength of CRs is often derived from the measurement of the ratio of primary to secondary nuclei. The latter are produced through spallation during propagation in the Galaxy. As an example, the measured boron-to-carbon ratio is shown in Fig. 3 (*top*) as function of energy Stephens & Streitmatter 1998. The energy dependence of the measured ratios is frequently explained in leaky box models by a decrease of the pathlength of CRs in the Galaxy $\Lambda(R) = \Lambda_0(R/R_0)^{-\delta}$, with typical values $\Lambda_0 \approx 10 - 15$ g/cm^2, $\delta \approx 0.5 - 0.6$, and the rigidity $R_0 \approx 4$ GV/c.

At higher energies such measurements are not feasible due to the limited mass resolution of air shower experiments. However, at these energies the large scale anisotropy is expected to reveal properties of the CR propagation. The Rayleigh formalism is applied to the right ascension distribution of extensive air showers measured by KASCADE Antoni *et al.* 2004a. No hints of anisotropy are visible in the right ascension distributions in the energy range from 0.7 to 6 PeV. This accounts for all showers, as well as for subsets containing showers induced by predominantly light or heavy primary nuclei. Upper flux limits are shown together with results from other experiments in Fig. 3 (*bottom*). It presents the Rayleigh amplitude as function of energy. The experimental results are compared to the anisotropy expected from calculations of the propagation of CRs in the Galaxy. Two versions of a leaky box model, with and without reacceleration, seem to be ruled out by the measurements Ptuskin *et al.* 1993. The data reflect a trend predicted by a diffusion model Candia *et al.* 2002. This indicates that leakage from the Galaxy and consequently a decreasing pathlength $\Lambda(E)$ plays an important part during CR propagation at high energies and most likely, also for the origin of the knee.

Most valuable to reveal the origin of the knee are measurements of the energy spectra for individual elements or at least elemental groups. The KASCADE group studied the influence of different hadronic interaction models used in the simulations to interpret the data on the resulting spectra Ulrich *et al.* 2004. Two sets of spectra, derived from the observation of the electromagnetic and muonic air shower components, applying an unfolding procedure based on the Gold algorithm and using CORSIKA Heck *et al.* 1998 with the hadronic interaction models QGSJET and SIBYLL are compiled in Fig. 4 for three elemental groups, namely protons, helium, and iron. As can be seen in the figure, the fluxes depend on the model used. The KASCADE group emphasizes that, at present, there are systematic differences between measured and simulated observables which cause the ambiguities of the spectra. These conclusions apply in a similar way also to other experiments. A correct deconvolution of energy spectra requires a precise knowledge of the hadronic interactions. Presently, they do not describe the measurements with a sufficiently high precision.

Fig. 4 also shows the spectrum of primary protons, which has been derived from the flux of unaccompanied hadrons measured with the KASCADE experiment Antoni *et al.* 2004c. The spectrum is compatible with the proton flux as obtained from the unfolding procedure when using the QGSJET model. The EAS-TOP experiment published two sets of spectra with different assumptions about the contribution of protons and helium nuclei derived from the measurements of the electromagnetic and muonic shower components Navarra *et al.*

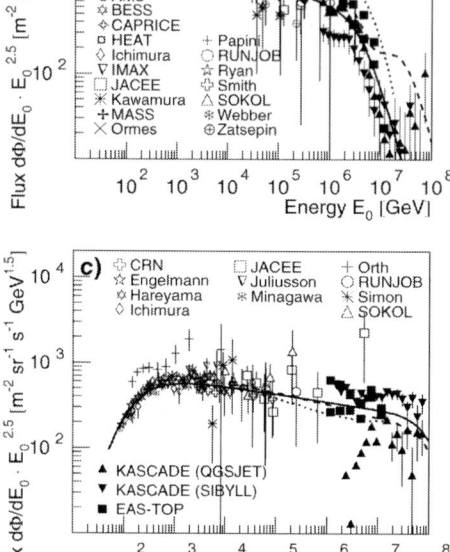
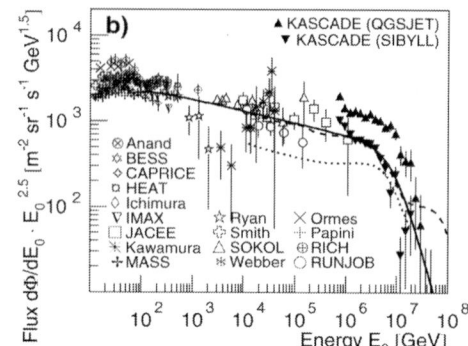

Figure 4. Energy spectra for elemental groups **a)** protons, **b)** helium, and **c)** iron. Open symbols give results of direct measurements, for references see Hörandel 2003a. Filled symbols represent data from air shower measurements: KASCADE electrons/muons interpreted with two interaction models Ulrich *et al.* 2004 (preliminary), KASCADE single hadrons Antoni *et al.* 2004c, and EAS-Top Navarra *et al.* 2003. The data are compared to calculations by Kalmykov & Pavlov 1999 ($\cdots$), Sveshnikova 2003 (- - -), and the Poly-Gonato model Hörandel 2003a (—).

2003. The resulting fluxes are indicated by two squares per primary energy. For comparison, also the results of direct measurements at the top of the atmosphere are presented at lower energies. To guide the eye, the solid lines indicate power law spectra with a cut-off at $Z \cdot 4.5$ PeV. For the iron spectrum at low energies the modulation due to the magnetic fields of the heliosphere causes the flux suppression.

The dashed lines represent calculations of energy spectra for nuclei accelerated in supernovae Sveshnikova 2003. It is assumed that the particles are accelerated in a variety of supernovae populations, each having an individual maximum energy to be attained during acceleration, which results in the bumpy structure of the obtained spectra. The dotted lines reflect calculations of the diffusive propagation of particles through the Galaxy Kalmykov & Pavlov 1999. The leakage of particles yields a rigidity dependent cut-off. Comparison with the data may suggest a *qualitative* understanding of the energy spectra. However, for a precise *quantitative* understanding, detailed investigations of the systematic errors of the measurements are necessary and the description of the interaction processes in the atmosphere needs to be improved.

Taking the cut-off behavior of the flux of individual elements as shown in Fig. 4 by the solid line, one obtains an increase of the mean logarithmic mass

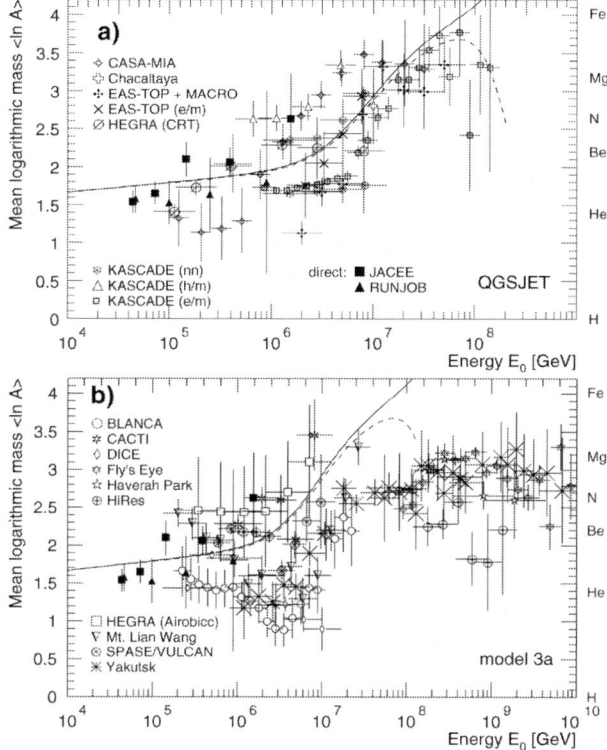

Figure 5 Mean logarithmic mass of cosmic rays reconstructed from **a)** experiments measuring electrons, muons, and hadrons at ground level interpreted with simulations based on CORSIKA/QGSJET, see also Hörandel 2003a and **b)** observations of the shower maximum with experiments registering Čerenkov and fluorescence light interpreted with a modification of QGSJET Hörandel 2003b. The lines indicate expectations according to the Poly Gonato model Hörandel 2003a.

with energy. This rise is plotted as solid line in both panels of Fig. 5. Its behavior is reflected by the results of various air shower experiments measuring electrons, muons, and hadrons at ground level, interpreted with simulations based on CORSIKA/QGSJET, compiled in Fig. 5a. Another class of air shower experiments reconstructs the average depth of the shower maximum X_{max} from the observation of Čerenkov and fluorescence light. Using the model QGSJET to derive the mean logarithmic mass from the data results in a light mass composition at high energies in contradiction to the values shown in Fig. 5a Hörandel 2003a. Introducing modifications to the model QGSJET, namely lowering the inelastic cross sections and slightly increasing the elasticity of hadronic interactions, this discrepancy can be reduced Hörandel 2003b. The mean logarithmic mass derived from X_{max} measurements using the modified version is shown in Fig. 5b. Also these measurements follow the trend of an increase of the mean logarithmic mass as function of energy.

The average experimental values from air shower measurements (Fig. 5) are combined in the light grey area shown in Fig. 6. The dark grey area represents the results of direct measurements above the atmosphere. This experimental situation will be compared to predictions of various models in the next section.

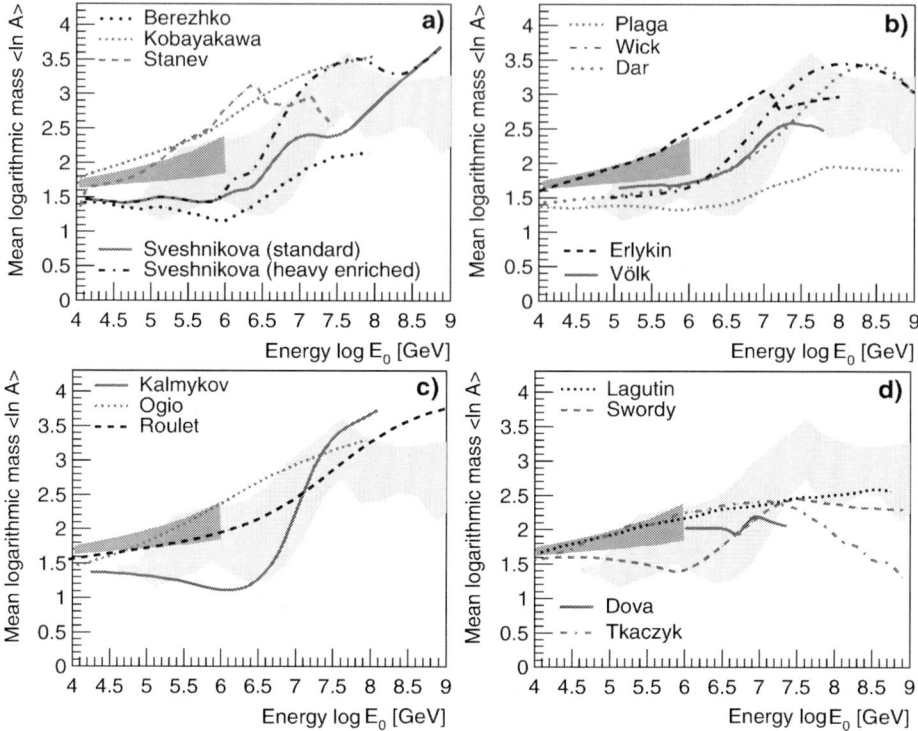

Figure 6.   Mean logarithmic mass as function of energy as obtained by direct observations (dark grey area) and air shower experiments (light grey area) compared with different models (lines). **a)** Acceleration in SNRs Berezhko & Ksenofontov 1999, Kobayakawa *et al.* 2002, Stanev *et al.* 1993, Sveshnikova 2003; **b)** Acceleration in GRBs Plaga 2002, Wick *et al.* 2004, Dar 2004, single source model Erlykin & Wolfendale 2001, reacceleration in the galactic wind Völk & Zirakashvili 2003; **c)** Diffusion in Galaxy Kalmykov & Pavlov 1999, Ogio & Kakimoto 2003, Roulet 2004; **d)** Propagation in the Galaxy Lagutin *et al.* 2001, Swordy 1995 and interaction with background photons Karakula & Tkaczyk 1993 and neutrinos Dova *et al.* 2001. For details see Hörandel 2004.

Discussion

To summarize the experimental results a different representation is given in Fig. 7. The average of the flux values shown in Fig. 1 is displayed by the data points. The spectra for elemental groups are presented according to a parameterization of the measurements, which corresponds to the solid lines in Fig. 4 and 5, where the agreement with the data has been demonstrated. Also shown is a proposed contribution of ultra-heavy elements ($Z \geq 28$), extrapolated from measurements at GeV energies. The individual spectra exhibit a cut-off at $E_Z = Z \cdot 4.5$ PeV. The cut-off for the heaviest elements agrees with the energy of the second knee at ~ 400 PeV, which is interpreted as the end of the galactic cosmic rays in this framework, while the knee is caused by the cut-off of the light elements. The sum spectrum of all elements is given by the

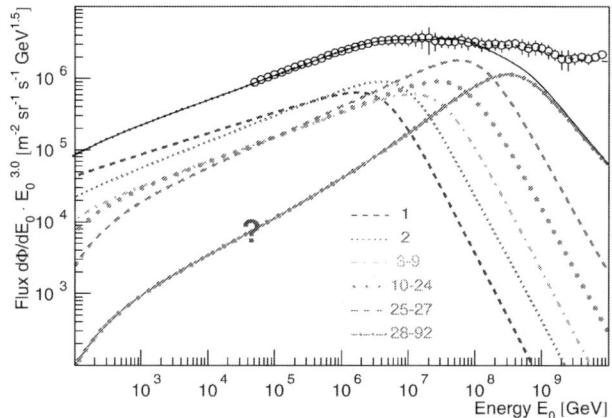

Figure 7 Average measured all-particle energy spectrum derived from many air shower experiments. Additionally, spectra for elemental groups with the indicated charge number range according to a parameterization of the measurements are depicted, including a proposed contribution of ultra-heavy elements ($Z \geq 28$), extrapolated from measurements at GeV energies Hörandel 2003a.

solid line, which fits nicely the average measured spectrum up to 100 PeV. For details see Hörandel 2003a.

How do the measured sequential cut-offs for the individual elements come about? Contemporary ideas will be discussed in the following.

The bulk of CRs is assumed to be accelerated in shock fronts of SNRs. The present understanding of acceleration in strong shock fronts has been initiated by Blanford & Ostriker 1978 which could demonstrate that at strong shocks particles are accelerated efficiently. The finite lifetime of a shock front ($\sim 10^5$ a) limits the maximum energy attainable to $E_{max} \sim Z \cdot (0.1 - 5)$ PeV for particles with charge Z.

Various versions of this scenario have been discussed, see e.g. Berezhko & Ksenofontov 1999, Stanev *et al.* 1993, Kobayakawa *et al.* 2002, Sveshnikova 2003. The models differ in assumptions of properties of the SNRs like magnetic field strength, available energy etc. This yields differences in $\langle \ln A \rangle$, as can be inferred from Fig. 6a. While older models limit the maximum energy to about 0.1 PeV Stanev *et al.* 1993, recent ideas, taking into account latest observations of SNR, achieve maximum energies above 1 PeV Sveshnikova 2003. In such a model sufficient energy is released from SNR to explain the observed spectra, see Fig. 4. A special case of SNR acceleration is the single source model Erlykin & Wolfendale 2001, which predicts pronounced structures in the all-particle energy spectrum in the knee region, caused by a single SNR. Such structures can not be seen in the compilation of Fig. 1. The SNR origin of CRs is supported by the detection of TeV γ-ray emission from Cassiopeia A as discussed.

With charged particles no point sources have been detected, but the similarity of the measured CR composition with the abundances in the Solar system indicate that CRs are accelerated out of a sample of well mixed interstellar matter Wiedenbeck *et al.* 2003.

In the literature also other acceleration mechanisms, like the acceleration of particles in γ-ray bursts, are discussed Plaga 2002, Wick *et al.* 2004, Dar 2004. They differ in their interpretation of the origin for the knee. The approach by Plaga, assuming Fermi acceleration in a "cannon ball" is not compatible with the measured $\langle \ln A \rangle$ values, see Fig. 6b. A different interpretation of acceleration in the cannonball model yields – at the source – a cut-off for individual elements proportional to their mass due to effects of relativistic beaming in jets. The present predictions of the model are compatible with recent data Dar 2004. However, it remains to be clarified how a detailed consideration of the propagation processes, e.g., in a diffusion model, effects the cut-off behavior observed at earth. Gamma-ray bursts as a special case of supernova explosions are proposed to accelerate CRs from 0.1 PeV up to the highest energies ($> 10^{20}$ eV) Wick *et al.* 2004. In this approach the propagation of cosmic rays is taken into account and the knee is caused by leakage from the Galaxy leading to a rigidity dependent cut-off behavior.

After acceleration, the particles propagate in a diffusive process through the Galaxy, being deflected many times by the randomly oriented galactic magnetic fields ($B \sim 3 \mu$G). From the measured abundance of radioactive nuclei in CRs a residence time of about $15 \cdot 10^6$ a for particles with GeV energies is derived Yanasak *et al.* 2001. The propagation is accompanied by leakage of particles from the Galaxy. With increasing energy it is more and more difficult to magnetically confine the nuclei to the Galaxy. As mentioned above, the pathlength decreases as $\Lambda \propto E^{-\delta}$. Such a decrease will ultimately lead to a complete loss of the particles, with a rigidity dependent cut-off of the flux for individual elements. Many approaches have been undertaken to describe the propagation process, see Ptuskin *et al.* 1993, Ogio & Kakimoto 2003, Roulet 2004, Swordy 1995, Lagutin *et al.* 2001. The leaky box model Swordy 1995 and the anomalous diffusion model Lagutin *et al.* 2001 yield cut-offs significantly weaker than the data shown in Fig. 4, see Hörandel 2004.

The propagation as described in diffusion models Kalmykov & Pavlov 1999, Ogio & Kakimoto 2003, Roulet 2004 yields $\langle \ln A \rangle$-values presented in Fig. 6c. The models are based on the approach by Ptuskin *et al.* 1993, but take into account different assumptions on details of the propagation process, like the structure of galactic magnetic fields etc. This results in a more or less strong cut-off for the flux at the individual knees and, consequently, in a more or less strong increase of $\langle \ln A \rangle$. The model by Kalmykov & Pavlov 1999 has been used to describe the observed spectra in Fig. 4.

During the propagation phase, reacceleration of particles has been suggested at shock fronts in the galactic wind. Also this mechanism yields a rigidity dependent cut-off Völk & Zirakashvili 2003.

Another hypothetical explanation for the knee are interactions of CRs with background particles like massive neutrinos Dova *et al.* 2001, Wigmans 2003 or photo disintegration in dense photon fields Karakula & Tkaczyk 1993, Candia *et al.* 2002. Such models appear to be excluded with a high level of confidence. The interactions would produce a large amount of secondary protons, which results in a light mass composition at high energies, not confirmed by the experiments, see Fig. 6d. Furthermore, a massive neutrino, proposed in Dova *et al.* 2001, Wigmans 2003 can be excluded by measurements of the WMAP and 2dFGRS experiments Hannestad 2004.

A completely different reason for the knee is the idea to transfer energy in nucleon-nucleon interactions into particles, like gravitons Kazanas & Nikolaidis 2001 or extremely high-energy muons Petrukhin 2003, which are not observable (or not yet observed) in air shower experiments. The latter proposal seems to be excluded by recent measurements of the Baikal experiment setting upper limits for the flux of muons above 10^5 GeV Wischnewski *et al.* 2004.

Conclusion

Comparing the present results to the status about one decade ago, our knowledge about high-energy CRs has significantly improved. In particular, the KASCADE experiment has provided a wealth of high-quality data. It could be shown that the knee is caused by sequential cut-offs for individual elements, starting with protons and helium nuclei and that the mean logarithmic mass increases as function of energy.

Summarizing the large number of experimental observations, there are indications for a standard picture of galactic CRs. At least a large fraction of them seems to be accelerated in supernova remnants up to energies of $Z \cdot (0.1 - 5)$ PeV. Higher energies may be reached in additional sources, such as γ-ray bursts. The particles propagate in a diffusive process through the Galaxy. With rising energy the pathlength decreases and the particles escape easier from the Galaxy. This brings about the knee in the energy spectrum. The main shape of the elemental spectra should be determined by the propagation process, maybe slightly modulated by properties of the source spectra.

Acknowledgments It was a pleasure to participate in an interesting and inspiring school and to experience the great hospitality in Erice. I would like to thank my colleagues from the KASCADE-Grande and TRACER experiments for stimulating scientific discussions.

References

Aharonian F. *et al.* (2001), *Astronomy & Astrophys.* **370**, 112.
Antoni T. *et al.* (2004a), *Astroph. J.* **604**, 687.
Antoni T. *et al.* (2004b), *Astroph. J.* **608**, 865.
Antoni T. *et al.* (2004c), *Astroph. J.* **612**, 914.

Berezhko E.G. & Ksenofontov L.T. (1999), *JETP* **89**, 391.

Berezhko E.G. *et al.* (2003), *Astronomy & Astrophys.* **400**, 971.

Blanford R.D. & Ostriker J.P. (1978), *Astroph. J.* **221**, L29.

Candia J. *et al.* (2002), *Astrop. Phys.* **17**, 23.

Candia J. *et al.* (2003), *J. Cosmol. Astropart. Phys.* **5**, 3.

Chilingarian A. *et al.* (2003), *Astroph. J.* **597**, L129.

Chilingarian A. (2004), poster at the 19th ECRS, Florence.

Dar A. (2004), *preprint* astro-ph/0408310.

Drescher H.J. *et al.* (2004), *Astrop. Phys.* **21**, 87.

Dova M.T. *et al.* (2001), *preprint* astro-ph/0112191.

Erlykin A.D. & Wolfendale A.W. (2001), *J. Phys. G: Nucl. Part. Phys.* **27**, 1005.

Hannestad S. (2004), *New Journal of Physics* **6**, 108.

Heck D. *et al.* (1998), Report FZKA 6019, Forschungszentrum Karlsruhe;
    and http://www-ik.fzk.de/~heck/corsika.

Hörandel J.R. (2003a), *Astrop. Phys.* **19**, 193.

Hörandel J.R. (2003b), *J. Phys. G: Nucl. Part. Phys.* **29**, 2439.

Hörandel J.R. (2003c), *Nucl. Phys. B (Proc. Suppl.)* **122**, 455.

Hörandel J.R. (2004), *Astrop. Phys.* **21**, 241.

Kalmykov N.N. & Pavlov A.I. (1999), *Proc. 26th ICRC, Salt Lake City*, vol. **4**, 263.

Karakula S. and Tkaczyk W. (1993), *Astrop. Phys.* **1**, 229.

Kazanas D. & Nikolaidis A. (2001), *Gen. Rel. Grav.* **35**, 1117.

Kobayakawa K. *et al.* (2002), *Phys. Rev. D* **66**, 083004; and astro-ph/0008209.

Lagutin A.A. *et al.* (2001), *Nucl. Phys. B (Proc. Suppl.)* **97**, 267.

Milke J. *et al.* (2004), *Acta Physica Polonica B* **35**, 341.

Navarra G. *et al.* (2003), *Proc. 28th ICRC, Tsukuba* vol. **1**, 147; Valchierotti S. *et al.* (2003), *Proc. 28th ICRC, Tsukuba*, vol. **1**, 151.

Ogio S. & Kakimoto F. (2003), *Proc. 28th ICRC, Tsukuba*, vol. **1**, 315.

Petrukhin A.A. (2003), *Phys. Atom. Nucl.* **66**, 517.

Plaga R. (2002), *New Astronomy* **7**, 317.

Ptuskin V.S. *et al.* (1993), *Astronomy & Astrophys.* **268**, 726.

Ptuskin V.S. (1997), *Adv. Space Res.* **19**, 697.

Roulet E. (2004), *Int. J. Mod. Phys. A* **19**, 1133.

Stanev T. *et al.* (1993), *Astronomy & Astrophys.* **274**, 902.

Stephens S.A. & Streitmatter R.E. (1998), *Astroph. J.* **505**, 266.

Sveshnikova L.G. (2003), *Astronomy & Astrophys.* , **409**, 799.

Swordy S.P. (1995), *Proc. 24th ICRC, Rome*, vol. **2**, 697.

Ulrich H. *et al.* (2004), *Europ. Phys. J. C* DOI: 10.1140/epjcd/s2004-03-1632-2.

Völk H.J. & Zirakashvili V.N. (2003), *Proc. 28th ICRC, Tsukuba*, vol. **4**, 2031.

Wick S.D. *et al.* (2004), *Astrop. Phys.* **21**, 125.

Wiedenbeck M.E. *et al.* (2003), *Proc. 28th ICRC, Tskuba*, vol. **4**, 1899.

Wigmans R. (2003), *Astrop. Phys.* **19**, 379.

Wischnewski R. *et al.* (2004), *Int. J. Mod. Phys. A* in press.

Yanasak N.E. *et al.* (2001), *Astroph. J.* **563**, 768.

THE KASCADE-GRANDE EXPERIMENT

M. Brüggemann[5], M. Stümpert[1], J. van Buren[2], T. Antoni[1], W.D. Apel[2],
F. Badea[2,9], K. Bekk[2], A. Bercuci[3], M. Bertaina[4], H. Blümer[2,1], H. Bozdog[2],
I.M. Brancus[3], P. Buchholz[5], C. Büttner[1], A. Chiavassa[4], K. Daumiller[1],
F. di Pierro[4], P. Doll[2], R. Engel[2], J. Engler[2], F. Fessler[2], P.L. Ghia[6], H.J. Gils[2],
R. Glasstetter[7], A. Haungs[2], D. Heck[2], J.R. Hörandel[1], K.-H. Kampert[7],
H.O. Klages[2], Y. Kolotaev[5], G. Maier[2], H.J. Mathes[2], H.J. Mayer[2],
J. Milke[2], B. Mitrica[3], C. Morello[6], M. Müller[2], G. Navarra[4], R. Obenland[2],
J. Oehlschläger[2], S. Ostapchenko[2,10], S. Over[5], M. Petcu[3], S. Plewnia[2],
H. Rebel[2], A. Risse[8], M. Roth[1], H. Schieler[2], J. Scholz[2], T. Thouw[2],
G. Toma[3], G.C. Trinchero[6], H. Ulrich[2], S. Valchierotti[4], W. Walkowiak[5],
A. Weindl[2], J. Wochele[2], J. Zabierowski[8], S. Zagromski[2], D. Zimmermann[5]

[1] *Institut für Experimentelle Kernphysik, Universität Karlsruhe, 76021 Karlsruhe, Germany*

[2] *Institut für Kernphysik, Forschungszentrum Karlsruhe, 76021 Karlsruhe, Germany*

[3] *National Institute of Physics and Nuclear Engineering, 7690 Bucharest, Romania*

[4] *Dipartimento di Fisica Generale dell'Università, 10125 Torino, Italy*

[5] *Fachbereich Physik, Universität Siegen, 57072 Siegen, Germany*

[6] *Istituto di Fisica dello Spazio Interplanetario, CNR, 10133 Torino, Italy*

[7] *Fachbereich Physik, Universität Wuppertal, 42097 Wuppertal, Germany*

[8] *Soltan Institute for Nuclear Studies, 90950 Lodz, Poland*

[9] *on leave of absence from Nat. Inst. of Phys. and Nucl. Engineering, Bucharest, Romania*

[10] *on leave of absence from Moscow State University, 119899 Moscow, Russia*

Abstract   The KASCADE-Grande experiment measures extensive air showers induced by primary cosmic rays in the energy range $10^{**} - 10^{**}$ eV. The major motivation for KASCADE-Grande is the investigation of the so called "knee" in the energy spectrum of cosmic rays and its presumed rigidity dependence. A short overview of the experimental setup with focus on the Grande array and its new data acquisition system is given. As an example of analysis the reconstruction of the total muon number is presented.

Keywords: KASCADE-Grande, EAS, air showers, cosmic rays

M. M. Shapiro et al. (eds.), Neutrinos and Explosive Events in the Universe, 377–382.
© 2005 *Springer. Printed in the Netherlands.*

Introduction

The major goal of the KASCADE-Grande experiment [Haungs et al. 2003] is to study the knee in the primary cosmic ray energy spectrum [Hrandel J. 2004] and the composition around the knee. Therefore KASCADE-Grande measures air showers caused by primary cosmic ray particles hitting the earth's atmosphere. Results so far by the smaller area KASCADE experiment indicate a rigidity dependence of the individual components. Thus one could expect a knee corresponding to iron at $\approx 10^{17}$ eV. KASCADE-Grande uses a multi-detector layout (Fig. 1) to measure as many air shower observables as possible, to achieve a high-quality reconstruction of extensive air showers.

A brief description of the experiment, focusing on the new Grande array is given in section 1. In section 1 we describe an extension to the already existing data acquisition system. It will allow time dependent measurements of the arriving particles in the shower disk, enabling an estimator for the electron to muon ratio for high energies above 10^{17} eV. In section 2, the reconstruction of the total muon number is described, necessary to infer the mass of the primary cosmic ray particle with unfolding techniques using the electron and muon size spectra.

The KASCADE-Grande experiment

The KASCADE-Grande experiment is located at the Forschungszentrum Karlsruhe, Germany, at 110m above sea level, corresponding to an average

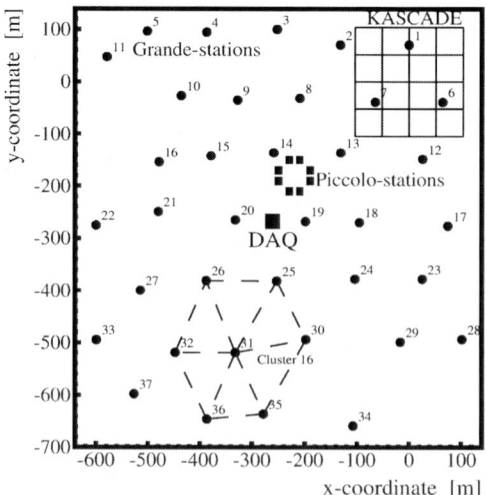

Figure 1.   The arrangement of the KASCADE-Grande detectors

vertical atmospheric depth of 1022 g cm^{-2}. For studies of the hadronic core of extensive air showers KASCADE-Grande uses a central detector containing a large hadron sampling calorimeter. Further components of the central detector measuring the muonic part of the core of EAS are also available. A muon tracking detector north of the central detector building measures muon tracks and densities outside the core. The original KASCADE array of 252 detector stations located around the central detector measures densities and arrival times of electrons and muons. See table 1 and Ref. [Antoni et al. 2003] for further details.

Table 1. The detectors of the KASCADE-Grande experiment with the detected particle types and their sensitive area.

Detector	Particles	Sensitive area (m$^\cdot$)
Grande array	e/μ	370
Piccolo	e/μ	80
Muon tracking detector	μ, $E_{thresh} = 800$ MeV	3 × 128
KASCADE array		
Liquid scintillators	e	490
Plastic scintillators	μ, $E_{thresh} = 230$ MeV	622
Central detector		
Trigger plane	μ, $E_{thresh} = 490$ MeV	208
MWPCs/LSTs	μ, $E_{thresh} = 2.4$ GeV	3 × 129
Calorimeter	Hadrons, $E_{thresh} = 10 - 20$ GeV	9 × 304

With its extension the KASCADE-Grande experiment is able to measure primary particles up to 10^{18} eV. This is achieved with 37 detector stations from the former EAS-Top experiment forming the Grande array. These stations are arranged in a 0.5 km^2 hexagonal grid with an average distance of 137 m.

Each station houses 16 scintillation detectors arranged in a 4 × 4 grid. These detectors consist of plastic scintillators ($80 \times 80 \times 4$ cm^3) read out by photomultiplier tubes. The signals of the 16 PMTs are summed up, amplified (high gain channel, ~ 1.6 pC/m.i.p.) and shaped before they are sent to the central data acquisition. The four central scintillators are equipped with additional PMTs with a 20 times lowered gain (low gain channel, ~ 0.32 pC/m.i.p.). Therefore the Grande array reaches a large dynamic range of $0.03 - 3000$ m.i.p./m^2. In the central data acquisition the 37 stations are connected to 18 trigger hexagons. Programmable trigger conditions are used to start TDC-measurements which are stopped by the individual stations. The energy deposit is determined by digitizing the signals of the stations using peak sensing ADCs. The obtained arrival times and particle densities are used to reconstruct the shower core, the total number of charged particles, the arrival direction and the shape of the shower front.

The KASCADE-Grande FADC-DAQ-System

Apart from the data acquisition (DAQ) system taking data since November 2003 the collaboration decided to build a Flash-ADC (FADC) based DAQ system for the Grande array. This system will sample the full pulse shape created by the photomultiplier tubes. Having the complete pulse shape recorded a correction for noise in order to improve the data quality is possible and new shower observables to be used in the analysis can be derived. In particular an intrinsic electron to muon separation at individual detector stations will be possible. Since the data will be transmitted optically, it will be resistant against pickup noise.

The FADC system is a modular system of custom made electronics. The main parts of the system are described in the following paragraphs.

- **Digitizer board:** One digitizer board will be installed in each station. It digitizes continuously two analog photomultiplier signals by two channels with 4 FADCs each. The FADCs work with 12 bit resolution and an effective sampling frequency of 250 MHz. A comparator compares each signal sample with a programmable threshold. If the current sample exceeds the threshold, a 1 μs long digitization period is triggered. If necessary this period is extended to prevent dead-time. The signals arriving at the FADCs are delayed compared to the ones seen by the comparator. This allows to record 112 ns before the signal. The data of 1 μs sampling cycle together with the timestamp of the KASCADE-Grande experiment forms one data package. This data package is transmitted over an optical fibre to the central DAQ station, where it is received by the receiver board. A scheme of the concept is depicted in Fig. 2.

- **Receiver board:** The counterpart of the digitizer board is a receiver board located in the DAQ station receiving data packages from up to 8 detector stations. Five receiver boards will be installed in the DAQ station. The task of the board is to translate the incoming optical signals back into electronic signals and to forward them to the PCI-interface which forms the connection of the FADC system to the computers.

- **PCI-Interface:** The PCI-Interface writes the data via direct memory access (DMA) into the PC-memory. The data transfer rate achieved is 85 MB/s which has to be compared with an expected data rate of 20 MB/s.

- **Event building:** In addition to the PCs buffering the digitized detector signals, a master PC will be used to scan the data packages for coincident timestamps and request all data within a programmable time window around the coincidence. With this procedure the data rate is reduced

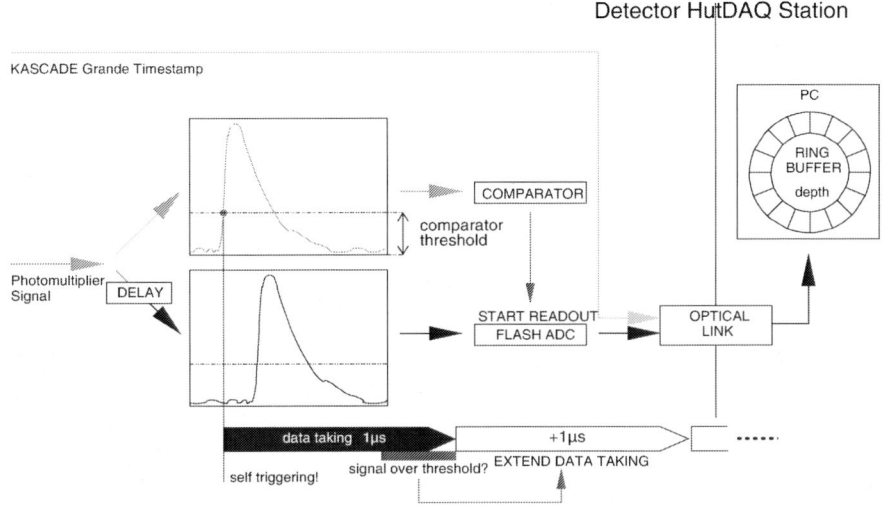

Figure 2. Schematic overview of the FADC concept [Over 2004]

from initial 37×2.5 MB/s to final 100 KB/s. In this estimation 2.5 MB/s correspond to the data rate of one detector station with an average event rate of 2.5 kHz and a FADC single event size of 1kB.

Reconstruction of Total Muon Number

The new Grande array measures the charged component, enabling to reconstruct the energy of air showers in the range 30 PeV to 1 EeV. To deduce the mass composition of cosmic rays, one is interested in the muon component of extensive air showers measured by 192 shielded scintillation detectors in the 200×200 m^2 sized KASCADE array. Together with the shower core and arrival direction provided by the Grande array, a lateral distribution function is fitted to the measured muon densities to obtain the total muon number of the extensive air shower. To reduce the effect of misreconstructed cores, a fiducial area of 600×600 m^2 is defined, which translates in a measurement of the muon density in a radial distance of $275 - 625$ m to the shower core for $\approx 68\%$ of the showers.

In the energy range $E_0 > 3 \times 10^{16}$ eV where Grande has 100% trigger efficiency, the systematic error of the reconstructed total muon number is constant at around 20% (see Fig. 3). Though it depends, like the muon number itself, on the primary particle. The statistical error decreases as expected with increasing primary energy. Furthermore the systematic error shows a dependency on the core distance to the KASCADE array, ranging from 30% for showers close to the array to 10% at larger radii.

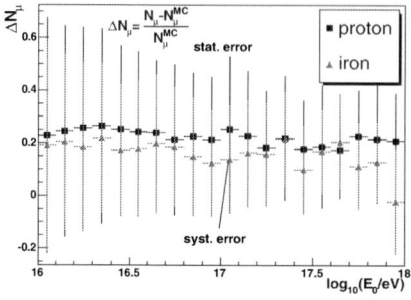 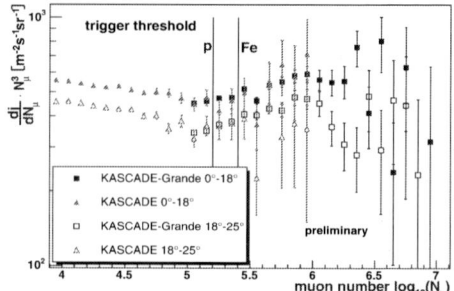

Figure 3.   Uncertainties in reconstruction   *Figure 4.*   Reconstructed muon size spectrum
of total muon number

The shown differential muon size spectrum in Fig. 4 is based on a data set of 5.4×10^{6} events, taken from December 2003 to April 2004. The effective time of combined data taking of the Grande array and the KASCADE array was approximately 117 days. As one can see there is a good agreement between the two measured fluxes in the overlap area in both shown zenith angle ranges. Furthermore also the spectral structure shows reasonable continuation. Around a total muon number of $\log_{10} N_{\mu} \approx 5$ one sees a decrease and then a re-increase of the index of the spectrum, corresponding to the knee in the light component of the energy spectrum and the relative increase of the heavy component. Further investigation of reconstruction systematics is necessary for both the reconstruction of high energetic showers with the KASCADE array and for showers with a reconstruction combining the information of both arrays.

Summary

The Grande array as an extension of the KASCADE-Grande experiment is taking data for nearly one year now. A new data acquisition system will give further improvements on the reconstruction of air shower observables. First analysis show a good agreement between data from the KASCADE and the Grande array and provide enhanced capabilities of the Grande array.

References

Antoni, T. et al. - KASCADE Coll., *Nucl. Instr. and Meth.* **A513**, 490-510, (2003)

Haungs, A. et al. - KASCADE-Grande Coll., ICRC (Tsukuba) **2**, 985, (2003)

Hrandel J., 2004 Proc. 14th ISCRA, these proceedings

S. Over - *Development and comissioning of data acquisition systems for the KASCADE-Grande experiment*, diploma thesis, Siegen University (2004)

CALIBRATION OF THE KASCADE-GRANDE HADRON CALORIMETER AT AN ACCELERATOR

S. Plewnia[1], Th. Berghöfer[2], H. Blümer[1], P. Buchholz[2], J. Engler[1], J.R. Hörandel[1], R. Lixandru[2], J. Milke[1], W. Walkowiak[2], J. Wochele[1]

[1]*Forschungszentrum and University of Karlsruhe, P.O. 3640, 76021 Karlsruhe, Germany*

[2]*University of Siegen, Emmy-Noether-Campus, Walter-Flex-Str. 3, 57072 Siegen, Germany*
plewnia@ik.fzk.de

Abstract   An iron sampling calorimeter with warm-liquid ionization chambers has been tested at the CERN SPS in order to study the signal development and to verify the energy calibration of the hadron calorimeter in the KASCADE-Grande air shower experiment. The signal calibration of the detectors is discussed. First results of the analysis of the longitudinal shower development in the calorimeter are presented and compared with results from simulations based on the GEANT/FLUKA code.

Keywords: air shower, hadron calorimeter

Introduction

The astrophysical interpretation of air shower data requires a precise understanding of the interaction processes in the atmosphere. The shower development is driven both, by the high-energy hadronic interactions and their particle production in the extreme kinematical forward region as well as by the low energy interactions influencing mostly the lateral particle density distributions. Unfortunatelly, both processes are only poorly known, e.g. Drescher et al. 2004. In order to improve the contemporary understanding of hadronic interaction processes the multi detector set-up of the KASCADE experiment Antoni et al. 2003 is operated, providing detailed insight into air shower development. Emphasize is given to the hadronic backbone of air showers which is investigated with a large hadron calorimeter.

The detector Engler et al. 1999 is an iron sampling calorimeter with the lateral dimensions 16×20 m^2. It consists of 9 layers of ionization chambers and a layer of plastic scintillation counters to provide fast trigger signals interspaced

M. M. Shapiro et al. (eds.), Neutrinos and Explosive Events in the Universe, 383–387.

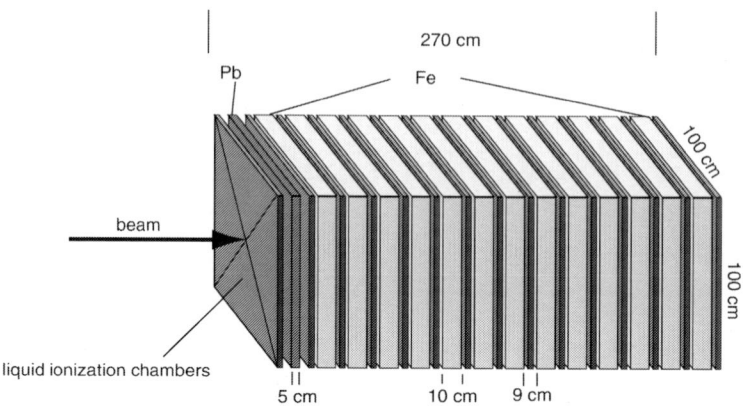

Figure 1.   The calorimeter set-up at the CERN SPS secondary beam line.

between 4000 t iron absorber. A lead filter above the iron absorber serves to suppress the electromagnetic component of air showers.

The energy calibration of the ionization chambers, filled with a "warm liquid" Engler 1996, is based on measurements with single muons Engler et al. 1999. The conversion Milke et al. 2000 from energy measured in the calorimeter to the energy of the incident hadron is based on simulations with the GEANT package GEANT 1993 using the FLUKA92 code Aarnio et al. 1990. The signal damping of strongly ionizing particles in liquid ionization chambers is taken into account in the code, based on measurements at the Karlsruhe cyclotron Engler et al. 1992, 1993.

Goal of the present investigations is to study the signal development in the chambers for the highest accelerator energies available to complement earlier measurements at lower energies Engler et al. 1992 and to check the Monte-Carlo based energy calibration applied so far. First results of the analysis are presented.

Experimental Set-Up

A set-up similar to the longitudinal structure of the KASCADE-Grande calorimeter has been chosen as sketched in Fig. 1. 60 ionization chambers are arranged in 15 layers (each divided in 16 segments), forming a detector with $1\ m^2$ active area. The absorber consists of a lead layer (5 cm thick) followed by 13 layers of iron slabs, each 10 cm thick, corresponding to 8.2 hadronic interaction lengths. The read-out electronics is identical to the one used in the KASCADE-Grande calorimeter Engler et al. 1999. Two plastic scintillators where used in coincidence in front of the calorimeter as trigger, a third one was installed behind the detector to select muons.

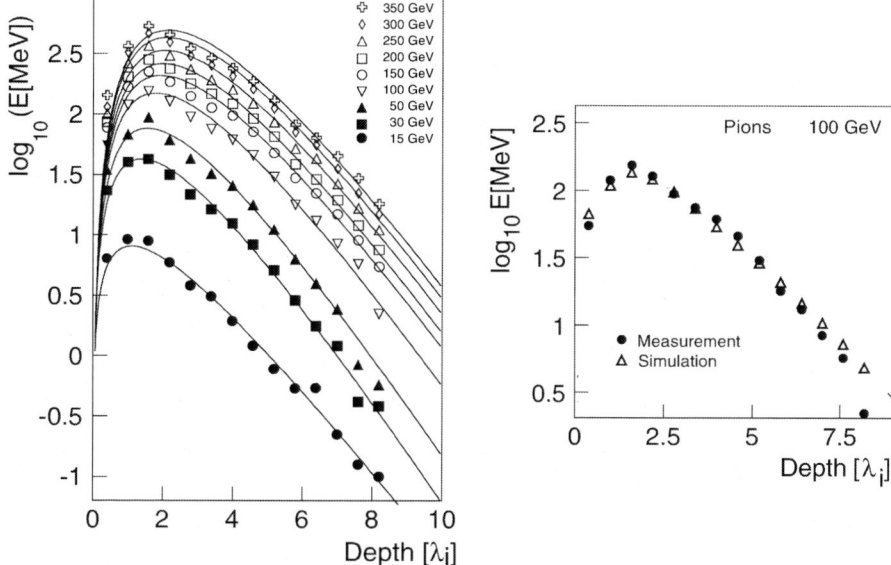

Figure 2.   Energy deposition in the calorimeter as function of depth in hadronic interaction lengths as measured for pions/protons with energies from 15 to 350 GeV (left) and measurements for 100 GeV compared to simulations with GEANT/FLUKA (right).

The calorimeter was set up at the H4 beamline of the Super Proton Synchrotron (SPS) at CERN and was tested with beams of protons, pions, electrons, and muons with energies between 15 and 350 GeV Plewnia et al 2004. Protons and pions could not be distinguished, they are treated as hadrons, as in the air shower experiment. To identify electrons, a lead plate (15 mm thick, corresponding to three radiation lengths) has been placed in front of the first layer of ionization chambers and the signal in this chambers was used to select primary electrons. Contaminations of muons and pions in the "electron" beam could be efficiently rejected.

Results

Firstly, the calibration of the individual chambers has been verified with muon runs. The charge yield obtained agrees well with the cosmic-ray muon calibrations performed several years ago. This is an indication for the long-term signal stability of these detectors.

The response of the detector for hadrons from 15 to 350 GeV is depicted in Fig. 2 (*left*), where the energy deposition in the ionization chambers is plotted as function of the depth in the calorimeter, measured in hadronic interaction lengths λ_i. To guide the eye, the measurements are parameterized according to

$$E_{dep}(x) = A\, x^B\, e^{-Cx} \tag{1}$$

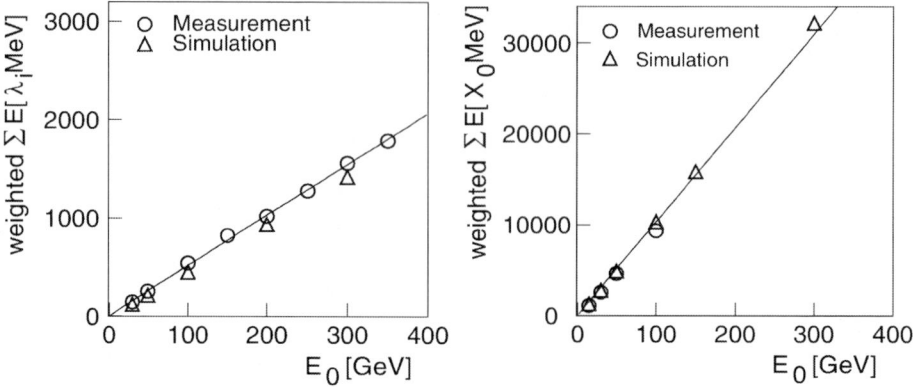

Figure 3. Weighted energy sum in the calorimeter as function of the incident particles energy for hadrons (left) and electrons (right). Measured values are compared with GEANT/FLUKA simulations.

giving the energy deposition as function of depth x in the detector. This function has been derived for electromagnetic showers, but apparently, describes hadronic cascades as well. The showers penetrate deeper into the calorimeter with increasing energy, and the maxima shift proportional to $\ln(E)$ as expected.

The measurements are compared to results from simulations in Fig. 2 (*right*) for pions of 100 GeV. The preliminary results indicate a good agreement between data and calculations. This verifies the description of signal generation for ionization chambers in the GEANT code.

The integral under the curves shown in Fig. 2 is expected to be proportional to the incident particle energy. This weighted energy sum (the energy measured in the chambers is weighted with the amount of absorber in front of each layer of ionization chambers) is shown in Fig. 3 as function of the incident particle energy for hadrons (*left*) and electrons (*right*). One recognizes a linear correlation as indicated by the straight lines in both diagrams. Also results of GEANT/FLUKA calculations are shown, which are found to agree with the measurements.

The RMS value of the frequency distribution of the weighted energy sum for a particular energy is taken as the energy resolution of the calorimeter. The present results can be approximated by

$$\frac{\sigma(E)}{E} = \frac{130\%}{\sqrt{E \text{ [GeV]}}} + 14\% \quad . \tag{2}$$

A good value for a large-area detector in a cosmic-ray experiment, which measures hadrons with energies up to beyond 10 TeV.

Conclusion

A hadron sampling calorimeter equipped with warm-liquid ionization chambers has been tested at a CERN SPS test beam in order to verify the energy calibration of the large hadron calorimeter in the KASCADE-Grande experiment. The longitudinal development of the energy deposition of hadrons and electrons in the calorimeter has been investigated. First results are in agreement with Monte Carlo calculations based on GEANT/FLUKA. More detailed analyses are in progress. They will increase the precision of the KASCADE data, and help to improve the understanding of high-energy interactions, necessary for an astrophysical interpretation of air shower data.

Acknowledgments We would like to thank all colleagues at CERN who made this work possible, in particular I. Efthymiopoulos, M. Hauschild, and L. Linssen. We acknowledge the valuable technical contributions by N. Bechtold, H. Bolz, W. Paulus, and M. Riegel of Forschungszentrum Karlsruhe as well as the assistance of N.O. Hashim from Siegen University during data acquisition.

References

Aarnio P.A. et al. (1990), Fluka user's guide. Technical Report TIS-RP-190, CERN.

Antoni T. et al. (2003), *Nucl. Instr. and Meth.* A **513**, 490.

Drescher H.J. et al. (2004), *Astrop. Phys.* **21**, 87.

Engler J. et al. (1992a), *Nucl. Instr. and Meth.* A **311**, 479.

Engler J. et al. (1992b), *Nucl. Instr. and Meth.* A **320**, 460.

Engler J. et al. (1993), *Nucl. Instr. and Meth.* A **327**, 128.

Engler J. (1996), *J. Phys. G: Nucl. Part. Phys.* **22**, 1.

Engler J. et al. (1999), *Nucl. Instr. and Meth.* A **427**, 528.

GEANT 3.15 (1993), Detector Description and Simulation Tool, CERN Program Library Long Writeup W5013, CERN.

Milke J. et al. (2000), in Calorimetry in High Energy Physics, G. Barreira, B. Tome, Eds., World Scientific p. 803, ISBN 981-02-4304-9

Plewnia S. et al. (2004), in Calometry in High Energy Physics, World Scientific, in press.

THE AMS EXPERIMENT

A. S. Torrentó Coello*
Dpto. Fusión y Partículas Elementales. CIEMAT
Avda. Complutente, 22. 28040 Madrid, Spain
ana.torrento@ciemat.es

Abstract   AMS is a particle physics detector which will operate on the International Space Station (ISS) for at least 3 years, to search for cosmic antimatter and dark matter signatures, to make a deep study of cosmic ray composition and energy spectrum and to do some gamma ray astrophysics. The detector design and its placement for a long period of time in a background-free environment will make possible a high precision and high statistics data collection, allowing a significant improvement of present results on that issues.

Keywords: Antimatter, cosmic ray, dark matter, gamma ray, magnetic spectrometer, neutralino.

AMS-02 detector

Since the discovery of cosmic rays balloon-borne experiments have been widely used to study both their composition and energy spectrum. However, they present some limitations which result in poor statistics and hence in a few-percent reconstruction efficiency.

AMS-02 is a large acceptance (0.45 m²sr) particle detector which will be installed on the ISS at 400-km altitude, where background is negligible, for a long period of time, to measure cosmic ray flux with unprecedented statistics and precision. Its design principles focus on detection of faint signals among intense backgrounds and meet the requirements imposed by operation on orbit for several years [1].

AMS-02 is a magnetic spectrometer which consists of a superconducting magnet and eight planes of Silicon Tracker (STD). It will measure the particle rigidity with 1.5% precision for protons at 10 GV. It is complemented by several subdetectors i.e. a Transition Radiation Detector (TRD), a Time of

---

*On behalf of the AMS collaboration.

M. M. Shapiro et al. (eds.), Neutrinos and Explosive Events in the Universe, 389–392.
© 2005 *Springer. Printed in the Netherlands.*

Flight system (TOF), a Ring Imaging Čerenkov (RICH) and an Electromagnetic Calorimeter (ECAL) (Figure 1).

This configuration provides redundant measurements of the particle velocity (TOF and RICH, with 3.5% and 0.1% precision for protons with $\beta=1$) and charge modulus (STD and TOF up to $Z \sim 20$, RICH up to $Z \sim 26$). ECAL measures $e^{\pm}$ energy with 3% resolution at 100 GeV. Either STD or ECAL measure γ energy with 5% and 3% precision in the range 1 GeV - 1 TeV and incoming direction with 0.1° and 1° angular resolution at 10 GeV. TRD and ECAL perform the lepton/hadron separation with rejection factors of 10^2-10^3 in 1.5-300 GeV and 10^4 in 1.5 GeV-1 TeV. The velocity and rigidity measurements given by RICH and STD will allow light isotope separation for $E \leq 10$ GeV/n.

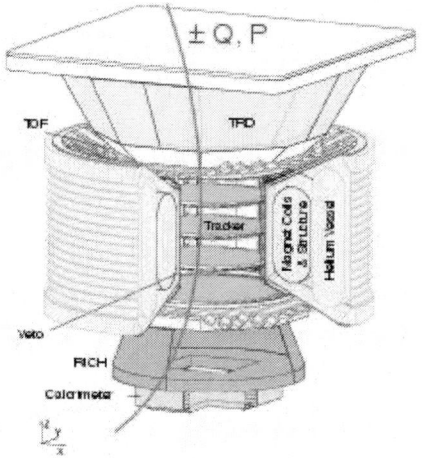

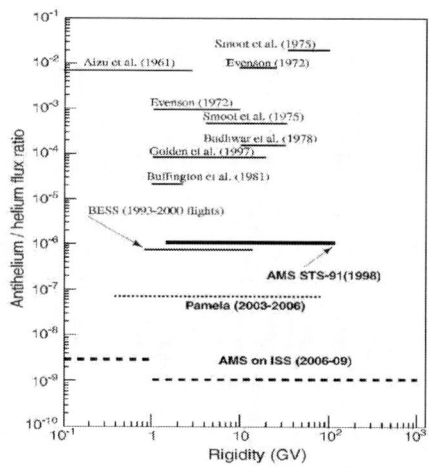

Figure 1.   AMS-02 layout.

Figure 2.   He/He ratio provided by several experiments and AMS-02 expected limit after a 3-year exposure.

Antimatter search

The Big Bang, the most accepted model to explain the origin and evolution of the Universe, requires matter and antimatter to be equally abundant at the very beginning, and a subsequent dominance of matter over antimatter (baryogenesis). But there is no evidence of large antimatter regions in our cluster of galaxies [2], and the baryogenesis mechanism proposed by several theories are based on baryon number and strong C and CP violations which have not been observed yet. Thus, more experimental data are necessary to enlighten this puzzle.

Although e^+ and $\overline{p}$ are widely produced in the interstellar medium, heavier antinuclei could only be produced in anti-stars. Up to now, no $\overline{He}$ nucleus has been detected in cosmic ray flux yet. AMS-02 detection capabilities will make possible to collect 10^9 He with $E \leq 1$ TeV in 3 years, improving the present $\overline{He}$/He ratio upper limit in 3 to 4 orders of magnitude (Figure 2).

Dark matter and gamma ray astrophysics

Some SUSY extensions of the Particle Physics Standard Model propose non-baryonic particles to be the constituents of dark matter [3], e.g. the neutralino, whose annihilation in the galactic halo would produce an excess in e^+, $\overline{p}$, $\overline{D}$ and γ fluxes.

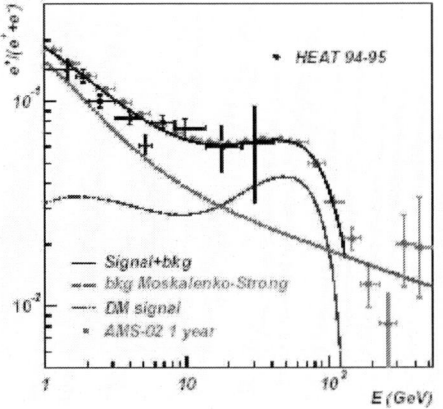

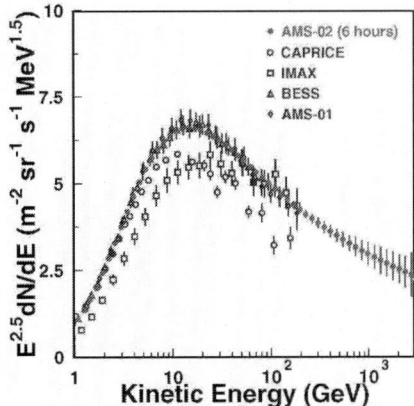

Figure 3. Positron excess due to neutralino annihilation, HEAT and AMS-02 data after a 1-year exposure.

Figure 4. AMS-02 proton flux expected after a 6-hour exposure compared to data from other experiments.

AMS will search for dark matter through antimatter from these annihilation channels. Electron and positron fluxes will be detected up to $\sim$300 GeV and antiproton flux up to $\sim$400 GeV. After 3 years, AMS-02 will collect $\sim 10^7$ e^- with $E \leq 10$ GeV, $\sim 10^6$ e^+ with $E \leq 5$ GeV and $\sim 10^6$ $\overline{p}$ with $E \leq 5$ GeV (Figure 3). AMS will also contribute to the study of gamma diffuse background, both galactic and extragalactic and gamma source location [4].

Cosmic rays

Cosmic rays are particles that arrive in Earth coming from the Sun, our Galaxy or other galaxies. Primary species fluxes give a constraint in source type and composition and primary acceleration mechanisms. AMS-02 will precisely determine individual element fluxes with $1 \leq Z \leq 26$ in the energy range 0.1 GeV/n $\leq E \leq$ 1 TeV/n. After 3 years, it will collect $\sim 10^8$ H, $\sim 10^7$ He and $\sim 10^5$ C with $E > 100$ GeV/n, improving both precision and statistics as compared with previous experiments (Figure 4).

Propagation models can be tested with secondary cosmic ray flux and the ratio of secondary to primary. The confinement time in the galaxy and thus source distances can be estimated with the ratio of unstable to stable isotope flux. AMS-02 will measure light element spectra in a wide energy range. In particular, it will identify $\sim 10^4$ B nuclei with $E < 100$ GeV/n and $\sim 10^5$ ^{10}Be after 3 years, and hence determine the B/C ratio for $E \leq 1$ TeV and separate ^{10}Be from ^{9}Be in the energy range 0.15 - 10 GeV/n (Figure 5).

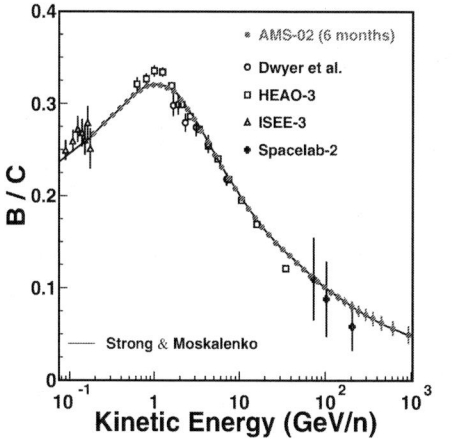 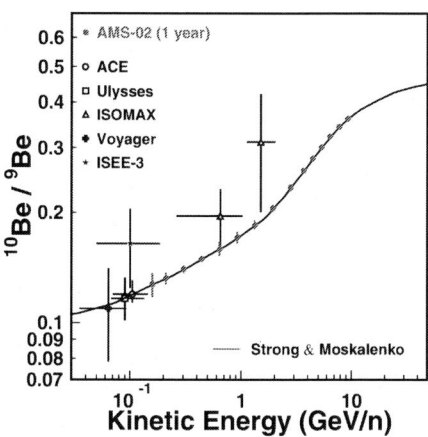

Figure 5. B/C (left) and $^{\cdot\cdot}$Be/$^{\cdot}$Be (right) ratios from several experiments, theoretical models and AMS-02 expected data after a 6-month and 1-year exposure.

References

[1] AMS collaboration. " AMS on ISS. Construction of a particle physics detector on the International Space Station". To be published in Nucl. Instr. Methods A.

[2] Cohen A., de Rujula A. and Glashow S. "A matter-antimatter universe?" Astrophysical Journal, 495 (1998), 539.

[3] Jungman G., Kamionkowski M. and Griest K. "Supersymmetric dark matter". Physics Reports 267 (1996), 195-373.

[4] Lamanna G. " High-energy gamma-ray detection with the Alpha Magnetic Spectrometer on board the International Space Station". Nuclear Physics B (Proc. Suppl.) 113 (2002) 177-185.

SOLAR NEUTRON OBSERVATION
AT GROUND LEVEL AND FROM SPACE

M. R. Moser[1], L. Desorgher[1], E. O. Flückiger[1], R. S. Miller[2], J. M. Ryan[3],
J. R. Macri[3], M. L. McConnell[3]

[1]*Physikalisches Institut, University of Bern, Sidlerstrasse 5, CH-3012 Bern, Switzerland*
[2]*Department of Physics, University of Alabama in Huntsville & National Space Science and Technology Center (NSSTC), Huntsville, AL 35899, USA*
[3]*Space Science Center, University of New Hampshire, Durham, NH 03824, USA*
Michael.Moser@phim.unibe.ch

Abstract   The study of particle acceleration, particle transport and interaction mechanisms associated with high-energy phenomena at the Sun is of general astrophysical relevance. In this context an exemplary overview is given of ground-based and spaceborne instruments to measure the energy spectrum of solar neutrons, and of the significance of such observations.

Introduction

Nuclear and pion related γ-rays provide important information about the spectra of protons and ions accelerated in solar flares [e.g. Hua and Lingenfelter, 1987; Murphy et al., 1987; Lockwood et al., 1997; Hua et al., 2002]. However, nuclear γ-ray lines probe the proton spectrum only up to $\sim$40 MeV, while γ-rays from pion decays are only observed in the most intense flares. In addition, any spectral break in the proton spectrum is likely to lie below the pion production threshold. Neutrons produced at the solar surface over a wide range of energies may provide important information from the 50–300 MeV regime, complementing γ-ray observations. Due to the long neutron thermalization time ($\sim$100 s) the 2.223 MeV neutron capture line is only a limited measure of neutron production. The spectrum of accelerated and interacting protons can be deduced more reliably from direct neutron measurements.

Fig. 1 illustrates the importance of the various energy ranges of the energetic flare particles, in particular for the neutron production. A typical solar proton production spectrum $\mathrm{d}J/\mathrm{d}E \sim E^{-2.5}$ (A) is folded with the cross-sections for the neutron (B), γ-ray line (C and D) and pion production (E), and weighted by the elemental abundances of the target material [Lockwood et al., 1997].

M. M. Shapiro et al. (eds.), Neutrinos and Explosive Events in the Universe, 393–397.

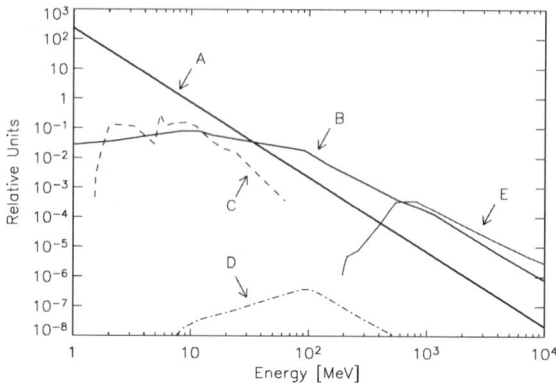

Figure 1.   Schematic representation of solar neutron and γ-ray production: proton energy spectrum (A), neutron production (B), 4.438 MeV γ-ray line production (C), 2.223 MeV γ-ray line production (D), and pion production (E). [Lockwood et al., 1997]

Ground-based Observation

The first identification of solar neutrons at Earth took place on June 3, 1982 (Fig. 2a), by neutron monitor measurements at Jungfraujoch, Lomnicky Stit, and Rome [Debrunner et al., 1983; Chupp et al., 1987]. This was two years after the discovery of solar neutrons in near-Earth space by the Gamma Ray Spectrometer (GRS) aboard the Solar Maximum Mission (SMM) satellite [Chupp et al., 1982]. Thereafter, standardized neutron monitors were set up at favorable observational locations at Earth, such as Haleakala, Hawaii [Pyle and Simpson, 1991]. Additionally, new ground-based detectors with enhanced sensitivity to solar neutrons were developed [e.g. Shibata et al., 1991; Muraki et al., 1993].

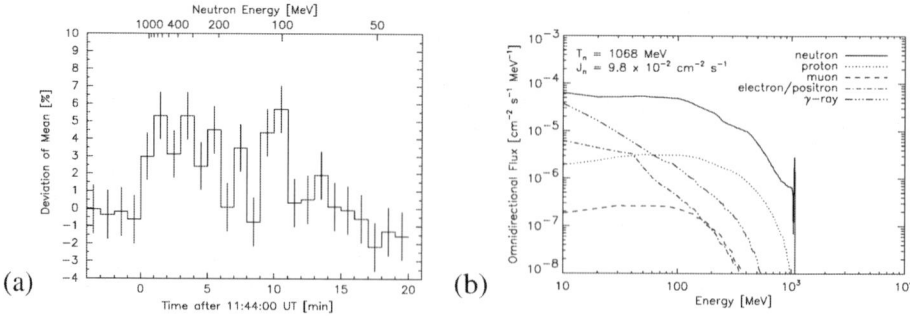

Figure 2.   June 3, 1982, solar neutron event: (a) Count rate enhancement of the Jungfrau-joch 18-IGY neutron monitor. (b) Simulated omnidirectional spectra of secondary particles at $700\,\mathrm{g\,cm^{-\bullet}}$ induced by 1068 MeV solar neutrons at onset time [Moser et al., 2003].

Initiated by the Solar-Terrestrial Environment Laboratory of the Nagoya University in Japan, a new global network of solar neutron telescopes was set up. Seven locations well distributed in longitude enable the observation of solar neutrons during 24 hours a day. The detectors are capable to determine the incident neutrons' energies and directions. Observations based on this network are reported e.g. by Sako et al., 2003.

Limitations of ground-based observations. In addition to the fact that only solar neutrons above a few hundred MeV survive in significant numbers until the Earth's orbit, the atmosphere prevents most of the primary neutrons from reaching ground-level without interacting with atmospheric nuclei [Moser et al., 2003; Shibata, 1994]. Fig. 2b illustrates that the majority of the particles reaching ground-level are secondaries that no longer contain the original information about energy and arrival direction. The deconvolution of the primary neutron spectrum from ground-based observations is even more difficult, since in any case most of the detected particles are induced by galactic cosmic rays. The atmospheric limitations can be overcome by placing the instruments on balloons or even better on spacecraft.

Spaceborne Instruments

The Compton Telescope (COMPTEL) on the Compton Gamma Ray Observatory (CGRO) was the first instrument built to detect γ-rays and solar neutrons in space [Ryan et al., 1993a]. COMPTEL observations of the June 15, 1991, solar neutron event were the first that allowed to generate an image of the Sun in neutrons [Ryan et al., 1993b; Nieminen, 1997].

Since the CGRO mission ended in 2000, leaving a lack of spaceborne neutron detectors, we will present two new projects of solar neutron detectors to be operated on future space missions.

Fast Neutron Imaging Telescope (FNIT). The FNIT detector [Moser et al., 2004a] is sensitive to solar neutrons in the energy range 3–100 MeV, and is intended to be operated on inner heliosphere missions.

The detection principle is based on multiple elastic neutron-proton scatterings (Fig. 3a) in parallel plastic scintillator plates (Fig. 3b). The scintillation light is read out by wavelength shifting fibers arranged in orthogonal directions on the upper and lower side of each layer. By reconstructing the event locations and measuring the recoil proton energies, the energy and event circle of an incident neutron can be determined. By superimposing the event circles of multiple events, the neutron source can be located in the sky. This is the same technique as used with the COMPTEL instrument [Ryan et al., 1992; de Boer et al., 1992].

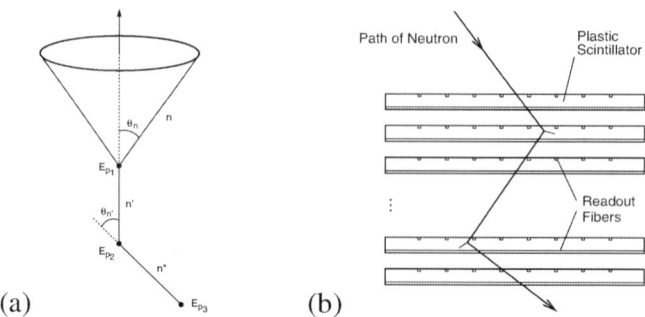

Figure 3.   (a) Event circle reconstruction of a triple scatter event. (b) Schematic view of the FNIT concept with its scintillator layers and readout fibers.

As a possible enhancement of FNIT to improve the response in the energy range 1–10 MeV a concept for an advanced telescope based on the addition of a phoswich-based detector is currently being studied [McKibben et al., 2004].

In order to optimize the instrument's performance, a Monte Carlo simulation package has been developed that allows us to optimize the design with respect to scintillator composition, plate thickness, and fiber pitch, as well as to develop appropriate data reconstruction algorithms [Moser et al., 2004a].

Solar Neutron Tracking and Imaging Spectrometer (SONTRAC).   The SONTRAC instrument [Miller et al., 2003] is being developed to detect energy and direction of neutrons in the 20–250 MeV energy range, and thus is optimized for near-Earth missions.

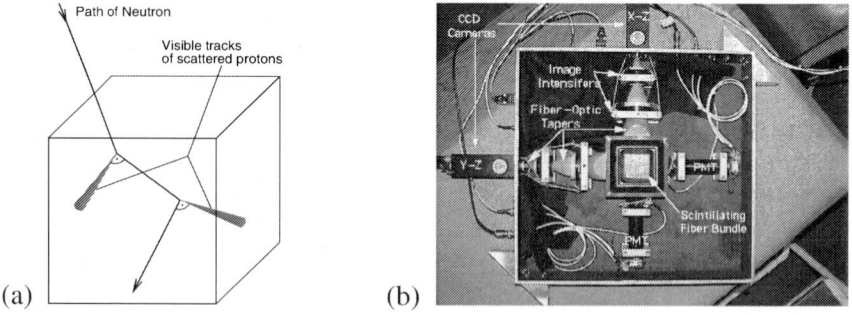

Figure 4.   (a) Diagram of a non-relativistic double scatter neutron event in a plastic scintillator block. (b) SONTRAC science model assembly (cover removed). [Miller et al., 2003]

The detection principle is also based on non-relativistic double scatterings of neutrons off ambient protons within a block of scintillating fibers. Using this doublescatter mode it is possible to uniquely determine neutron energy and direction on an event-by-event basis. Contrary to FNIT, the proton energy is determined by measuring the track length instead of the pulse height (Fig. 4a).

In the science model (Fig. 4b) two orthogonal imaging chains using CCD cameras for fiber readout allow full 3D reconstruction of scattered proton particle tracks. By using 500 μm fibers, the instrument has a measured energy resolution of 4.8% (2.1%) for 35 MeV (67.5 MeV) protons, and an angular resolution (1σ) of 4.6° (2.3°) [Miller et al., 2003].

In order to reduce off-line data space and transmission time, laboratory measurements and Monte Carlo simulations of the instrument are currently being used to develop data reconstruction algorithms based on three orthogonal 1D projections of fiber output signals [Moser et al., 2004b].

Acknowledgments

This research is supported in Switzerland by the Swiss National Science Foundation (grants 20-067092.01 and 200020-105435/1); and in the United States by NASA's Living With a Star Targeted Research and Technology program (grant NAG5-13519) and NASA's Sun-Earth Connection Instrument Development program (grant NAG5-13178). M.R.M. acknowledges the support by the Swiss Commission for Astronomy for participation at the ISCRA.

References

Chupp, E. L. et al. (1982). *Astrophys. J.*, 263:L95–L99.
Chupp, E. L. et al. (1987). *Astrophys. J.*, 318:913–925.
de Boer, H. et al. (1992). *Data Analysis in Astronomy*, IV:241–249.
Debrunner, H. et al. (1983). In *Proc. 18th ICRC., Bangalore*, volume 4, pages 74–78.
Hua, X.-M. et al. (2002). *Astrophys. J.*, 140:563–579.
Hua, X.-M. and Lingenfelter, R. E. (1987). *Astrophys. J.*, 323:779–794.
Lockwood, J. A. et al. (1997). *Sol. Phys.*, 173:151–176.
McKibben, R. B. et al. (2004). *Adv. Space Res.*, in press.
Miller, R. S. et al. (2003). *Nucl. Instr. and Meth. A*, 505:36–40.
Moser, M. R. et al. (2003). In *Proc. 28th ICRC, Tsukuba*, pages 3215–3218.
Moser, M. R. et al. (2004a). *Adv. Space Res.*, in press.
Moser, M. R. et al. (2004b). In *35th COSPAR Scientific Assembly, Paris*. D2.4/E3.4-0032-04.
Muraki, Y. et al. (1993). In *Proc. 23th ICRC, Calgary*, volume 3, pages 171–174.
Murphy, R. J. et al. (1987). *Astrophys. J. Suppl.*, 63:721–748.
Nieminen, P. (1997). PhD thesis, University of Bern, Bern, Switzerland.
Pyle, K. R. and Simpson, J. A. (1991). In *Proc. 22th ICRC, Dublin*, volume 3, pages 53–56.
Ryan, J. M. et al. (1992). *Data Analysis in Astronomy*, IV:261–270.
Ryan, J. M. et al. (1993a). In *AIP Conf. Proc. 280, St. Louis, MO, USA*, pages 631–642.
Ryan, J. M. et al. (1993b). *Adv. Space. Res.*, 13(9):255–258.
Sako, T. et al. (2003). In *Proc. 28th ICRC, Tsukuba*, pages 3175–3178.
Shibata, S. (1994). *J. Geophys. Res.*, 99(A4):6651–6665.
Shibata, S. et al. (1991). In *Proc. 22th ICRC, Dublin*, volume 3, pages 788–791.

MONTE CARLO SIMULATIONS AND SEMIANALYTICAL PARAMETERISATIONS OF THE ATMOSPHERIC MUON FLUX

Bogdan Mitrica[1], Iliana Brancus[1], Gabriel Toma[1], Juergen Wentz[1,2], Heinigerd Rebel[2], Alexandru Bercuci[1], Cristina Aiftimiei[1]

[1]*IFIN-HH, RO-76900 Bucharest, POB MG-6, Romania*
bogdan.mitrica@ifin.nipne.ro

[2]*Forschungszentrum Karlsruhe, POB 3640, 76021 Karlsruhe, Germany*
rebel@ik.fzk.de

Abstract   The atmospheric muon flux have been simulated using the CORSIKA code for two different geographical positions (Bucharest: $44°N$, $26°E$ and Hiroshima: $34°N, 132°E$). The simulations have been done for different angles of incidence between $0°$ and $70°$. The comparison between the simulations and the experiment have been done using the measurements of the muon charge ratio with the WILLI detector in Bucharest. The results of the Monte Carlo simulations of the muon flux for the geographical positions of Hiroshima and Bucharest are compared with the semi-analytical formulae of Judge and Nash, and of Gaisser for different angles of incidence between $0°$ and $70°$ and with experimental results of the Bess experiment (vertical incidence). Various sensitivities of the approach of Judge and Nash, in particular to variations of the pion and kaon production spectra have been studied.

Keywords: Muon, flux, simulation

Introduction

The muon belongs to the family of elementary particles known as leptons. Like the electron it may be positively or negatively charged and has a spin $\frac{1}{2}$. However its mass is about 100 MeV, more than two orders of magnitude larger than that of the electron, and about one order of magnitude less than of the proton. It is produced mainly by the decay of pions and kaons generated by high-energy collisions of cosmic rays with the atoms of the Earth atmosphere. Muons are unstable decaying to electrons and positrons and neutrinos (electron (ν_e) and muon (ν_μ) neutrinos) with a half - life of $\tau_\mu = 2.2\mu s$.

M. M. Shapiro et al. (eds.), Neutrinos and Explosive Events in the Universe, 399–402.
© 2005 *Springer. Printed in the Netherlands.*

The air shower simulation program CORSIKA

The simulation tool CORSIKA has been originally designed for the four dimensional simulation of extensive air showers with primary energies around 10^{15} eV. The particle transport includes the particle ranges defined by the life time of the particle and its cross-section with air. The density profile of the atmosphere is handled as continuous function, thus not sampled in layers of constant density.

Ionization losses, multiple scattering, and the deflection in the local magnetic field are considered. The decay of particles is simulated in exact kinematics, and the muon polarization is taken into account.

In contrast to other air shower simulations tools, CORSIKA offers alternatively six different models for the description of the high energy hadronic interaction and three different models for the description of the low energy hadronic interaction. The threshold between the high and low energy models is set by default to $E_{Lab} = 80$ GeV/n.

Calculation of atmospheric muon flux

The calculation of muon flux proceeds by a full 3D-simulation (CORSIKA). The simulations have been done using for the primary particle's spectrum the expresion: $J_p(E) \sim E^{-2.78}$.

The differential particle flux **?**

$$J_\mu = \frac{dN}{dt \cdot dA \cdot d\Omega \cdot dP} \qquad (cm^{-2} \cdot s^{-1} \cdot sr^{-1} \cdot (GeV/c)^{-1}) \qquad (1)$$

resulting from the simulation was calculated by deviding the number of particles detected by the surface of the particle collection area (cm^2), solid angle, momentum bin size, and equivalent sampling time of the CR flux.

Semi-analytical approaches

There are several empirical approximations describing the fluxes in by analytical expressions like power-law distributions (see P.Grieder). Recent approaches by T.K. Gaisser display explicitly the dependence on primary energy, but with complicated mathematical procedures and valid only for muon energies above 10 GeV. This holds also for the simplication given in Gaisser's Book:

$$\phi_\mu = \frac{0.14}{cm^2 \cdot s \cdot sr \cdot GeV} \cdot (E/GeV)^{-2.7} \left[\frac{1}{1 + \frac{E \cdot cos\theta}{110 GeV}} + \frac{0.37}{1 + \frac{E \cdot cos\theta}{760 GeV}} \right] \qquad (2)$$

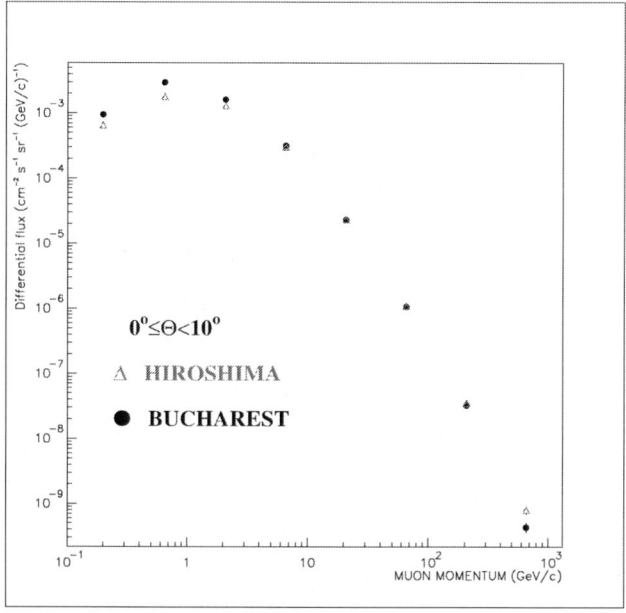

Figure 1. Differential flux of the muons for $0° \leq \theta < 10°$

used for example by Unger. In Fig.2 this formula is compared with the the results of the Monte Carlo simulations , displaying the disaggreement in particular at lower energies.

The approach by Judge and Nash uses as input the production spectra of parent pions and kaons and calculates the flux resulting from pion and kaon decay by:

$$D_\pi(E_\mu, \theta) = \frac{A_\pi \cdot W_\mu \cdot E_\pi^{-\gamma_\pi} \cdot H_\pi}{E_\pi \cdot \cos\theta + H_\pi} \tag{3}$$

$$D_k(E_\mu, \theta) = \frac{A_k \cdot W_\mu \cdot E_k^{-\gamma_k} \cdot H_k}{E_k \cdot \cos\theta + H_k} \tag{4}$$

There $H_{\pi,k}$ and H_μ are parameters accounting for tte propagation of the particles in the atmosphere. The parameters $A_{\pi,k}$ are the normalisations of the pion and kaon production spectra. There are several other parameters entering in the approximation: the absorption lengths of the primary particle λ_p, of the pions λ_π and kaons λ_k. There is clearly some influence, but in the present investigation the values have fixed along the original proposal. Only the $A_{\pi,k}$ values have been changed in order to adjust the calculated fluxes to the the results of the Monte Carlo simulations and to experimental data from the BESS experiment.

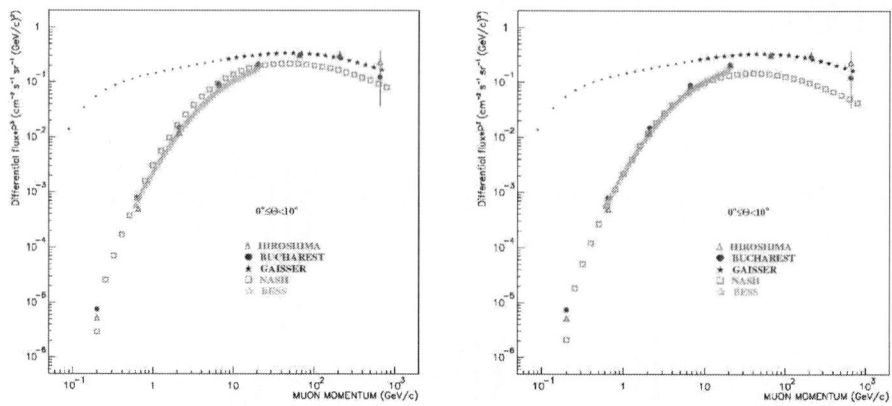

Figure 2. Comparison of the results of Monte Carlo simulations and BESS data with predictions of simplified semi-analytical formulae. In the approach of Jugde and Nash $A_\pi = 0.373$ and $A_k = 1.0$ (Fig. 1.a) and $A_\pi = 0.373$ and $A_k = 0,373$ (Fig. 1.b) is used.

Concluding remarks

Semi-analytical approaches are able to reproduce globally the results of Monte Carlo simulations and experimental data, and in particular the approach of Judge and Nash does account for muon energies $< 10 GeV$. However, these approaches can be hardly modified in order to take into account also finer effect like the influence of the geomagnetic field. For that detailed Monte Carlo simulations have to be invoked.

References

J.O.H. Stone et al, Geochimica et Cosmochimica Acta 62(1998)433

G.Huber et al. Annual Report Beschleunigerlaboratorium der Universitaet

T.K. Gaisser, Cosmic Rays and Particle Physics, Cambridge University Press, Cambridge (1992)

T.K. Gaisser, Astropart. Phys. 16(2002)285

R.J.R.Judge and W.F.Nash, Il Nuovo Cimento, XXXV-4(1965)999

M. Motoki et al, Astropart. Phys. 19(2003)113-126

P.K.Grieder, Cosmic Rays at Earth, Researcher's Reference Manual and Data Book, Elsevier (2001)354-454

M.Unger, PhD thesis, Humboldt-Universitat zu Berlin, (2003)

G.Toma, Laboratory Report (2003), Institut fur Kernphysik, Forschungszentrum Karlsruhe, unpublished

B.Mitrica, Master Thesis, University Bucharest (2004)

MEASUREMENT OF THE FLUORESCENCE YIELD IN AIR WITH THE AIRLIGHT EXPERIMENT

T. Waldenmaier[1], J. Blümer[1], H. Klages[1], S. Klepser[2]

[1]*Forschungszentrum Karlsruhe, Institut für Kernphysik*
P.O. Box 3640, 76021 Karlsruhe, Germany

[2]*Universität Karlsruhe (TH), Institut für Experimentelle Kernphysik*
P.O. Box 6980, 76128 Karlsruhe, Germany

Abstract   For the detection of ultra-high energy cosmic rays, many experiments rely on the fluorescence technique to measure the longitudinal development of extensive air showers in the atmosphere. The number of emitted fluorescence photons is related to the energy deposited by the shower in the air and therefore can be used to estimate the energy of the primary particle. The aim of the AirLight Experiment is to measure this relation for electrons in the energy range between ~ 500 keV and 2 MeV for different pressures, temperatures and air compositions with an accuracy of about 10 %.

Introduction

Due to the very steep energy spectrum of cosmic rays, very-high energy cosmic rays can only be detected by the indirect method of observing extensive air showers (EAS). Apart from their detection by large detector arrays on the ground, very-high energy EAS can also be detected by the measurement of their fluorescence light emissions. This technique utilizes the atmosphere as a scintillator with the advantage of being able to directly access fundamental shower parameters, as the longitudinal development of the total electromagnetic energy deposit along the shower axis, without relying on theoretical interaction models. Furthermore, it is possible to investigate the whole longitudinal development of an EAS instead of measuring "just" the lateral particle distributions at a certain shower stage as it is done by large detector fields. Challenging is the need of a very good understanding of the entire fluorescence detector, including the atmosphere! The relation between the number of observed fluorescence photons in the detector N_γ and the deposited energy E_{dep} of an EAS per unit of traversed matter X is assumed to be:

M. M. Shapiro et al. (eds.), Neutrinos and Explosive Events in the Universe, 403–408.
© 2005 *Springer. Printed in the Netherlands.*

$$\frac{dN_\gamma}{dX} = \frac{dE_{\text{dep}}}{dX} \int y(\lambda, T, p) \cdot \varepsilon_{\text{atm}}(\lambda) \cdot \varepsilon_{\text{FD}}(\lambda) \, d\lambda \qquad (1)$$

where ε_{atm} and ε_{FD} are the integral efficiencies of the atmosphere and the fluorescence detector which have to be monitored very carefully. The quantity $y(\lambda, T, p)$ is the fluorescence yield, which depends on the wavelength λ of the emitted fluorescence light as well as on the temperature T and the pressure p of the air at the position of emission. Eq. (1) is only applicable if $y(\lambda, T, p)$ does not depend on the energy of the ionizing particles which implies the number of emitted fluorescence photons at the shower axis N_γ^0 per wavelength and unit of traversed matter to be [2; 3]:

$$\frac{d^2 N_\gamma^0}{dX \, d\lambda} = y(\lambda, T, p) \cdot \frac{dE_{\text{dep}}}{dX} \qquad (2)$$

Fluorescence Emission in Air

Almost all the fluorescence emissions in air in the wavelength range between 300 nm and 400 nm originate from transitions of excited N_2 and N_2^+ molecules [1]. In an EAS most of the excitations are caused by electrons and positrons with energies below 1 GeV [2]. Once the nitrogen states are excited they will return to the ground state after their mean lifetime. Since the nitrogen molecules may suffer collisions with other molecules in the air, some of the excited states will lose their energy radiationless and therefore less fluorescence photons will be emitted. This effect is called collisional quenching. According to kinetic gas theory the mean time between molecular collisions τ_c is decreasing with increasing pressure and temperature and it depends also on the type of the colliding molecules (i.e. oxygen, water vapor). The relaxation rate $\frac{dN}{dt}$ and the effective lifetime τ of an exited state are therefore:

$$\frac{dN}{dt} = -N(t)\left(\frac{1}{\tau_0} + \frac{1}{\tau_c}\right), \qquad \tau = \frac{\tau_0 \tau_c}{\tau_0 + \tau_c} \qquad (3)$$

with the number of excited states $N(t)$, their intrinsic lifetime τ_0 and the effective lifetime τ. The fluorescence yield $y(\lambda, T, p)$ therefore, is expected to depend also on temperature, pressure and the gas composition. Since an EAS usually develops from its first interaction point at high altitudes down towards sea level, the fluorescence yield has to be known for pressures between 10 hPa and 1000 hPa and temperatures in the range from -60°C to 20°C. The effect of water vapor in the atmosphere needs to be investigated, too.

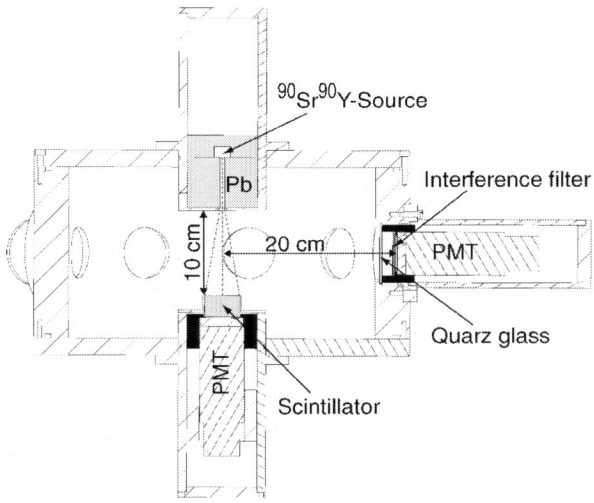

Figure 1.   Experimental setup of the AirLight Experiment.

The AirLight Experiment

The aim of the AirLight Experiment is to verify the relation of the fluorescence photon production to the ionization energy deposit, Eq. (2), and to measure the fluorescence yield $y(\lambda, T, p)$ of electrons in air around their minimum ionizing energy between $\sim$ 500 keV and 2 MeV. The fluorescence yield will be studied in several wavelength bands for different air pressures, temperatures and water vapor contaminations.

As shown in Fig. 1, the experiment consists of a cylindrical aluminum chamber with seven photomultipliers (PMT) mounted perpendicular to the chamber axis at a distance of 20 cm. Six PMTs are equipped with narrow band interference filters (FWHM $\sim$ 10 nm) to measure the fluorescence light just of certain nitrogen transitions. One PMT has a M-UG6 absorption filter as it is used in the fluorescence detectors of the Pierre Auger Observatory [4] to measure the integral fluorescence spectrum between 320 nm and 400 nm. The chamber is black anodised to suppress photons scattered off the chamber walls which would bias the acceptance of the PMTs. The electrons are emitted from a ^{90}Sr-^{90}Y beta source with a maximum energy of 2.3 MeV. The source has an activity of 37 MBq and is located behind a lead collimator of 6 cm length at the top of the chamber. Once the electrons have passed the collimator they traverse 10 cm of the test gas (normally air or pure nitrogen) and are finally stopped in a scintillation detector in the bottom of the chamber. The scintillator measures the energy of the electrons and was calibrated by the measurement of the two

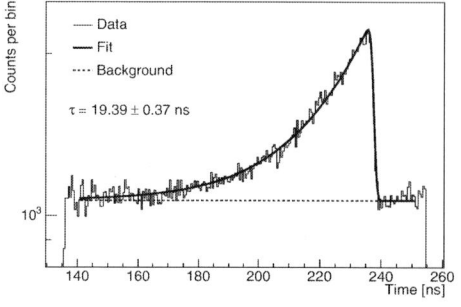
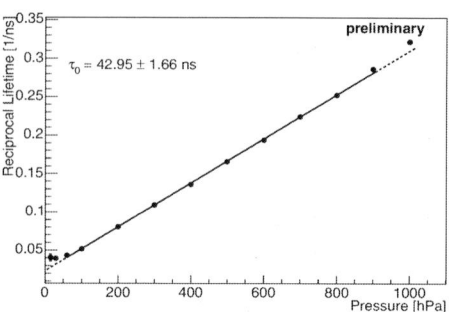

Figure 2.   Fluorescence signal of the 337 nm line in pure nitrogen at a pressure of 100 hPa and 20 °C.

Figure 3.   Reciprocal lifetimes of the 337 nm line in pure nitrogen vs. pressure at 20 °C (Error bars are covered by the data points).

well-known compton-edges in the energy spectrum of a ^{22}Na gamma emitter. The scintillation detector has a typical resolution of about $\dfrac{10\%}{\sqrt{E[\text{MeV}]}}$.

First Data

The measuring procedure takes advantage of the coincidence between the electron signal of the scintillator and the induced fluorescence photons. If any of the PMTs detects a signal within a coincidence interval of 120 ns after an electron has been detected in the scintillator, the event is accepted. Accidental coincidences can be subtracted as background from the fluorescence signals because of their missing time correlation with the scintillator signal.

Fig. 2 shows an example of a time spectrum of the prominent 337 nm line in pure nitrogen at a pressure of 100 hPa. It illustrates the time differences between the electron signal and the photon signal. For technical reasons the time scale is inverted. Random coincidences, mainly due to thermal noise of the PMT, lead to a constant offset which has to be subtracted. The fluctuations of the background play a crucial role for the determination of the signal especially for low pressure measurements. According to Eq. (3) the spectrum is fitted with a gauss-convoluted exponential function to take into account also the 0.8 ns time resolution of the detector. The fit provides the number of fluorescence photons above the background as well as the value of the effective lifetime τ at a certain pressure. According to kinetic gas theory the mean collision time τ_c should be inversely proportional to the pressure [1]. Therefore a linear behavior of the reciprocal lifetime with the pressure, as it is shown in Fig. 3 for the 337 nm line, is expected of Eq. (3). At low pressure even slight gas impurities are able to influence the molecular quenching, resulting in a sys-

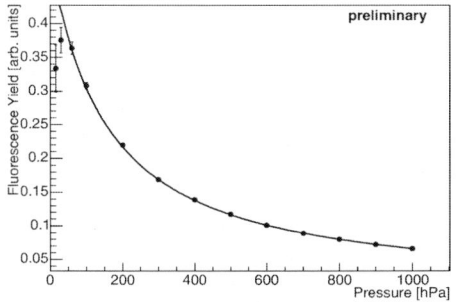 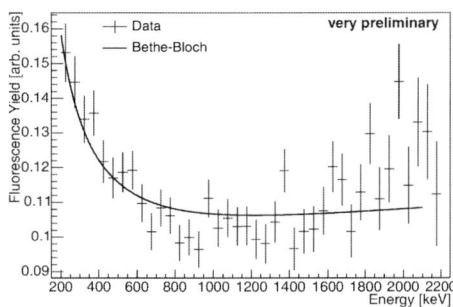

Figure 4. Fluorescence yield vs. pressure of the 337 nm line in pure nitrogen at 20 °C.

Figure 5. Fluorescence yield vs. energy of the 337 nm line in pure nitrogen at 400 hPa and 20 °C.

tematic shift to smaller lifetimes. The linear fit between 60 hPa and 900 hPa leads to an intrinsic lifetime τ_0 of the molecular state of 42.95 ± 1.66 ns which is in good agreement with other experiments [5].

The pressure dependence of the fluorescence yield for the 337 nm line is shown in Fig. 4. The fluorescence yield is decreasing with increasing pressure. This is due to the fact that the mean time between molecular collisions is decreasing at higher pressure and therefore the quenching of the excited states is getting stronger.

Fig. 5 illustrates the relative dependence of the fluorescence yield of the 337 nm line on different electron energies at a pressure of 400 hPa. The drawn curve corresponds to the Bethe-Bloch function [6] for ionization energy loss which was fitted to the data by a constant factor. The statistics in this plot is still very limited but the number of emitted fluorescence photons indeed seems to be proportional to the energy loss as it is suggested by Eq. (2).

Status and Outlook

The experimental setup of the AirLight Experiment at Forschungszentrum Karlsruhe has almost been completed. First measurements have been performed in pure nitrogen at 20 °C. The data analysis has been started recently and some of the first results were presented and are quite encouraging. Future work will concentrate on the absolute calibration of the PMTs, the full use of different wavelength and the reduction of background. In addition GEANT4-simulations [7] of the experiment are under progress.

Acknowledgments:
The authors thank Günter Wörner who constructed the chamber and provided excellent expertise in technical matters.

References

[1] A.N. Bunner: Cosmic Ray Detection by Atmospheric Fluorescence. PhD Thesis, Cornell University, Ithaca, New York (1967)

[2] M. Risse, D. Heck: Astropart. Phys. **20**, 661 (2004)

[3] M. Nagano et al.: Astropart. Phys. **20**, 293 (2003)

[4] Auger Collab., J. Abraham et al.: Nucl. Inst. and Meth. A **523**, 50 (2004)

[5] L.W. Dotchin, E.L. Chupp, D.J. Pegg: J. Chem. Phys. **59**, 3960 (1973)

[6] S.M. Seltzer, M.J. Berger: Int. J. Appl. Radiat. Isot. **33**, 1189 (1982)

[7] Geant4 Collab., S. Agostinelli et al.: Nucl. Inst. and Meth. A **506**, 250 (2003)

Participants

Natalia Agafonova
Institute for Nuclear Research RAS,
60th October Anniversary, 7a
117312 Moscow
Russia
agafonova@vaxmw.tower.ras.ru

Felix Aharonian
Max-Planck Institut fuer Kernphysik
Saupfercheckwerg 1
69117 Heidelberg
Germany
flex.aharonian@mpi-hd.mpg.de

Eun-Joo Ahn
Dep.t of Physics and Astronomy
University of Chicago
5640 S. Ellis Ave.
Chicago, IL 60637, USA
e-mail: sein@oddjob.uchicago.edu

Ester Aliu
Institut de Fisica d'Altes Energies
Edifici Cn. Universitat Autonoma
de Barcelona
08193 Bellaterra (Barcelona)
Spain
e-mail: aliu@ifae.es

Alexandre Armada
Institut de Fisica d'Altes Energies
Edifici Cn. Universitat Autonoma
de Barcelona
08193 Bellaterra (Barcelona)
Spain
e-mail: armada@ifae.es

Luisa Arruda
LIP – Laboratorio de Instrumentacao
e Fisica Experimental de Particular
Av. Elias Garcia, 14, 1 andar
1000-149 Lisboa
Portugal
e-mail: luisa@lip.pt

Sezgin Aydin
Nigde University
Department of Physics
Nigde
Turkey
e-mail: sezgin@nigde.edu.tr

Marek Bartosik
Pomorska Street 149/153
90-236 Lodz
Poland
e-mail: bartosik@uni.lodz.pl

Jacob Bekenstein
Kaplun Hall, Room 106
RACAH Institute of Physics
Hebrew University
Jerusalem 91904
Israel
e-mail: berkenste@vms.huji.ac.il

Vadim Boyarkin
Institute for Nuclear Research RAS,
60th October Anniversary, 7a
117312 Moscow
Russia
e-mail: boyarkin@vaxmw.tower.
ras.ru

410

Lukasz Bratek
Department of Physics
Reymonta 4
30-059 Krakow
Poland
lukasz.bratek@durham.ac.uk

Marc Brüggemann
Universität Siegen
Fachbereich 7 Physik,
Walter-Flex-Str. 3
57068 Siegen, Germany
e-mail: brueggemann@hep.physik.
uni-siegen.de

Bai Cai
Department of Physics
University of Minnesota
116 Church St.
Minneapolis, MN 55455, USA
e-mail: cai@physics.umn.edu

Marco Cavaglia
Dept. of Physics and Astronomy
105 Lewis Hall, Univ. of Mississippi
P.O. Box 1848
University, MS 38677-1848, USA
e-mail: cavaglia@phy.olemiss.edu

Poonam Chandra
Theoretical Astrophysics Group
Tata Instit. of Fundamental Research
Homi Bhabha Road
Mumbai 400 005, India
e-mail: poonam@tifr.res.in

J. Taylor Childers
School of Physics and Astronomy
University of Minnesota
116 Church St. SE
Minneapolis, MN 55455
USA
e-mail: childers@physics.umn.edu

Andrew Clough
Department of Physics
University of Minnesota-Duluth
10 University Dr.
Duluth, MN 55812, USA
e-mail: clou0011@d.umn.edu

Rosa Coniglione
INFN-LNS
Via S. Sofia 62
95123 Catania
Italy
e-mail: coniglione@Lns.infn.it

Buckner Creel
Physics and Astronomy Department
University of New Mexico
Albuquerque, NM 87131
USA
e-mail: creelbm@unm.edu

Roger Firpo Curcoll
Institut de Fisica d'Altes Energies
Edifici Cn [UAB]
08193 Cerdanyola del Valles
Spain
e-mail: rfirpo@ifae.es

Giulia DeBonis
Physics Department
University "LaSapienza" Roma-1
Piazzale Aldo Moro
2 00185 Rome
Italy
e-mail: giulia.debonis@roma1.
infn.it

Carla Distefano
LNS-INF
Via S. Sofia 62
95123 Catania
Italy
e-mail: distefano_c@lns.infn.it

Lev Dorman
P.O. Box 2217
Qazrin 12900
Israel
e-mail: lid@physics.technion.ac.il

Ioana Dutan
Faculty of Physics
University of Bucharest
Bucharest
Romania
e-mail: idutan@mpifr-bonn.mpg.de

Michal Dyrda
Jagiellonian University
M. Smoluchowski Inst. of Physics
Ul. Raymonta 4,
30-059 Krakow
Poland
e-mail: dyrda@th.if.uj.edu.pl

Veronique van Elewyck
Instituto de Ciencias Nucleares,
UNAM
Dept. de Fisica de Atlas Energias
Apartado Postal 70-543
04510 Mexico, D.F.
e-mail: vero@nuclecu.unam.mx

Andrij Elyiv
Astronomical Observatory
Kyiv National University
3 Observatorna Str.
Kiev, 04053
Ukraine
e-mail: elyjiw@ukr.net

Anton Empl
CERN
BAT 544/R-030
CH 1211 Geneva 23
Switzerland
e-mail: anton.empl@cern.ch

Giovanni Fazio
Harvard Smithsonian Center
for Astrophysics, MS/65
60 Garden St.
Cambridge, MA 02138, USA
e-mail: gfazio@cfa.harvard.edu

Thomas Gaisser
Bartol Research Institute
University of Delaware
217 Sharp Laboratory
Newark, DL 19716, USA
e-mail: gaisser@brivs2.bartol.udel.edu

412

Heiko Geenen
University of Wuppertal
Physics Department
Gauss Str. 20,
D-42119 Wuppertal
Germany
e-mail: geenen@physik.uni-
wuppertal.de

Florian Goebel
Max-Planck-Institute fur Physik
Foehringer Ring 6
80805 Muenchen
Germany
e-mail: fgoebel@mppmu.mpg.de

Jordan Goodman
Department of Physics
University of Maryland
College Park, MD 20742-4111
USA
e-mail:
Goodman@umdgrb.umd.edu

Dariusz Gora
Institute of Nuclear Physics
Radzikowskiego 152
31-342 Krakow
Poland
e-mail: gora@auger5.ifj.edu.pl

Christian Hededal
Astronomisk Observatorium
Rockefeller V108
Juliane Maries Vej 30
DK-2100 Copenhagen
Denmark
e-mail: hededal@astro.ku.dk

Joerg Hoerandel
Universitaet und
Forschungszentrum Karlsruhe
Insitut fuer Kernphysik
Hermann-von-Helmholz-Platz 1
76344 Eggenstein-Leopoldshafen
Germany
e-mail: joerg@ik.fzk.de

Piotr Homola
Institute of Nuclear Physics PAS
Ul. Radzikowskiego 152
31-342 Krakow
Poland
e-mail: Piotr.Homola@ifj.edu.pl

Alexei Illarionov
Laboratory of High Energy
Joint Institute for Nuclear Research
141980 Dubna, Moscow Region
Russia
e-mail: Alexei.Illarionov@jinr.ru

Joanna Jalocha
Jagiellonian University
Smoluchowski Institute of Physics
Ul. Raymonta 4,
30-059 Krakow
Poland
e-mail: jalocha@amun.ifj.edu.pl

Yuki Kaneko
University of Alabama in Huntsville
National Space Science and Technology
Center
320 Sparkman Dr.
Huntsville, AL 35805, USA
e-mail: Yuki.Kaneko@msfc.nasa.gov

Koray Karaca
Department of Physics
Middle East Technical University
06531 Ankara
Turkey
e-mail: karacak@metu.edu.tr

Karl Kosack
Physics Department
Washington University
St. Louis, MO 63130
USA
e-mail: kosack@hbar.wustl.edu

George Keros
Photon Physics
18 N Main St. Room 208
Concord, NH 03301
USA
e-mail:
photonphysics@mcttelecom.com

Daria Kosenko
Relativist Astrophysics Department
Sternberg Astronomical Institute
13, Universitetskij pr.
Moscow 119992
Russia
e-mail: lisett@xray.sai.msu.ru

Maxim Yu. Khlopov
Center for Cosmoparticle Physics
Keldysh Inst. Of Applied
Mathematics
Russian Academy of Sciences
Moscow
Russia
maxim.khlopov@roma1.infn.it

Alexandra Kozyreva
Sternberg Astronomical Institute
Moscow State University
Universitetski av. 13
119992 Moscow
Russia
e-mail: sasha@sai.msu.su

Vladimir Kuzmin
Institute for Nuclear Research
Russian Academy of Sciences
60th October Anniversary Prosp. 7a
Moscow 117312
Russia
e-mail: kuzmin@ms2.inr.ac.ru

Hans Klages
Forchungszentrum Karlsruhe
Institute of Nuclear Physics
P.O. box 3640
76021 Karlsruhe
Germany
e-mail: klages@ik.fzk.de

Alexandrer V. Kuznetsov
Department of Physics & Technology
Kharkiv National University
4 Svobody Sq.
Kharkiv 61077
Ukraine
e-mail:
alex_kuznetsov2002ua@yahoo.com

Marcin Kolonko
Institute for Nuclear Physics
Polish Academy of Sciences
Ul. Radzikowskiego 152
31-342 Krakow
Poland
e-mail: Marcin.Kolonko@ifj.edu.pl

414

Gwenaelle Lefeuvre
PCC – College de France
11 place Marcelin Berthelot
F75231 Paris Cedex 05
France
e-mail: lefeuvre@cdf.in2p3.fr

Smadar Levy
Hebrew University
Edmond Safra Campus
Givat Ram
91904, Jerusalem
Israel
e-mail: smadar@phys.huji.ac.il

Chune Yang Lum
Physics Department, Blk S12
Faculty of Science
National University of Singapore
2 Science Drive 3
Singapore 117542
e-mail: lightinsky@hotmail.com

Alexandr Malinovsky
Astro Space Center
Lebedev Physical Institute
117997 Moscow
Russia
e-mail: amalin@lukash.asc.rssi.ru

Volodymyr Marchenko
Astronomical Observatory of Taras
Shevchenko
Kyiv National University
3 Observatorna Str.
Kiev, 04053
Ukraine
e-mail: marv@observ.univ.kiev.ua

Ioana Maris
Faculty of Physics
University of Bucharest
Bucharest
Romania
e-mail: ioanamaris@yahoo.com

Kirill Martianov
Institute of Applied Physics of the
    Russian Academy of Science
46 Ulyanov St.,
603950 Nizhny Novgorod
Russia
e-mail: mca1@appl.sci-nnov.ru

Ivan Masnyak
Astronomical Observatory of Taras
Shevchenko
Kyiv National University
3 Observatorna Str.
Kiev, 04053
Ukraine
e-mail: masnyak@ukr.net

Daniel Mazin
Max-Planck Institute for Physics
Foehringer Ring 6
80805 Munich
Germany
e-mail: mazin@mppmu.mpg.de

Christine Meurer
Forschungszentrum Karlsruhe
Institut fuer Kernphysik
Postfach 3640
76021 Karlsruhe
Germany
e-mail: Christine.Meurer@cern.ch

Emilio Migneco
Laboratori Nazionali del Sud
dell'I.N.F.N.
University of Catania
Viale V. Doria 8
95129 Catania
Italy
e-mail: migneco@Lns.infn.it

Aleksei Mikhailov
Institute of Cosmophysical Research
and Aeronomy
Lenin ave. 31
Yakutsk 667891
Russia
e-mail: mikhailov@ikfia.ysn.ru

Stanislav Mikheyev
Institute for Nuclear Research
Russian Academy of Sciences
60th October Anniversary Prosp.
7a
Moscow 117312
e-mail:
mikheyev@pcbai10.inr.ruhep.ru

Katarina Miljkovic
Faculty of Mathematics
University of Belgrade
11000 Belgrade
Serbia and Montenegro
e-mail: mkaja@ptt.yu

Jamal Mimouni
Physics Institute
Mentouri University
Constantine, 25000
Algeria
e-mail: jamalm@wissal.dz

Bogdan Mitrica
National Institute of Physics and
   Nuclear Engineering (IFIN-HH)
Str. Atomistilor no. 407
P.O. Box MG-6
76900 Bucharest
Romania
bogdan.mitrica@ifin.nipne.ro

Satoko Mizobuchi
Max-Planck-Institute fur Physik
Werner-Heisenbert-Institute
Foehringer Ring 6
80805 Muenchen
Germany
e-mail: satoko@icrr.u-tokyo.ac.jp

Michael Moser
Physikalisches Institut
University of Bern
Sidlerstrasse 5
CH-3012 Bern
Switzerland
e-mail: Michael.moser@phim.unibe.ch

Dietrich Müller
Enrico Fermi Institute
University of Chicago
933 E. 56th St.
Chicago, IL 60637
e-mail: muller@ulysses.uchicago.edu

Jacek Niemiec
Institute of Nuclear Physics
Polish Academy of Sciences
Ul. Radzikowskiego 152
31-342 Krakow
Poland
e-mail: niemiec@crab.ifj.edu.pl

Victor Olmos-Gilbaja
Department of Particle Physics
University of Santiago de
Compostela
15706 Santiago de Compostela
A Coruna,
Spain
e-mail: volmos@fpaxp1.usc.es

Emma de Oña-Wilhelmi
Dpto. Fisica Atomica,
Molecular & Nuclear
Universidad Complutense de
Madrid (U.C.M.)
Avda. Complutense sn/n
28040 Madrid
Spain
e-mail: emma@gae.ucm.es

Michal Ostrowski
Astronomical Observatory
Jagiellonian University
Ul. Orla 171
30-244 Krakow
Poland
e-mail: mio@oa.uj.edu.pl

Sacit Ozdemir
Gazi University
Kirsehir Science and Arts
Faculty
Dept. of Physics,
Kirsehir
Turkey
e-mail: ozdemir@comu.edu.tr

Jan Pekala
Institute of Nuclear Physics
Polish Academy of Sciences
Ul. Radzikowskiego 152
31-342 Krakow
Poland
e-mail: Jan.Pekala@ifj.edu.pl

Jeremy Perkins
Physics Department
Washington University
One Brookings Dr.
St. Louis, MO 63130, USA
e-mail: jperkins@freyda.wustl.edu

Moshe Pessing
Hebrew University
Givat Ram, Levin 16c
Jerusalem
Israel
e-mail: mop@phys.huji.ac.il

Paolo Piatelli
INFN-LNS
Via S. Sofia 62
95123 Catania, Italy
e-mail: piattelli@Lns.infn.it

Stefan Plewnia
Forschungszentrum Karlsruhe
P.O. Box 3640
76021 Karlsruhe
Germany
e-mail: plewnia@ik.fzk.de

Sergei Popov
Sternberg Astronomical Institute
Relativistic Astrophysics
Universitetskii pr. 13
119992 Moscow
Russia
e-mail: popov@pd.infn.it

Konstantin Postnov
Sternberg Astronomical Institute
Moscow State University
119899 Moscow
Russia
e-mail: pk@sai.msu.ru

Wolfgang Priester
Institut fuer Astrophysik
Universitaet Bonn
Auf dem Huegel 71
53121 Bonn
Germany
e-mail: priester@astro.uni-
bonn.de

Vladimir Ptuskin
IZMIRAN
Russian Academy of Sciences
Troitsk, Moscow Region
142120, Moscow
Russia
e-mail: vptuskin@hotmail.com

Milan Raicevic
Faculty of Mathematics
University of Belgrade
11000 Belgrade
Serbia and Montenegro
e-mail: rakhan@eunet.yu

Martin Raue
Universitaet Hamburg
Institut fuer Experimental Physik
Luruper Chaussee 149
D-22761 Hamburg
Germany
e-mail: mraue@mail.desy.de

Paul Rebillot
Physics Department
Washington University
One Brookings Dr.
St. Louis, MO 63130, USA
e-mail: rebillot@hbar.wustl.edu

Raquel de los Reyes
Dpto. Fisica Atomica, Molecular &
Nuclear
Universidad Complutense de Madrid
Avda. Complutense sn/n
28040 Madrid, Spain
e-mail: reyes@gae.ucm.es

Giorgio Riccobene
Univ. of Catania
LNS-INFN
Via S. Sofia 44
I-95123 Catania
Italy
e-mail: riccobene@lns.infn.it

Javier Rico
Institut de Fisica d'Altes Energies
Edifici Cn. Universitat Autonoma de
Barcelona
08193 Bellaterra (Bracelona)
Spain
e-mail: jrico@ifae.es

Cecile Roucelle
LPNHE
4 Place Jussieu
T33 RDC
75252 Paris Cedex 05
France
e-mail: roucelle@lpnhep.in2p3.fr

Benjamin Morales Ruiz
Instituto de Fisica, UNAM
Departamento de Fisica Teorica
Mexico D.F., A.P. 20-364
Del. Alvaro Obergon, 01000
Mexico
e-mail: bamr@fisica.unam.mx

Piera Sapienza
INFN-LNS
Via S. Sofia 62
95123 Catania
Italy
e-mail: sapienza@Lns.infn.it

Hisham Sayed
Astronomy Department
Faculty of Science
Cairo University
Giza
Egypt
e-mail: hisham@mailer.eun.eg

Villi Scalzotto
Dipartimento di Fisica G.
Galilei
Via Marzolo, 8
35131 Padova
Italy
e-mail: scalz8@pd.infn.it

Viviana Scherini
University of Wuppertal
Physics Department
Gauss Str. 20
D-42119 Wuppertal
Germany
e-mail: scherini@physik.uni-wuppertal.de

Martin Schroedter
FLWO
P.O. Box 97
Amado, AZ 85645
USA
martins@physics.arizona.edu

Lauren Scott
Department of Physics
Washington University
One Brookings Drive
St. Louis, MO 63130
USA
e-mail: lscott@hbar.wustl.edu

Maurice M. Shapiro
5809 Nicholson Lane
Apt. 801
Rockville, MD 20852
USA
e-mail: mmshapiro@mailaps.org

Nuria Sidro
Institut de Fisica d'Altes Energies
Edifici Cn. Universitat Autonoma de
Barcelona
08193 Bellaterra (Barcelona)
Spain
e-mail: nsidro@ifae.es

Andrea Silvestri
University of California - Irvine
Department of Physics and
Astronomy
4129 Federick Reines Hall
Irvine, CA 92697-4575
USA
e-mail: silvestri@cosmic.ps.uci.edu

Lawrence R. Sulak
Antares Neutrino Observatory
Centre for Particle Physics
University of Marseille
163 av. De Luminy, Case 907
13288 Marselle, CEDEX 09
France
e-mail: sulak@bu.edu

Arthur Smith
University of Oxford
Clarendon Laboratory
Park Road
Oxford
United Kingdom
e-mail:
A.Smith1@physics.ox.ac.uk

Sebastian Szybka
Jagiellonian University
Institute of Physics
Dept. of General Relativity and Astr.
Ul. Raymonta 4
30-059 Krakow
Poland
e-mail: szybka@if.uj.edu.pl

Todor Stanev
Bartol Research Institute
University of Delaware
Newark, DE 19711
USA
e-mail:
stanev@muon.bartol.udel.edu

Oana Tascau
BU Wuppertal
Gausstrasse 20
42097 Wuppertal
Germany
e-mail: andorada@physik.uni-
wuppertal.de

Elisabetta Strazzeri
LNS-INFN
Via S. Sofia, 44
I-95123 Catania
Italy
e-mail: strazzeri@lns.infn.it

Masahiro Teshima
Max-Planck-Institute for Physics
Foehringer Ring 6
80805 Munchen
Germany
e-mail: mteshima@mppmu.mpg.de

Mathias Stuempert
Forschungszentrum Karlsruhe
P.O. Box 3640
76021 Karlsruhe
Germany
Mathias.Stuempert@ik.fzk.de

Alessandro Thea
Dipartimento di Fisica dell'Universita
INFN, Sezione di Genova
Via Dodecaneso 33
1-16146 Genova, Italy
e-mail: thea@ge.infn.it

420

Omar Tibolla
Departimento di fisica G. Galilei
Via Marzdo, 8
Padova
Italia
e-mail: omar.tibolla@pd.infn.it

Inez Valino-Rielo
Department of Particle Physics
University of Santiago de Compostela
15706 Santiago de Compostela
A Coruna, Spain
e-mail: inesvr@usc.es

Irina Titkova
Joint Institute of Nuclear Research
Joliot-Curie 6
141980 Dubna, Moscow Region
Russia
e-mail: kitti@nusun.jinr.ru

Jurriaan Van Buren
Forschungszentrum Karlsruhe
P.O. Box 3640
76021 Karlsruhe
Germany
e-mail: vanburen@ik.fzk.de

Nadia Tonello
Max-Planck Institute for Physics
Foehringer Ring, 6
D-80805 Munich
Germany
e-mail: tonello@mppmu.mpg.de

Vincenzo Vitale
Max Planck Institute for Physics
Foehringer Ring, 6
D-80805 Munich
Germany
e-mail: vitale_tmp_2003@yahoo.it

Ana Torrento-Coello
CIEMAT
Dpto. Fusion y Paticulas Elementales
Avda. Complutense, 22
28040 – Madird
Spain
e-mail: Ana.Torrento@ciemat.es

Tilo Waldenmaier
Institut fuer Kernphysik
Forschungszentrum Karlsruhe
P.O. Box 3640
76021 Karlsruhe
Germany
e-mail: waldenmaier@ik.fzk.de

Virginia Trimble
Physics Department
University of California, Irvine
Irvine, CA 92697
USA
e-mail: vtrimble@astro.umd.edu

John P. Wefel
Department of Physics and Astronomy
Louisiana State University
Baton Rouge, LA 70803
USA
e-mail: wefel@phunds.phys.lsu.edu

Motohiko Yoshimura
Okayama University
Faculty of Science
3-1-1 Tsushima-naka
Okayama 700-8530
Japan
e-mail: yoshim@icrr.u-tokyo.ac.jp
yoshim@astro.hep.okayama-u.ac.jp

Jochen Zonnchen
Institut fur Astrophysik und
Extraterrestrische Physik
Forschung der Univeritat Bonn
Auf dem Hugel 71
53121 Bonn
Germany
e-mail: zoenn@astro.uni-
bonn.de
jzoenn@yahoo.de

Index